PHYLOGENOMIC DATA ACQUISITION

PHYLOGENOMIC DATA ACQUISITION

PRINCIPLES AND PRACTICE

W. BRYAN JENNINGS

CRC Press is an imprint of the
Taylor & Francis Group, an **informa** business

The cover image shows a computer model of a Tn5 synaptic complex, which is comprised of a Tn5 transposase enzyme (blue) bound to the ends of a DNA transposon (red and green). These enzymes play a vital role in some Next Generation Sequencing methods. (image credit: Laguna Design, Science Photo Library/Getty Images.)

CRC Press
Taylor & Francis Group
6000 Broken Sound Parkway NW, Suite 300
Boca Raton, FL 33487-2742

First issued in paperback 2019

ISBN-13: 978-1-4822-3534-0 (hbk)
ISBN-13: 978-0-367-86980-9 (pbk)

Library of Congress Cataloging-in-Publication Data

Names: Jennings, W. Bryan, author.
Title: Phylogenomic data acquisition : principles and practice / author, W. Bryan Jennings.
Description: Boca Raton : Taylor & Francis, 2017. | Includes bibliographical references.
Identifiers: LCCN 2016032138| ISBN 9781482235340 (hardback : alk. paper) | ISBN 9781482235357 (ebook)
Subjects: | MESH: Sequence Analysis, DNA--methods | Polymerase Chain Reaction--methods | Phylogeny | Biological Evolution | Data Collection
Classification: LCC QP624.5.D726 | NLM QU 500 | DDC 572.8/6--dc23
LC record available at https://lccn.loc.gov/2016032138

Visit the Taylor & Francis Web site at
http://www.taylorandfrancis.com

and the CRC Press Web site at
http://www.crcpress.com

CONTENTS

PREFACE

Phylogenomics intersects and unites many areas in evolutionary biology including molecular and genomic evolution, systems biology, molecular systematics, phylogeography, conservation genetics, DNA barcoding, and others. Although these disciplines differ from each other in their study questions and methods of data analysis, they all use DNA sequence datasets. Phylogenomics is moving forward at a dizzying pace owing to advances in biotechnology, bioinformatics, and computers. This is reminiscent of what occurred two decades ago when the field of molecular systematics was coming of age. Another major factor that undoubtedly helped spur the growth of molecular systematics was the arrival of *Molecular Systematics*, 2nd edition (Hillis et al. 1996), a book that allowed me (along with countless others) to jump into this exciting field. As phylogenomics has grown substantially since the dawn of the genomics era—in large part due to the advent of next generation sequencing—the time is right for a book that presents the principles and practice of obtaining phylogenomic data.

This book enables beginners to quickly learn the essential concepts and methods of phylogenomic data acquisition so they can confidently and efficiently collect their own datasets. Directed at upper level undergraduate and graduate students, this book also benefits experienced researchers. The inference of gene trees from DNA sequence data represents one of the fundamental aspects of phylogenomic analysis. Accordingly, because robust gene tree inferences are generally made using longer DNA sequences (e.g., ~200–2,000 base pairs long), this book focuses on methods for obtaining sequences in this length range.

This book is organized as follows. Chapter 1 introduces phylogenomics within a historical context, points out connections between DNA sequence data and gene trees, discusses gene trees versus species trees, and provides an overview of the methods used today to acquire phylogenomic datasets. Chapter 2 describes the landscapes of eukaryotic genomes followed by discussion of molecular processes that govern the evolution of DNA sequences. Chapter 3 continues the discussion about properties of DNA sequence loci by reviewing six common assumptions that pertain to data characteristics before describing the different types of DNA sequence loci used in phylogenomic studies. Chapter 4 covers DNA extraction methods including high-throughput methods. Chapter 5 reviews PCR theory, discusses applications in phylogenomics, and considers high-throughput workflow. Chapter 6 describes Sanger sequencing including high-throughput sequencing. Chapter 7 explains Illumina sequencing technology and how it is used to obtain phylogenomic datasets. Chapter 8 reviews theory and methods for designing novel DNA sequence loci. Finally, Chapter 9 offers a vision of the future in phylogenomic data acquisition.

Most of the information contained in this book can be found elsewhere, but it is worthwhile to bring it together. This synthesis provides detail including reference to the foundational papers. I hope these discussions will stimulate and direct the reader—especially students—to study these classic papers. Not only will these extra readings provide additional details about the subject at hand, but should also evoke feelings of admiration for those works and thereby generate inspiration and excitement about phylogenomics research. This book is biased toward eukaryotic organisms because of my research experience and interests in vertebrates. Therefore, an apology is in order to my colleagues who study microorganisms though they may still find at least some parts of this book useful.

I am grateful to a number of colleagues who reviewed earlier versions of chapters and provided helpful comments. For their help and encouragement I thank David Blackburn, Steve Donnellan, Andrew Gottscho, Fábio Raposo, Sean Reilly, Todd Schlenke, and especially Ryan Kerney who read four of the chapters. Any remaining errors are my own. I thank my editor Chuck Crumly at Taylor & Francis for suggesting that I undertake this project and for his constant encouragement and patience. I would also like to thank Cynthia Klivecka of Taylor & Francis as well as Mohamed Hameed and Karthick Parthasarathy of Novatechset for their tremendous help to produce the final book. Special thanks go to my doctoral advisor Eric Pianka who first encouraged me to apply molecular data to study evolutionary questions. I am indebted to all my mentors—Samuel Sweet, Jonathan Campbell, Eric Pianka, and Scott Edwards—who have helped me with my career and provided inspiration. I would also like to thank some colleagues from my graduate school days: Mark and Kris Holder, Tom Wilcox, Marty Badgett, and Todd Schlenke taught me many things in the laboratory and provided much assistance while David Hillis and James Bull generously allowed me to work in their laboratory. One of the pleasures of being a professor is having excellent graduate students who help to inspire me. I thus want to thank my students Andrew Gottscho, Sean Reilly, Igor Rodrigues da Costa, Carla Quijada, and Piero Ruschi. I would like to thank my Humboldt colleagues especially Jacob Varkey, Sharyn Marks, Brian Arbogast, Jeffrey White, John Reiss, Anthony Baker, Julie Davy, Leslie Vandermolen, and Christopher Callahan for their support of my teaching and research. I am also very grateful to my colleagues at the Federal University of Rio de Janeiro for their support and wonderful collaborations. In particular I wish to thank Ronaldo Fernandes, Paulo Buckup, Marcelo Weksler, Jose Pombal Jr, Paulo Passos, Ulisses Caramaschi, Marcos Raposo, Marcelo Britto, and Francisco Prosdocimi. This work was made possible through support from HHMI, Humboldt State University, Dean of Graduate Studies and Research (PR-2) at the Federal University of Rio de Janeiro, and the Brazilian funding agencies CAPES, CNPq, and FAPERJ. My wife Vivian Menezes Leandro provided me with constant support and encouragement, which allowed me to complete this book. I dedicate this book to her.

W. BRYAN JENNINGS
Rio de Janeiro, Brazil
December 2016

AUTHOR

W. Bryan Jennings is a foreign visiting professor in the Department of Vertebrates and Post-Graduate Program in Zoology at the National Museum of Brazil and Federal University of Rio de Janeiro. He earned his BA in zoology from the University of California at Santa Barbara; MS in biology from the University of Texas at Arlington; and PhD in ecology, evolution, and behavior from the University of Texas at Austin. He was a postdoctoral fellow in the Department of Biology at the University of Washington, and in the Museum of Comparative Zoology and Department of Organismic and Evolutionary Biology at Harvard University. He was then appointed teaching fellow for one year in the Department of Molecular and Cellular Biology at Harvard before becoming an assistant professor of biology at Humboldt State University. At Humboldt, he taught genetics labs, bioinformatics, biogeography, introductory molecular biology, and introductory biology for nonbiology majors. In 2010, he moved to the National Museum of Brazil to accept a CAPES foreign visiting professorship. At the National Museum, he cofounded the Molecular Laboratory of Biodiversity Research, teaches a graduate course in phylogeography, and mentors masters students, doctoral students, and postdocs. Studies in his lab are focused on phylogenomics of vertebrates with an emphasis on phylogeographic and conservation genetics studies.

CHAPTER ONE

Introduction

The great evolutionary geneticist, Theodosius Dobzhansky, famously wrote "*Nothing in biology makes sense except in the light of evolution*" (Dobzhansky 1973). One area in evolutionary biology that has shed much light on biological phenomena is the field of molecular phylogenetics. Phylogenetic trees inferred from molecular genetic data have led to quantum leaps in our understanding about molecular evolution and the *Tree of Life*. The Tree of Life Project is a worldwide collaboration of evolutionary biologists that aims to elucidate the evolutionary history for all life found on Earth (Maddison et al. 2007; http://tolweb.org/tree/). Another important initiative is the Open Tree of Life (http://opentreeoflife.org/). Advances in DNA sequencing capability, computers, and bioinformatics from the late 1970s through the 1990s spurred the rapid growth of molecular phylogenetics (Hillis et al. 1996). An outgrowth of this field, which began slowly in the 1990s but later blossomed into its own field due to the emergence and explosive growth of genomics, is the discipline of *phylogenomics*. Given that substantial overlap obviously exists between molecular phylogenetics and phylogenomics, we should ask the following questions: *What is phylogenomics and how does it differ from molecular phylogenetics?*

1.1 WHAT IS PHYLOGENOMICS?

Before we further consider a definition for phylogenomics, let's first examine a traditional definition of molecular phylogenetics. The field of molecular phylogenetics can be defined as follows: *molecular phylogenetics is the discipline concerned with using phylogenetic methodology on molecular genetic data to infer evolutionary phylogenies or "trees" to elucidate the evolutionary relationships and distances or divergence times among DNA sequences, amino acid sequences, populations, species, or higher taxa*. The vast majority of molecular phylogenetic studies have been based on DNA sequence data, typically representing one to several genes, though other types of molecular genetic data such as amino acid sequences are also used. Molecular phylogenies, especially those inferred for a single gene, are commonly called *gene trees*.

In contrast to molecular phylogenetics, "phylogenomics" is more difficult to define for two reasons. First, some methodological and conceptual overlap exists between them—namely both fields rely on phylogenetic methodology to infer phylogenies from molecular data. Secondly, researchers have used the term phylogenomics to characterize different types of studies. We will now take a closer look at how researchers have used the term phylogenomics before we settle on a definition to follow in this book.

1.1.1 The Early View of Phylogenomics

Eisen (1998a) originally coined the term phylogenomics and defined this discipline as the prediction of gene function and study of gene and genome evolution using molecular phylogenies in conjunction with modern comparative methods. For example, in an early phylogenomic study, Eisen (1998b) first inferred the gene tree among members (amino acid sequences) of the MutS family of proteins, a group of proteins important for recognition and repair of DNA mismatches caused by errors during DNA replication. He then used this tree to investigate the evolutionary diversification of this gene family by looking at MutS homologs found within and among genomes across the Tree of Life. Phylogenetic-based methods are not only superior to similarity-based methods for predicting the functions of unknown genes, but they allow a researcher to split genes into orthologous and paralogous subfamilies and identify key events in the histories of gene families such as gene divergences, lateral gene transfers, and gene losses (Eisen et al. 1995; Eisen 1998a,b).

Shortly thereafter, O'Brien and Stanyon (1999) used the term phylogenomics differently, as they mentioned "comparative phylogenomics" to describe studies using comparative gene maps for a number of closely related species combined with cladistic analysis to reconstruct ancestral genomes (e.g., Haig 1999). Although these two uses of the term phylogenomics were both applied to the study of molecular or genomic evolution, these studies nonetheless differed from each other in terms of data, analytical methods, and study goals.

1.1.2 An Expanded View of Phylogenomics

The purview of phylogenomics broadened further during the early 2000s soon after the genome era commenced, as the rapidly increasing volumes of genomic data—including some fully sequenced eukaryotic genomes such as the human genome—allowed researchers to dramatically scale up sizes of their datasets in phylogenetically based evolutionary studies. It was during this time that researchers could begin using phylogenetic methodology to analyze enormous genome-wide datasets for addressing problems ranging from genome evolution to reconstructing the Tree of Life (Eisen and Fraser 2003; Rokas et al. 2003; Delsuc et al. 2005; Philippe et al. 2005).

The viewpoint that phylogenomics is a discipline comprised of two main areas of inquiry—one concerned with questions in molecular and genomic evolution and the other focused on the evolutionary history of organisms—was subsequently reinforced at the first phylogenomics symposium (Philippe and Blanchette 2007) and in the first book focused on phylogenomics (Murphy 2008). Aside from the dramatic growth in phylogenomic studies since that time, little has changed regarding this dichotomy of research goals. Thus, at the present time we can think of phylogenomics as comprising two major subdisciplines: *molecular phylogenomics* and *organismal phylogenomics*. Accordingly, we may broadly define "phylogenomics" as *the field of study concerned with using genome-wide data to infer the evolution of genes, genomes, and the Tree of Life*. What primarily differentiates molecular phylogenetics from phylogenomics is that the latter field often uses much larger or "computer bursting" datasets and gene trees as independent units of analyses in evolutionary studies.

1.2 ANATOMY OF GENE TREES

Gene trees that have been inferred from DNA sequence data represent fundamental units of analysis in various types of phylogenomic studies. The basic anatomy of a gene tree is illustrated in Figure 1.1. In this figure we see the genealogical relationships among a sample of five DNA sequences labeled a through e for a single gene. Except for some specialized cases (e.g., studying microorganisms in a laboratory), the *true* gene tree cannot be known for a given set of DNA sequences. Instead, the genealogical history must be *inferred* or reconstructed using phylogenetic methodology.

The structure of an inferred gene tree is defined by its branching pattern or "topology" and length of each branch. For example, the tree in Figure 1.1 has four *nodes* (labeled 1 through 4). The bottom-most node (node 4) represents the *root* of the gene tree. The root node is particularly important because it represents the *most recent common ancestor* or "MRCA" of the five DNA sequences and therefore provides directionality or time's arrow along the tree (Figure 1.1). Similarly, we can describe node 1 as the MRCA of a and b, node 2 is the MRCA of a, b, and c, and node 3 is the MRCA of d and e. The other major structural features of gene trees are its *branches*, which represent *lines of descent*.

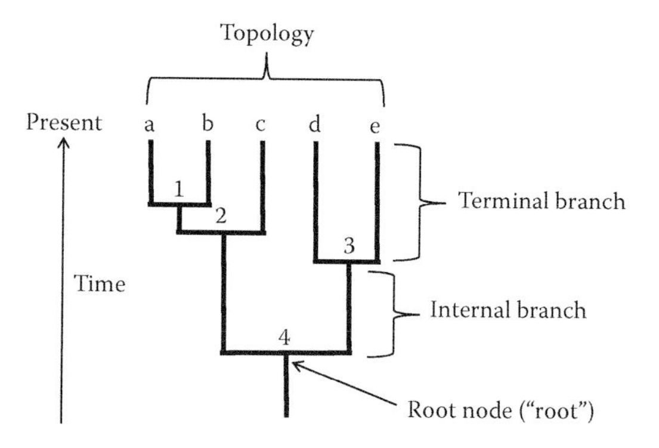

Figure 1.1. An example of a rooted gene tree for five DNA sequences labeled (a–e). In this tree there are four nodes, five terminal branches, and four internal branches. The placement of the root in this tree (node 4) gives the tree a direction with respect to time and thus ancestor–descendant relationships can be inferred.

Branches (i.e., the vertical lines in the tree) can be subdivided into two categories: *terminal branches*, which connect the tips (observed DNA sequences a–e) with nodes below them. For example, the branches connecting node 1 to sequences a and b represent two terminal branches. Likewise, the branch between node 2 and sequence c is also a terminal branch. The other type of branch is known as an *internal branch*. There are a total of four internal branches in this tree and they are found between nodes 1 and 2, nodes 2 and 4, nodes 3 and 4, and below the root node. The lengths of the branches in gene trees can indicate rates of molecular evolution or evolutionary time depending on how the tree is constructed. Note that the tree in Figure 1.1 represents one of the 105 possible different topologies that could be generated for five labeled tips and a tree that is completely bifurcating (i.e., each node has exactly two descendent branches connected to it) and has a root (Felsenstein 2004). The numbers of unique rooted tree topologies becomes shockingly high as the number of sequences increases. For example, a perusal of Table 3.1 in Felsenstein (2004) shows that for only 10 sequences there are more than 34 million different rooted bifurcating trees and a mind boggling 2.75×10^{76} different trees for 50 sequences! As the focus of this book is on acquiring phylogenomic data, we will not delve into the details on how these trees are made. Readers wanting to learn about methods for inferring phylogenetic trees using molecular data should consult the following references: Hillis et al. (1996), Felsenstein (2004), and Lemey et al. (2009).

1.3 GENE TREES VERSUS SPECIES TREES

When a molecular biologist reconstructs a gene tree for a particular gene family, the interpretations of the resulting tree are clear-cut: the tree shows the inferred evolutionary relationships among the *molecules* (amino acid or DNA sequences) used to make the tree and may also display the rates or timing of lineage divergences. In other words, what is gleaned from a gene tree in this type of phylogenomic study is the evolution of the molecular sequences themselves. In contrast, when a single gene tree is used to infer a phylogeny of populations or species, as has been done innumerable times in traditional molecular phylogenetics, then the researcher is extrapolating an *organismal phylogeny* from a molecular phylogeny (Maddison 1995, 1997). In Tree of Life studies, an organismal phylogeny is more commonly referred to as a *species tree* because it shows the branching relationships and times of divergence among populations or species (Maddison 1995, 1997). Although a gene tree may to some extent mirror a species tree, it is important to realize that gene trees and organismal trees are not the same thing. Thus, a researcher who uses a single gene tree to infer a species tree hopes that they match each other (i.e., are congruent).

Schematic examples of gene and species trees are shown side-by-side in Figure 1.2. Let's first consider the meaning of the topology in each evolutionary tree. The gene tree in Figure 1.2a shows the evolutionary relationships between two DNA sequences (haplotypes) sampled from one gene in two species, whereas the species tree in Figure 1.2b shows the evolutionary relationships between the two species. In addition to the topologies, the divergence times in each type of tree are also fundamentally different from each other. Divergence times derived from a gene tree represent *gene divergence times*, whereas in the species tree the divergence times represent *population* or *species divergence times* (Figure 1.2). The MRCA for the two haplotypes (i.e., gene divergence) corresponds to a single individual in the ancestral population, while the MRCA for the two species is the ancestral species. Thus the gene tree shows the inferred genealogical history of the sampled DNA sequences, whereas the latter exhibits the evolutionary history for populations or species. It can be a perilous practice to naively equate a gene tree with a species tree because, even if a gene tree is reconstructed without error, its topology may be incongruent with the corresponding species tree (Hudson 1983, 1992; Tajima 1983; Maddison 1995, 1997; Rosenberg 2002; Felsenstein 2004).

Notice in Figure 1.2 that the widths of the branches differ between gene and species trees. The branches of a gene tree are thin lines of constant width (Figure 1.2a) while the branches of a species tree are much wider branches. In some cases, the branch widths are drawn in this manner simply to distinguish the schematic of a gene tree from a species tree. In other cases, the widths of each branch in a species tree are drawn to be proportional to the *effective population sizes* for that population at particular points in time (Figure 1.2b). Thus wider branches represent larger effective

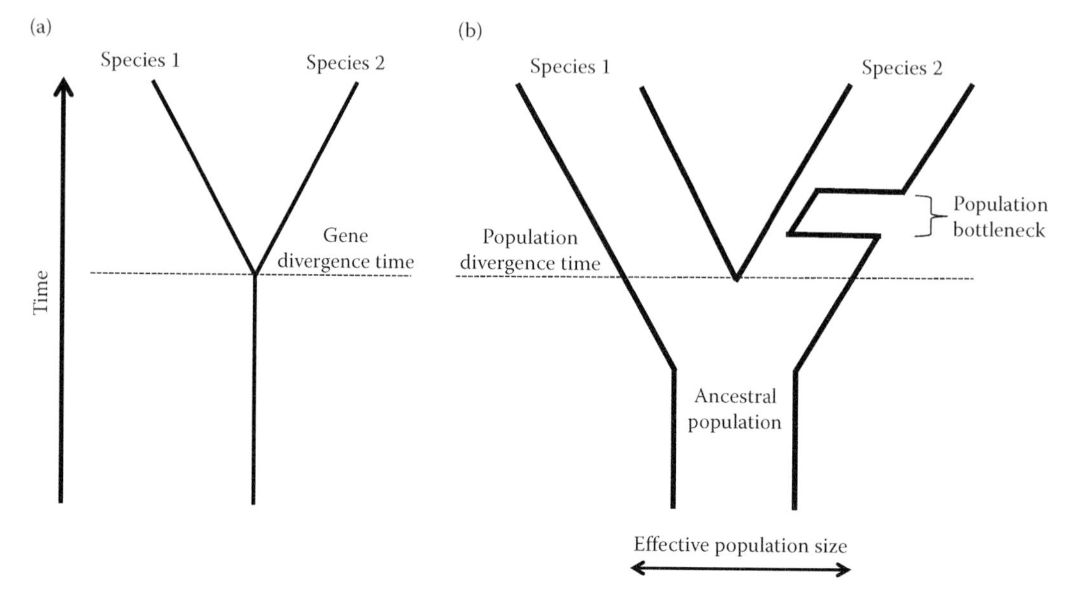

Figure 1.2. A comparison between the anatomy of a gene tree versus a species tree. (a) Shown is a gene tree that depicts the evolutionary history of two haplotypic lineages from two species. The timing of divergence between lineages is called gene divergence time. (b) A species tree showing the evolutionary history of two sister species. The timing of divergence between lineages in a species tree is referred to as the population divergence or speciation time. The branch widths in a species tree are drawn in proportion to the effective population sizes, which may vary through time. In this example, the ancestral population of species 2 is shown to have suffered a population bottleneck.

population sizes than do thinner branches. For example, a species tree with a branch having the same width from node to node means that the long-term effective population sizes have remained constant in size (e.g., ancestral populations of Species 1 in Figure 1.2b). In contrast, the ancestral populations may have undergone fluctuations in size such as from a population bottleneck (e.g., ancestral populations of Species 2 in Figure 1.2b). In contrast, line width in a gene tree has no meaning. Despite the potential lack of congruence between gene and species trees, phylogenomic methods have been developed to address these issues, which in turn, are enabling researchers to generate more accurate and robust estimates of species trees than was ever possible using single gene trees (Knowles and Kubatko 2010).

1.4 PHYLOGENOMICS AND THE TREE OF LIFE

The task of reconstructing the Tree of Life represents a monumental undertaking. In the realm of phylogenomics, there are several levels of study that are contributing to this effort. First, at the shallowest levels in the Tree of Life (i.e., "recent" speciation events), the use of phylogenomic datasets in conjunction with phylogeographic methodology is helping to enumerate the true numbers of extant species as well as provide insights into the history of their formation. *Phylogeography* is the study of how past demographic processes and environmental forces have contributed to speciation and shaped the genetic structure of contemporary populations and species (Avise 2000). Thus, phylogeography provides insights about the temporal and geographical aspects of speciation in recent species radiations and therefore this field shares a close connection to the Tree of Life. On a larger scale, the application of phylogenomic methods for inferring species trees is assisting with the reconstruction of the topology and branch lengths of the Tree of Life (Knowles and Kubatko 2010). Lastly, the use of *DNA barcoding* (Hebert et al. 2003), whether based on single organellar genes, entire organellar genomes, or even vast numbers of nuclear loci, is providing biologists with a powerful and simple tool for identifying known and possible new species. We will now briefly introduce each of these approaches.

Phylogenomics and phylogeography—During the first two decades of phylogeography beginning in the 1980s, the vast majority of phylogeographic studies relied on gene trees that were inferred from mitochondrial DNA sequences. However, in later years two factors dramatically increased the sophistication of this discipline. First, the commencement of the genomics era provided researchers with far greater access to genomic sequences, which, in turn, enabled them to mine genomes for large numbers of DNA sequence-based loci (e.g., Chen and Li 2001). A second critical factor was the development of software implementing coalescent theory-based Bayesian and maximum likelihood statistical methods (e.g., Yang 2002; Hey and Nielsen 2004; Hey 2010). With the newfound availability of more voluminous datasets and bioinformatics tools, phylogeographers could for the first time conduct more robust multilocus analyses, which were capable of yielding more accurate and precise estimates of key phylogeographic parameters such as divergence times between populations (Yang 2002; Jennings and Edwards 2005).

Phylogenomic methods for estimating species trees—Various approaches for inferring species trees have been employed in phylogenomic studies, some of which make use of individual gene trees while others do not. These methods fall into two main categories: *DNA sequence-based approaches* and *whole-genome features* (Delsuc et al. 2005). By far, most phylogenomic studies have relied on DNA sequence datasets for inferring species trees.

Within the category of DNA sequence-based approaches, three primary methods have been used to estimate species trees: (1) the coalescent-theory or independent gene tree approach (Edwards et al. 2005; Knowles and Kubatko 2010); (2) the "supermatrix" approach (Delsuc et al. 2005; Edwards et al. 2005; Philippe et al. 2005); and (3) the "supertree" approach (Delsuc et al. 2005; Philippe et al. 2005). The independent gene tree approach directly estimates a species tree by analyzing the gene trees in a *coalescent theory* framework (Edwards et al. 2005; Edwards 2009). Coalescent theory is a population genetics-based theory that models, in a probabilistic manner, how historical demography (e.g., population sizes, gene flow) of ancestral populations can influence the distributions of gene trees found in contemporary populations and species (Felsenstein 2004; Wakeley 2009). The study of Australian grass finches by Jennings and Edwards (2005) provides an early example in which a species tree was inferred from a coalescent-based analysis of a genome-wide dataset.

The supermatrix approach basically involves joining together, end-to-end, multiple loci to generate long contiguous sequences with one sequence representing each species thereby resulting in a supermatrix of phylogenetic characters. This method, which is also commonly referred to as *concatenation*, can involve the joining together of tens, hundreds, or thousands of loci from throughout a genome. Once a supermatrix is assembled, the investigator can then use phylogenetic methods to infer a species tree. Two early studies on eukaryotes that used this approach with genome-wide datasets include the work on hominoids by Chen and Li (2001) and on yeasts by Rokas et al. (2003).

Lastly, in the supertree approach, the individual trees, which can be based on molecular data, morphological data, or both, are first independently estimated (usually these are already published) and then a single "supertree" is constructed by essentially stitching the overlapping gene trees into a single composite tree (e.g., Hinchliff et al. 2015).

The "whole-genome features" approach for inferring species trees includes a variety of different methods that rely on using architectural features of genomes or rare genomic changes as phylogenomic characters from which species trees can be inferred (Delsuc et al. 2005; Edwards et al. 2005). Phylogenomic datasets making use of rare genomic changes may include data on: (1) gene order or "synteny" (Delsuc et al. 2005; Edwards et al. 2005); (2) retroposon insertion sites (Shedlock and Okada 2000; Okada et al. 2004; Shedlock et al. 2004; Edwards et al. 2005); or (3) "DNA strings" or "genomic signatures" (Deschavanne et al. 1999; Karlin and Burge 1995; Edwards et al. 2002; Delsuc et al. 2005). Although these methods may help resolve particular deep branches in the Tree of Life, their use thus far has been limited.

DNA barcoding and mitogenomics—In DNA barcoding studies, a single gene sequence, which is typically a mitochondrial gene, is used to identify existing and possible new species (e.g., Hebert et al. 2003; Guarnizo et al. 2015; Jennings et al. 2016). The increase in DNA sequencing power and availability of bioinformatics tools now enables biologists to sequence and annotate entire mitochondrial genomes far faster and simpler than before. Moreover, many recently published mitogenomes essentially represent "byproducts" of next-generation sequencing (NGS) studies with a focus on nuclear DNA sequences (e.g., Souto et al. 2014; Amaral et al. 2015). This surge in mitogenomes is providing molecular biologists with a treasure trove of data from which to study mitochondrial diseases, evolution of mitochondria, and the Tree of Life. Thus, the nascent field of "mitogenomics" is expected to continue to grow and become a major area of inquiry in phylogenomics.

1.5 SEQUENCING WORKFLOWS TO GENERATE PHYLOGENOMIC DATA

At the present time there are two major workflows for generating phylogenomic datasets: *Sanger sequencing* and NGS. Although numerous methodological differences exist between these workflows, both utilize the same basic three steps: (1) "DNA extraction," which is the isolation and purification of genomic DNA from tissues; (2) mass-copying or "amplification" of *target DNA templates* for each locus; and (3) sequencing these target templates using a DNA sequencing platform. We will now briefly look at the differences between Sanger and NGS workflows.

1.5.1 Sanger Sequencing Workflow

The Sanger method for DNA sequencing, named in honor of the English biochemist Frederick Sanger, has been the primary method for sequencing DNA over the past three decades. Sanger and his coworkers published a groundbreaking paper (Sanger et al. 1977) that described a new method for sequencing DNA. For this accomplishment, Sanger was awarded his second Nobel Prize in Chemistry and the method has since gone on to revolutionize biology and medicine.

The three main steps of the Sanger sequencing workflow are (1) isolation and purification of genomic DNA; (2) amplification of target templates using the *polymerase chain reaction* or "PCR"; and (3) sequencing of the PCR products using fluorescently labeled chain-terminating dideoxynucleotides in an automated sequencing machine. In Chapter 4, we will learn how DNA is isolated and purified before we take a detailed look at PCR and Sanger methodologies in Chapters 5 and 6, respectively.

Through the years, various improvements to the Sanger workflow have enhanced its high-throughput capabilities while lowering the cost per sequence. "Throughput" simply refers to the number of DNA sequences that are essentially generated at the same time. These advances made it easier and cheaper for researchers to generate phylogenomic datasets of modest sizes (~1–10 loci). However, the ability of researchers to generate larger phylogenomic datasets such as hundreds of loci or entire genome sequences, is still beyond the reach of most laboratories that rely solely on Sanger sequencing mainly because of cost limitations (i.e., cost of individual NGS experiments is relatively expensive compared to Sanger sequencing).

1.5.2 NGS Workflow

During the mid- to late 2000s several different next-generation sequencing or "NGS" methods and sequencing machines appeared on the genomics scene (Mardis 2008; Shendure and Ji 2008), which set the stage for a dramatic change in the nature of phylogenomics data acquisition. In September of 2005, two papers unveiled the first two NGS methods each of which presented powerful new methods for sequencing genomes: the "polony sequencing" method of Shendure et al. (2005) and the "pyrosequencing" method of Margulies et al. (2005). The sequencing technology presented by Margulies et al. (2005) was incorporated into the first commercial NGS platform called the "Roche/454" (Mardis 2008). Illumina introduced another type of NGS platform a few years later, which was based on a "sequencing by synthesis" technology (Mardis 2008). Bentley et al. (2008) provided an early demonstration of the genome-sequencing ability of the Illumina platform. A fourth NGS method, which used a "sequencing by ligation" approach (McKernan et al. 2009), was incorporated into a sequencing platform called the "SOLiD" (Mardis 2008).

These NGS platforms are often referred to as *second-generation* sequencing platforms while Sanger sequencing is considered a *first-generation* platform (Glenn 2011; Niedringhaus et al. 2011; Hui 2014). However, as is the typical theme in evolution whether it is organismal- or consumer product-based, one to a few of the many novel forms survives the initial diversification event and thrives thereafter. Indeed, from among the flurry of early NGS platforms, the Illumina platform has emerged to be, by far, the dominant NGS platform used in phylogenomics (Videvall 2016). Newer NGS technologies are appearing on the horizon any one of which might become the future dominant sequencing platform. These include the *third-generation* NGS platforms such as the Ion Torrent by (Life Sciences), PacBio (Pacific Biosciences), and Complete Genomics while a so-called *fourth-generation* NGS platform by Oxford Nanopore is in development (Glenn 2011; Niedringhaus et al. 2011; Hui 2014). Given Illumina's current dominance in the NGS market, in Chapter 7 we will focus on how this platform is being used to generate phylogenomic datasets.

The Illumina platforms have provided a substantial boost to phylogenomic studies in a number of ways. Although this technology was originally developed for sequencing entire genomes in a more efficient manner than Sanger sequencing, researchers have devised various methods for isolating target templates for large numbers of loci, which could then be sequenced. Thus, instead of using PCR to directly generate the intermediary products for a small number of loci (~1–10), the researcher prepares a *sequencing library* that represents a targeted portion of a genome, which can potentially contain the templates for dozens, hundreds, or thousands of loci from an individual's genome. Researchers can further dramatically increase the throughput of single Illumina runs by pooling multiple libraries together for sequencing. This kind of brute sequencing power is producing phylogenomic datasets that are orders of magnitude larger than those generated using Sanger sequencing (e.g., Faircloth et al. 2012; Jarvis et al. 2014; Prum et al. 2015).

NGS technology has impacted phylogenomic data acquisition in at least two other important ways. First, researchers can now generate partial or complete genome sequence datasets for "non-model organisms" far faster and cheaper than using the traditional shotgun cloning-Sanger workflow—the approach that was used to sequence the human genome (Venter et al. 2001). This easier access to substantial amounts of genome data means that researchers can rapidly and easily design new phylogenomic loci using bioinformatics methods rather than having to resort to laborious DNA cloning methods (Thomson et al. 2010). In Chapter 8, we will learn more about developing loci for use in either the Sanger or NGS workflows.

1.5.3 Is Sanger Sequencing Still Relevant in Phylogenomics?

Despite the recent surge in popularity of NGS, Sanger sequencing is still the most widely used method today. Still, as the hurdles to acquiring phylogenomic data using NGS become fewer, at some point in upcoming years some form of NGS technology will inevitably replace the Sanger method as the primary sequencing method used by phylogenomics researchers. Given this eventuality, it is fair to ask the following question: *Is Sanger sequencing still relevant in phylogenomics?*

In the short term, say for at least another 5–10 years, the answer to this question is clearly "yes." Despite the dramatically increased sequencing power offered by NGS, the Sanger method still retains some advantages over NGS such as cheaper costs of data collection for small-scale projects, longer sequence read lengths, and simple computer analysis of raw sequence data. The high cost per run for an NGS sequencer, which generally runs in the thousands of dollars, means that Sanger sequencing still has an important role for smaller projects or, at the very least, for filling in sequence "gaps" caused by incomplete coverage in NGS sequences. Another current difficulty with using NGS to generate phylogenomic data is that the construction of some types of sequencing libraries requires considerable molecular biology skills and specialized equipment not normally found in a Sanger-equipped laboratory. Further, the excitement of these new NGS-based methods for acquiring phylogenomic data can quickly subside for those who are not accustomed to performing bioinformatics analysis of NGS data. Once obtained, NGS data poses a number of terrific challenges owing to difficulties arising from storage, manipulation, and filtering of raw sequence files. Not only that, but such data manipulation must be done using UNIX-based "super" computers that can handle *terabytes* of genomic data (McCormack et al. 2013). At the present time only *one* run of an Illumina sequencer can generate more than a one-half terabyte of sequence data! In contrast, even the highest throughput of Sanger sequence data cannot overwhelm the most basic laptop computer using a Mac or Windows operating system. Thus, while NGS offers tremendous promise for the future, it does not represent a practical sequencing solution for all researchers and projects at the present time.

Even in the longer term after NGS becomes the standard method of phylogenomics data acquisition, there will be good reasons for all phylogenomics researchers to have a solid understanding of Sanger sequencing. First, genomic and nucleotide databases such as GenBank already contain an enormous volume of DNA sequence data generated via this technology. Therefore, researchers who mine sequence databases for bioinformatics purposes should have some knowledge about how those Sanger sequences were generated and understand their characteristics and limitations, which we will review in Chapter 6. Given these considerations, Sanger sequencing will likely remain the primary method for obtaining DNA sequence data by the majority of small labs and perhaps many of the larger labs until the various pre- and postdata NGS hurdles are reduced or eliminated.

Regardless of when NGS relegates Sanger sequencing to a minor role in phylogenomics, this issue of "obsoleteness" touches upon a broader phenomenon in molecular biology that we will now consider. When one looks more closely at the history of molecular biology, one sees the pattern of new technologies (i.e., lab methods and reagents) helping to spur scientific revolutions only to be rendered obsolete some years later as they become replaced by newer higher-performing technologies. What is most interesting about this fact is that these supposedly obsolete methods or reagents are often later co-opted to play essential new roles in these "cutting-edge" methods. My favorite example of this is the Sanger method for sequencing DNA: Sanger et al. (1977) simply combined pre-existing reagents and methods into a novel methodology for determining the sequence of DNA molecules. Another example is the enzyme DNA ligase, which was independently discovered by five groups of researchers in 1967 (Lehman 1974) and subsequently became a key component in recombinant DNA technology throughout the 1970s and 1980s. Once PCR largely replaced molecular cloning as a means for generating target DNA templates for Sanger sequencing in the late 1980s (except for shotgun sequencing of genomic libraries), ligase was no longer relevant for routine DNA sequencing, correct? The answer is no! The SOLiD (Mardis 2008) NGS platform, works by a *sequencing-by-ligation* principle (McKernan et al. 2009), *and*, the method of ligating DNA molecules together using DNA ligase is an essential step in the construction of many types of NGS sequencing libraries! A third example comes from PCR. As the use of Sanger sequencing workflow declines, does this mean that PCR and the design of PCR primers will be less important in the era of NGS? Again, the answer is a resounding no! Achieving a thorough knowledge about PCR is not only essential for successful Sanger sequencing, but PCR has also found important new niches in many NGS-based methods as we will see in Chapter 7. This phenomenon of rediscovering methods is a recurring theme in molecular biology. The important message here is that it is worthwhile for you to learn all of these methods because future cutting-edge technologies may co-opt existing technologies to produce important new methods for acquiring phylogenomic data.

1.6 THE PHYLOGENOMICS LABORATORY

We will close this first chapter with a brief discussion about what the modern phylogenomics laboratory looks like and see, perhaps surprisingly, how easily they can be set up. Large universities, natural history museums, and other research institutes typically have spacious and expensive molecular genetic laboratories full of various machines (e.g., automated DNA sequencers), equipment, and lab technicians where phylogenomics projects are undertaken. Although historically these sophisticated well-funded facilities were a prerequisite for conducting technically challenging and costly projects that involved gene cloning, high-throughput Sanger sequencing, or NGS-based sequencing, today a small-budget lab can be established to handle such genome-scale projects. Speaking from my own experiences, even a dirty old small storage room can be converted into a clean, shiny, and high-throughput phylogenomics laboratory. All one needs is a small workspace and the minimal equipment. How is this possible?

Recall from the previous section that the process of acquiring phylogenomic data can be divided into three basic steps:

DNA extraction → template acquisition → DNA sequencing

Although DNA extraction is a method common to all labs, methods for acquiring DNA templates for sequencing vary. Laboratories only equipped for PCR and Sanger sequencing (Chapters 5 and 6) are easy to set up and can be done so at low cost. Figure 1.3 shows a layout of what a minimalistic phylogenomics lab would resemble. If space and funding allow, additional pieces of equipment to the lab can further enhance its capabilities. For example, adding a second thermocycler and electrophoresis box (to be used with the same voltage box) can increase the lab's throughput. If a second electrophoresis apparatus is placed within the pre-PCR area apparatus (with dedicated single-channel and multichannel pipettes), then purified DNA extracts can be evaluated within a "clean" part of the lab, which would further reduce the risk of contamination. Adding a *Qubit* fluorometer (Life Technologies), *BioAnalyzer* (Agilent Technologies), and qPCR machine would enable researchers to construct NGS libraries. Finally, a −80°F or "ultra-cold" freezer can be used for long-term storage of genetic samples (e.g., tissues and DNA extracts *but not PCR products due to contamination concerns*) and reagents (e.g., concentrated primer stocks, enzymes, etc.). We will discuss strategies for minimizing contamination risks (Chapters 4 and 5) and maximizing the sample throughput capability of labs (Chapters 4 through 6).

You may have noticed the absence of an automated sequencing machine as part of the equipment repertoire in Figure 1.3. This is because of a fantastic innovation—*outsourcing* of DNA sequencing to another laboratory. Once the sequencing templates—regardless whether they are PCR products or NGS libraries—are prepared, they can be outsourced to another laboratory that performs the sequencing using their Sanger or NGS machines. The advent of outsourcing of DNA sequencing obviates the need for all laboratories to have their own expensive sequencing machines and technicians to run them, which effectively reduces the costs of setting up a phylogenomics lab from hundreds of thousands (or more) dollars down to tens of thousands of dollars. In addition to the cost savings, this outsourcing option represents a great equalizer because the *per capita* output of high-quality research projects from a small

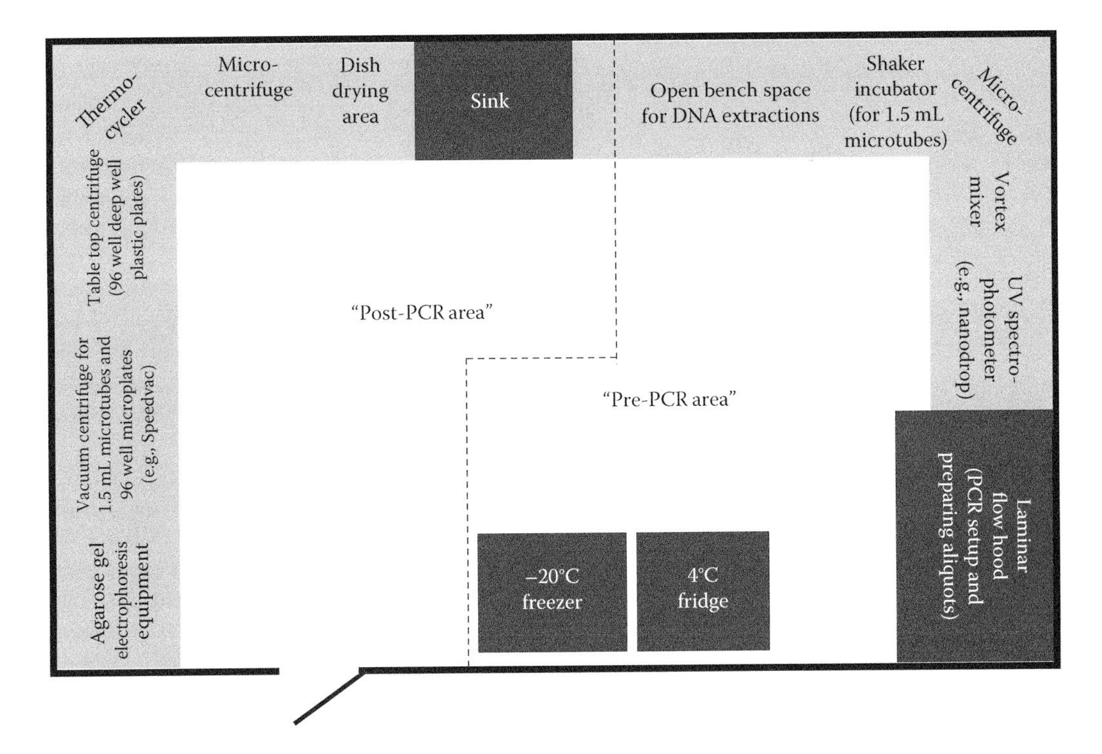

Figure 1.3. Diagram showing layout of a basic phylogenomics lab. This illustration is meant to give an idea of how a small room can be converted into a DNA extraction and PCR lab. The lab is divided into "pre-PCR" and "post-PCR" areas for purposes of minimizing the risk of contamination. Additional equipment can be included to further increase the capabilities of the lab. Shelves (not shown) for storing sterile plastics (kits, microtubes, microplates, pipette tips, etc.) can be placed on the walls above the work benches.

lab can now be more comparable with the larger well-established laboratories found in universities and research institutes.

REFERENCES

Amaral, F. R., L. G. Neves, M. F. Resende Jr et al. 2015. Ultraconserved elements sequencing as a low-cost source of complete mitochondrial genomes and microsatellite markers in non-model amniotes. *PloS One* 10:e0138446.

Avise, J. C. 2000. *Phylogeography: The History and Formation of Species*. Cambridge: Harvard University Press.

Bentley, D. R., S. Balasubramanian, H. P. Swerdlow et al. 2008. Accurate whole human genome sequencing using reversible terminator chemistry. *Nature* 456:53–59.

Chen, F. C. and W.-H. Li. 2001. Genomic divergences between humans and other hominoids and the effective population size of the common ancestor of humans and chimpanzees. *Am J Hum Genet* 68:444–456.

Delsuc, F., H. Brinkmann, and H. Philippe. 2005. Phylogenomics and the reconstruction of the tree of life. *Nat Rev Genet* 6:361–375.

Deschavanne, P. J., A. Giron, J. Vilain, G. Fagot, and B. Fertil. 1999. Genomic signature: Characterization and classification of species assessed by chaos game representation of sequences. *Mol Biol Evol* 16:1391–1399.

Dobzhansky, T. 1973. Nothing in biology makes sense except in the light of evolution. *Am Biol Teach* 35:125–129.

Edwards, S. V. 2009. Is a new and general theory of molecular systematics emerging? *Evolution* 63:1–19.

Edwards, S. V., B. Fertil, A. Giron, and P. J. Deschavanne. 2002. A genomic schism in birds revealed by phylogenetic analysis of DNA strings. *Syst Biol* 51:599–613.

Edwards, S. V., W. B. Jennings, and A. M. Shedlock. 2005. Phylogenetics of modern birds in the era of genomics. *Proc R Soc Lond B: Biol Sci* 272:979–992.

Eisen, J. A. 1998a. Phylogenomics: Improving functional predictions for uncharacterized genes by evolutionary analysis. *Genome Res* 8:163–167.

Eisen, J. A. 1998b. A phylogenomic study of the MutS family of proteins. *Nucleic Acids Res* 26:4291–4300.

Eisen J. A. and C. M. Fraser. 2003. Phylogenomics: Intersection of evolution and genomics. *Science* 300:1706–1707.

Eisen, J. A., K. S. Sweder, and P. C. Hanawalt. 1995. Evolution of the SNF2 family of proteins: Subfamilies with distinct sequences and functions. *Nucleic Acids Res* 23:2715–2723.

Faircloth, B. C., J. E. McCormack, N. G. Crawford, M. G. Harvey, R. T. Brumfield, and T. C. Glenn. 2012. Ultraconserved elements anchor thousands of genetic markers spanning multiple evolutionary timescales. *Syst Biol* 61:717–726.

Felsenstein, J. 2004. *Inferring Phylogenies*. Sunderland: Sinauer.

Glenn, T. C. 2011. Field guide to next-generation DNA sequencers. *Mol Ecol Res* 11:759–769.

Guarnizo, C. E., A. Paz, A. Muñoz-Ortiz, S. V. Flechas, J. Méndez-Narváez, and A. J. Crawford. 2015. DNA barcoding survey of anurans across the eastern cordillera of Colombia and the impact of the Andes on cryptic diversity. *PLoS One* 10:e0127312.

Haig, D. 1999. A brief history of human autosomes. *Philos Trans R Soc Lond B: Biol Sci* 354:1447–1470.

Hebert, P. D. N., A. Cywinska, S. L. Ball, and J. R. deWaard. 2003. Biological identifications through DNA barcodes. *Proc R Soc Lond B: Biol Sci* 270:313–321.

Hey, J. 2010. Isolation with migration models for more than two populations. *Mol Biol Evol* 27:905–920.

Hey, J. and R. Nielsen. 2004. Multilocus methods for estimating population sizes, migration rates and divergence time, with applications to the divergence of *Drosophila pseudoobscura* and *D. persimilis*. *Genetics* 167:747–760.

Hillis, D. M., B. K. Mable, and C. Moritz. 1996. *Molecular Systematics*, 2nd edition. Sunderland: Sinauer.

Hinchliff, C. E., S. A. Smith, J. F. Allman et al. 2015. Synthesis of phylogeny and taxonomy into a comprehensive tree of life. *Proc Natl Acad Sci USA* 112:12764–12769.

Hudson, R. R. 1983. Testing the constant-rate neutral allele model with protein sequence data. *Evolution* 37:203–217.

Hudson, R. R. 1992. Gene trees, species trees, and the segregation of ancestral alleles. *Genetics* 131:509–513.

Hui, P. 2014. Next generation sequencing: Chemistry, technology and applications. *Top Curr Chem* 336:1–18.

Jarvis, E. D., S. Mirarab, A. J. Aberer et al. 2014. Whole-genome analyses resolve early branches in the tree of life of modern birds. *Science* 346:1320–1331.

Jennings, W. B. and S. V. Edwards. 2005. Speciational history of Australian Grass Finches (*Poephila*) inferred from thirty gene trees. *Evolution* 59:2033–2047.

Jennings, W. B., H. Wogel, M. Bilate, R. O. L. Salles, and P. A. Buckup. 2016. DNA barcoding reveals species level divergence between populations of the microhylid frog genus *Arcovomer* (Anura: Microhylidae) in the Atlantic Rainforest of southeastern Brazil. *Mitochondrial DNA* Part A 27:3415–3422.

Karlin, S. and C. Burge. 1995. Dinucleotide relative abundance extremes: A genomic signature. *Trends Genet* 11:283–290.

Knowles, L. L. and L. S. Kubatko, eds. 2010. *Estimating Species Trees: Practical and Theoretical Aspects*. Hoboken: John Wiley and Sons.

Lehman, I. R. 1974. DNA ligase: Structure, mechanism, and function. *Science* 186:790–797.

Lemey, P., M. Salemi, and A. Vandamme, eds. 2009. *The Phylogenetic Handbook: A Practical Approach to Phylogenetic Analysis and Hypothesis Testing*, 2nd edition. Cambridge: Cambridge University Press.

Maddison, D. R., K. S. Schulz, and W. P. Maddison. 2007. The tree of life web project. *Zootaxa* 1668:19–40.

Maddison, W. P. 1995. Phylogenetic histories within and among species. In *Experimental and Molecular Approaches to Plant Biosystematics. Monographs in Systematic Botany* eds. P. C. Hoch, and A. G. Stephenson, 53:273–287. St. Louis: Missouri Botanical Garden.

Maddison, W. P. 1997. Gene trees in species trees. *Syst Biol* 46:523–536.

Mardis, E. R. 2008. Next-generation DNA sequencing methods. *Annu Rev Genomics Hum Genet* 9:387–402.

Margulies, M., M. Egholm, W. E. Altman et al. 2005. Genome sequencing in microfabricated high-density picolitre reactors. *Nature* 437:376–380.

McCormack, J. E., S. M. Hird, A. J. Zellmer, B. C. Carstens, and R. T. Brumfield. 2013. Applications of next-generation sequencing to phylogeography and phylogenetics. *Mol Phylogenet Evol* 66:526–538.

McKernan, K. J., H. E. Peckham, G. L. Costa et al. 2009. Sequence and structural variation in a human genome uncovered by short-read, massively parallel ligation sequencing using two-base encoding. *Genome Res* 19:1527–1541.

Murphy, W. J. 2008. *Phylogenomics: Methods in Molecular Biology*. New York: Humana Press.

Niedringhaus, T. P., D. Milanova, M. B. Kerby, M. P. Snyder, and A. E. Barron. 2011. Landscape of next-generation sequencing technologies. *Anal Chem* 83:4327–4341.

O'Brien, S. J. and R. Stanyon. 1999. Phylogenomics: Ancestral primate viewed. *Nature* 402:365–366.

Okada, N., A. M. Shedlock, and M. Nikaido. 2004. Retroposon mapping in molecular systematics. In *Mobile Genetic Elements: Protocols and Genomic Applications*, ed. P. Capy, 189–226. New York: Humana Press.

Philippe, H. and M. Blanchette. 2007. Overview of the first phylogenomics conference. *BMC Evol Biol* 7:1.

Philippe, H., F. Delsuc, H. Brinkmann, and N. Lartillot. 2005. Phylogenomics. *Annu Rev Ecol Evol Syst* 541–562.

Prum, R. O., J. S. Berv, A. Dornburg et al. 2015. A comprehensive phylogeny of birds (Aves) using targeted next-generation DNA sequencing. *Nature* 526:569.

Rokas, A., B. L. Williams, N. King, and S. B. Carroll. 2003. Genome-scale approaches to resolving incongruence in molecular phylogenies. *Nature* 425:798–804.

Rosenberg, N. A. 2002. The probability of topological concordance of gene trees and species trees. *Theor Popul Biol* 61:225–247.

Sanger, F., S. Nicklen, and A. R. Coulson. 1977. DNA sequencing with chain-terminating inhibitors. *Proc Natl Acad Sci USA* 74:5463–5467.

Shedlock, A. M. and N. Okada. 2000. SINE insertions: Powerful tools for molecular systematics. *Bioessays* 22:148–160.

Shedlock, A. M., K. Takahashi, and N. Okada. 2004. SINEs of speciation: Tracking lineages with retroposons. *Trends Ecol Evol* 19:545–553.

Shendure, J. and H. Ji. 2008. Next-generation DNA sequencing. *Nat Biotechnol* 26:1135–1145.

Shendure, J., G. J. Porreca, N. B. Reppas et al. 2005. Accurate multiplex polony sequencing of an evolved bacterial genome. *Science* 309:1728–1732.

Souto, H. M., P. A. Ruschi, C. Furtado, W. B. Jennings, and F. Prosdocimi. 2014. The complete mitochondrial genome of the ruby-topaz hummingbird *Chrysolampis mosquitus* through Illumina sequencing. *Mitochondrial DNA* 27:769–770.

Tajima, F. 1983. Evolutionary relationships of DNA sequences in finite populations. *Genetics* 105:437–460.

Thomson, R. C., I. J. Wang, and J. R. Johnson. 2010. Genome-enabled development of DNA markers for ecology, evolution and conservation. *Mol Ecol* 19:2184–2195.

Venter, J. C., M. D. Adams, E. Myers et al. 2001. The sequence of the human genome. *Science* 291:1304–1351.

Videvall, E. 2016. Results of the molecular ecologist's survey on high-throughput sequencing blog. *The Molecular Ecologist* blog. http://www.molecularecologist.com/2016/04/results-of-the-molecular-ecologists-survey-on-high-throughput-sequencing/ (accessed April 11, 2016).

Wakeley, J. 2009. *Coalescent Theory: An Introduction* (Vol. 1). Greenwood Village: Roberts & Company Publishers.

Yang, Z. 2002. Likelihood and Bayes estimation of ancestral population sizes in hominoids using data from multiple loci. *Genetics* 162:1811–1823.

CHAPTER TWO

Properties of DNA Sequence Loci: Part I

Genomes harbor a plethora of different loci types that can be used as evolutionary markers in phylogenomic studies. In a research project that aims to elucidate the evolutionary history of particular gene families, there is little or no ambiguity about which locus or loci will be the target(s) for DNA sequence acquisition. However, in organismal phylogenomic studies the researcher must choose the types and numbers of loci to use—a decision that may, if poorly made, adversely affect the quality of phylogenomic inferences. Indeed, some phylogenomic analyses require DNA sequence datasets to meet certain evolutionary assumptions and therefore it is essential to carefully select the type and quantity of data to acquire.

Ideally, locus or loci choice will be based on a researcher's sound knowledge about the properties of each locus type with an eye on the anticipated analyses and associated assumptions. What are these properties of DNA sequence loci? The characteristics of a locus can be influenced or determined by a number of factors. For example, a locus may contain sites that are highly conserved in an evolutionary sense or they may not be conserved at all. Also mutation rates may be constant for all sites within one type of locus or vary among sites in another locus. Thus patterns and processes of mutation may not only vary between loci but also within them as well. Loci may also differ from each other depending on whether they exist as an entity in the nuclear genome or are found in an organellar genome such as the mitochondrial genome. As we will see in this chapter, loci in the former type of genome evolve quite differently from loci in the latter and thus these differences must be taken into account in phylogenomic studies.

In order to better understand these and other loci properties regardless whether a study is focused on molecular/genomic evolution or organismal evolution, a researcher should be knowledgeable about the characteristics of genomes (e.g., size and composition) as well as how DNA sequences evolve over time. In this chapter, we will begin by reviewing aspects of organismal genomes across the tree of life with a focus on eukaryotes followed by a discussion about key aspects concerning the molecular evolution of DNA sequences. Included in this discussion will be a brief review on the biology of so-called repetitive DNA, an important class of genomic DNA comprised of simple repeat DNA (e.g., microsatellites), transposable elements, and various types of pseudogenes including mitochondrial pseudogenes or "numts." This chapter is important because it will prepare us for Chapter 3, the section of this book that considers the many standard assumptions of phylogenomic analyses and describes the major classes of DNA sequence-based loci used in tree of life studies. Prepared with this knowledge, the researcher can make informed decisions about which loci will be most appropriate for a given study and provide important context for phylogenomic analyses of those data.

2.1 GENOMIC BACKGROUND

2.1.1 Genome Types and Sizes

The three main types of genomes include *prokaryotic genomes*, *eukaryotic nuclear genomes*, and *eukaryotic organellar genomes*. Organellar genomes consist of two types—*mitochondrial genomes* and *chloroplast genomes*. Mitochondrial and chloroplast genomes reside within their namesake organelles in the cytoplasm and thus they are physically separated from the nuclear genome. Mitochondrial genomes are found in all different types of eukaryotes including protists, green alga, fungi, yeasts, animals, and land plants, but only photosynthetic eukaryotes also have chloroplast genomes (Brown 2007).

The genomes of prokaryotes and organellar genomes usually exist as single circular double-stranded DNA molecules though some prokaryotes have genomes subdivided into multiple linear chromosome-like molecules (Brown 2007). In contrast, all eukaryotes have their nuclear genomes apportioned among a number of linear chromosomes (Brown 2007). Interestingly, other similarities exist between prokaryotic and organellar genomes such as both having similar gene expression patterns and gene sequences (Brown 2007). The similarities between prokaryotic and eukaryotic organellar genomes may have an evolutionary basis according to the *endosymbiont theory*, which was developed by Lynn Margulis in the late 1960s. According to this theory, eukaryotic organellar genomes are descendants from free-living bacteria, which, long ago, not only lived inside primitive eukaryotic cells, but also had symbiotic relationships with their host cells (Sagan 1967; Margulis 1970).

Genome sizes are expressed in units of kilobases (kb), megabases (Mb), or gigabases (Gb), which are the equivalents to thousands, millions, and billions of bases, respectively. Prokaryotic genomes vary between 0.49 and 30 Mb (Table 2.1) with most being less than 5 Mb (Brown 2007). Organellar genomes range from about 6 kb to more than 11,000 kb (Table 2.2). A great amount of variation can be seen in the sizes of mitochondrial genomes. The protozoan that causes malaria (*Plasmodium falciparum*) has one of the smallest known mitochondrial genomes at 6 kb in size (Table 2.2). At the other end of the size spectrum, catchfly plants in the genus *Silene* have enormous mitochondrial genomes with one species attaining a size of 11,319 kb, which is >600 times larger than the human mitochondrial genome (Table 2.2)! The chloroplast genomes of photosynthesizing eukaryotes range in size from 120 kb for pea plants (*Pisum sativum*) to 195 kb in green alga (*Chlamydomonas reinhardtii*; Table 2.2).

From Table 2.2 it is obvious that the sizes of mitochondrial genomes are not closely related to organismal complexity, as some "simple" eukaryotes such as yeast and fungi have much larger mitochondrial genomes than "more complex" invertebrates or

TABLE 2.1
Sizes of some prokaryotic genomes

	Size of genome (Mb)
Archaea	
Nanoarchaeum equitans	0.49
Methanococcus jannaschii	1.66
Archaeoglobus fulgidus	2.18
Bacteria	
Mycoplasma genitalium	0.58
Streptococcus pneumoniae	2.16
Vibrio cholerae El Tor N16961	4.03
Mycobacterium tuberculosis H37Rv	4.41
Escherichia coli K12	4.64
Yersinia pestis CO92	4.65
Pseudomonas aeruginosa PA01	6.26
Bacillus megaterium	30

SOURCE: Data from Table 8.3, Page 234, in Brown, T. A. 2007. *Genomes 3*. New York: Garland Science/Taylor & Francis. With permission.

vertebrates but they are smaller than those found in flowering plants. Surprisingly, the mitochondrial genomes of *Chlamydomonas reinhardtii* (a green alga) and *Homo sapiens* are nearly the same size (Table 2.2). Even similar organisms such as the two protozoans in Table 2.2—*Plasmodium falciparum* and *Reclinomonas americana*—have mitochondrial genomes that differ from each by a factor of ten.

Eukaryotic nuclear genomes range from a minimum size of 12 Mb for the yeast (*Saccharomyces cerevisiae*) genome to the enormous genomes of *Fritillaria* lilies, some of which attain sizes up to 120,000–127,000 Mb (Table 2.3; Brown 2007; Ambrožová et al. 2010). To put this in perspective, the *Fritillaria* lily genomes are 40 times larger than the human genome and 10,000 times larger than the yeast genome! Despite the existence of NGS for more than a decade now, the complete sequences of these largest genomes have still not been fully sequenced. As of September 2015, the species with the largest complete genome sequence in Genbank is the White Spruce (*Picea glauca*), which has a genome size of 26.9 Gb (Table 2.3; Birol et al. 2013). The sizes of eukaryotic genomes are roughly correlated with organismal complexity, as, for example, yeast, fungi, and protists have the smallest genomes while vertebrates and plants have the largest genomes (Table 2.3; Brown 2007; Watson et al. 2014).

TABLE 2.2
Sizes of mitochondrial and chloroplast genomes

	Type of organism	Genome size (kb)
	Mitochondrial genomes	
Plasmodium falciparum	Protozoan (malaria parasite)	6
Chlamydomonas reinhardtii	Green alga	16
Mus musculus	Vertebrate (mouse)	16
Homo sapiens	Vertebrate (human)	17
Metridium senile	Invertebrate (sea anenome)	17
Drosophila melanogaster	Invertebrate (fruit fly)	19
Chondrus crispus	Red alga	26
Aspergillus nidulans	Ascomycete fungus	33
Reclinomonas americana	Protozoa	69
Saccharomyces cerevisiae	Yeast	75
Suillus grisellus	Basidiomycete fungus	121
Viscum scurruloideum	Flowering plant (mistletoe)	66
Brassica oleracea	Flowering plant (cabbage)	160
Arabidopsis thaliana	Flowering plant (vetch)	367
Zea mays	Flowering plant (maize)	570
Cucumis melo	Flowering plant (melon)	2,500
Silene noctiflora	Flowering plant (catchfly)	6,728
Silene conica	Flowering plant (catchfly)	11,319
	Chloroplast genomes	
Pisum sativum	Flowering plant (pea)	120
Marchantia polymorpha	Liverwort	121
Oryza sativa	Flowering plant (rice)	136
Nicotiana tabacum	Flowering plant (tobacco)	156
Chlamydomonas reinhardtii	Green alga	195

SOURCE: Data from Table 8.5, Page 240, in Brown, T. A. 2007. *Genomes 3*. New York: Garland Science/Taylor & Francis, with permission, except for *Viscum scurruloideum*, *Silene noctiflora*, and *S. conica*, which were obtained from Figure 2 in Skippington, E. et al. 2015. *Proc Natl Acad Sci USA* 112:E3515–E3524.

2.1.2 Composition of Eukaryotic Organellar Genomes

As of October 2015, Genbank contained the complete sequences for over 6,000 mitochondrial and 700 chloroplast genomes. These data are providing us with a clearer picture about variation in the composition of these organellar genomes. There is also at least one peer-reviewed journal (i.e., *Mitochondrial DNA*) that is dedicated to publishing papers focused on the biology of mitochondrial genomes.

Tremendous variation in the composition of mitochondrial genomes exists among eukaryotes. The numbers of genes (RNA- and protein-coding) varies from a low of five in *Plasmodium falciparum* up to at least 92 genes in *Reclinomonas americana* (Table 2.4). Notice that these two protozoans bracket all other eukaryotes in terms of their gene numbers listed in Table 2.4. Thus, the number of genes found in mitochondrial genomes does not relate to the complexity of an organism. Mitochondrial genomes include genes that code for proteins involved in the respiratory complex, ribosomal RNAs, transfer RNAs, and a control region (Table 2.4; Randi 2000). In contrast to nuclear genomes, mitochondrial genomes generally contain few or no introns (Table 2.4). One of the two strands of the mitochondrial genome has a higher G + T

TABLE 2.3

Sizes of various eukaryotic genomes

Species	Type of organism	Genome size (Mb)
Human (*Homo sapiens*)	Vertebrate	3,260
Mouse (*Mus musculus*)	Vertebrate	2,804
Anolis lizard (*Anolis carolinensis*)	Vertebrate	1,799
Chicken (*Gallus gallus*)	Vertebrate	1,047
Zebrafinch (*Taeniopygia guttata*)	Vertebrate	1,232
Zebrafish (*Danio rerio*)	Vertebrate	1,412
Pufferfish (*Takifugu rubripes*)	Vertebrate	392
Fruit fly (*Drosophila melanogaster*)	Invertebrate	164
Nematode worm (*Caenorhabditis elegans*)	Invertebrate	100
Thale cress (*Arabidopsis thaliana*)	Land plant	127
Black cottonwood (*Populus trichocarpa*)	Land plant	417
Maize (*Zea mays*)	Land plant	2,068
White spruce (*Picea glauca*)	Land plant	26,936
Fritillary lily (*Fritillaria assyriaca*)	Land plant	120,000
Yeast (*Saccharomyces cerevisiae*)	Yeast	12
Leishmaniasis parasite (*Leishmania major*)	Protist	33
Malaria parasite (*Plasmodium falciparum*)	Protist	27

SOURCE: Data obtained from Genbank except for the Fritillary lily, which was obtained from Table 7.2, Page 208, Brown, T. A. 2007. *Genomes* 3. New York: Garland Science/Taylor & Francis.

TABLE 2.4

Composition of mitochondrial genomes

Feature	*Plasmodium falciparum*	*Chlamydomonas reinhardtii*	*Homo sapiens*	*Saccharomyces cerevisiae*	*Arabidopsis thaliana*	*Reclinomonas americana*
Total number of genes	5	12	37	35	52	92
Types of genes						
Protein-coding genes	3	7	13	8	27	62
Respiratory complex	3	7	13	7	17	24
Ribosomal proteins	0	0	0	1	7	27
Transport proteins	0	0	0	0	3	6
RNA polymerase	0	0	0	0	0	4
Translation factor	0	0	0	0	0	1
Functional RNA genes	2	5	24	27	25	30
Ribosomal RNA genes	2	2	2	2	3	3
Transfer RNA genes	0	3	22	24	22	26
Other RNA genes	0	0	0	1	0	1
Number of introns	0	1	0	8	23	1
Genome size (kb)	6	16	17	75	367	69

SOURCE: Reproduced from Table 8.6, Page 241, in Brown, T. A. 2007. *Genomes* 3. New York: Garland Science/Taylor & Francis. With permission.

content than its complementary strand. Thus, the high G + T strand is referred to as the "heavy" or "H-strand," while the low G + T strand is called "light" or "L-strand" (Randi 2000).

The mitochondrial genomes of metazoans are relatively small (<21 kb) and thus many species in this group have had their mitochondrial genomes fully sequenced and annotated (see Boore 1999). With some exceptions (e.g., jellyfish), the composition of the metazoan mitochondrial genome is also remarkably stable. A typical metazoan mitochondrial genome is illustrated by the one found in the hummingbird *Chrysolampis mosquitus* shown in Figure 2.1. The gene composition in this mitochondrial genome includes: 13 protein-coding genes, 2 ribosomal RNA (rRNA) genes, and 22 transfer RNA (tRNA) genes (Figure 2.1). Additionally, metazoan mitochondrial genomes have a noncoding sequence called the *control region*. In vertebrates, this control region is called the "D-loop," while in invertebrates it is called the "AT-rich region" (Palumbi 1996; Randi 2000).

In addition to (usually) being circular, containing few genes and introns, there are other ways that mitochondrial genomes differ from their nuclear counterparts. While each cell only has one to several copies of a nuclear genome present in a nucleus depending on the ploidy level of the species, the same cells have an enormous number of mitochondrial genome copies. For example, human cells contain around 8,000 copies of the mitochondrial genome per cell (Pakendorf and Stoneking 2005). The number of mitochondrial genome copies per cell varies from 1,000 to 10,000 (Brown

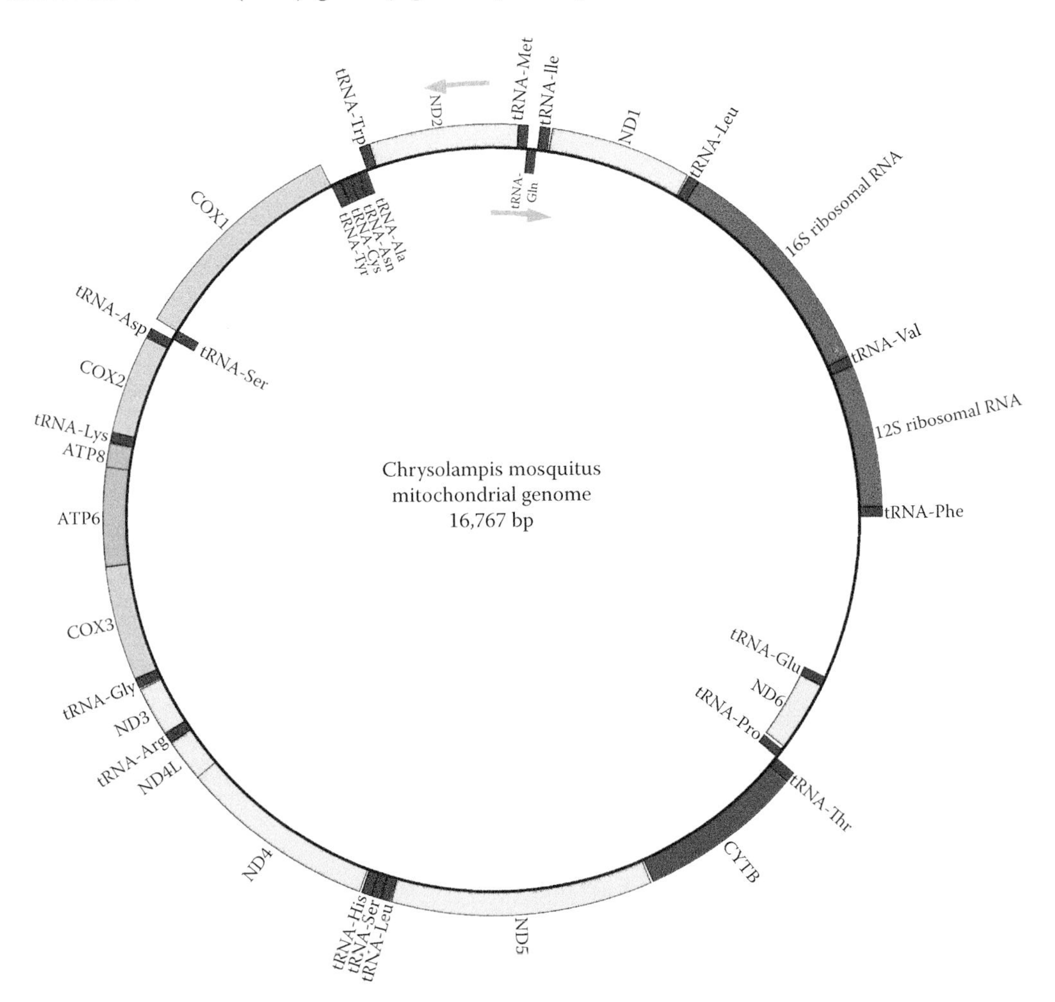

Figure 2.1. Complete mitochondrial genome for the hummingbird *Chrysolampis mosquitus* (Genbank # KJ619585; Souto, H. M. et al. 2014. *Mitochondrial DNA* 27:769–770) Figure was made using the software *OGdraw*. (From Lohse, M. et al. 2013. *Nucleic Acids Res*, doi: 10.1093/nar/gkt289. http://ogdraw.mpimp-golm.mpg.de/.)

2007). The mode of inheritance of mitochondrial genomes also differs from nuclear genomes. For most eukaryotes mitochondrial DNA is inherited strictly via the maternal line though some exceptions have been noted for marine and freshwater mussels that have biparental inheritance (Randi 2000; see review in Galtier et al. 2009). Owing to a high mutation rate, a cell can have a population of mitochondrial genomes that are not all identical in sequence. When more than two distinct types of mitochondrial genome sequences are present in a cell, this condition is called *heteroplasmy* (Randi 2000). However, only the copy present in highest abundance in a cell will be passed to the offspring (Randi 2000). Nuclear and mitochondrial genomes also differ regarding recombination. While recombination is a major feature of nuclear genomes, it is not yet clear if this evolutionary force plays any role in mitochondrial DNA evolution (Wilson et al. 1985; Pakendorf and Stoneking 2005; see review in Galtier et al. 2009). Evidently, though, the effects of any recombination appear to be negligible in mitochondrial genomes and thus researchers have assumed that these molecules are effectively passed from one generation to the next as a single nonrecombined "supergene" or linkage group (Wilson et al. 1985; Randi 2000). Excellent reviews of the properties of mitochondrial DNA can be found in Wilson et al. (1985), Moritz et al. (1987), Randi (2000), Ballard and Rand (2005), Pakendorf and Stoneking (2005), and Galtier et al. (2009).

In addition to not varying much in overall size (i.e., 100–200 kb), the composition of chloroplast genomes also appears to not vary much either (Brown 2007). Each chloroplast tends to show the same set of ~200 genes needed for photosynthesis, which includes proteins, ribosomal RNAs, tRNAs, and some introns (Brown 2007).

2.1.3 Composition of Eukaryotic Nuclear Genomes

Thanks to genome sequencing efforts over the past two decades, we not only have a better idea about the sizes of many nuclear genomes, but our knowledge of the genomic landscapes found across the tree of life has also dramatically improved. Many of these sequenced genomes have been well annotated and therefore we know far more now about the various elements and their representation in genomes. These elements include protein-coding genes and introns, regulatory regions for gene expression (e.g., promoters, enhancers, and noncoding RNAs), transfer and ribosomal RNAs, pseudogenes, various types of repetitive DNA, and intergenic DNA with presumably no function.

2.1.3.1 *Gene Numbers and Densities among Nuclear Genomes*

The total number of functional genes (i.e., RNA- and protein-coding sequences) per prokaryote genome ranges between 500 and 5,700 genes per genome (Brown 2007). In the human genome, the actual protein-coding genes (exons) occupy only about 48 Mb (1.5%) of the genome (from Figure 7.13 in Brown 2007). In addition to these functional genes, genomes also contain large numbers of gene-related DNA sequences that likely have little or no function such as introns, pseudogenes, and gene fragments. In the human genome, these gene-related elements are found in about 1,152 Mb (36%) of the genome (from Figure 7.13 in Brown 2007).

Within eukaryotes, the size and composition of genomes varies in different ways. As one might expect, both genome size and numbers of genes per genome increase with organismal complexity as the yeast, fruit fly, and human data show in Table 2.5. The numbers of introns also increases in this way. For example, the yeast genome has a total of 239 introns, whereas the human genome has 300,000 (Brown 2007). Similarly, the density of introns per gene is lowest in yeast, a little higher in the fruit fly genome, and highest in the human genome (Table 2.5). The percentage of genomes comprised of *genome-wide repeats*, a type of *repetitive DNA*, also increases with organismal complexity when low complexity eukaryotes are compared to high complexity eukaryotes (Table 2.5). However, when the gene count is examined in light of genome size, gene density shows the opposite trend with respect to organismal complexity. The yeast genome shows the highest gene density while the human genome the lowest (Table 2.5). This phenomenon can be explained by the huge amounts of noncoding DNA such as introns and especially repetitive DNA (Brown 2007).

2.1.3.2 *Intergenic DNA*

Located in the genomic spaces between RNA- and protein-coding genes is a type of DNA collectively called *intergenic DNA*. Intergenic DNA is a

TABLE 2.5
Genome statistics for three eukaryotes of varying complexity

	Yeast	Fruit fly	Human
Genome size (Mb)	12.1	180	3,200
Approximate number of genes	6,100	13,600	30,000–40,000
Gene density (average number of genes/Mb)	496	76	~11
Average number of introns/gene	0.04	3	9
Percentage of genome with interspersed repeats	3.4%	12%	44%

SOURCE: Data for yeast (*Saccharomyces cerevisiae*), fruit fly (*Drosophila melanogaster*), and human (*Homo sapiens*) are from Tables 7.3 and 7.4, Pages 210–211, Brown, T. A. 2007. *Genomes* 3. New York: Garland Science/Taylor & Francis. With permission.

heterogeneous class of DNA comprised of various types of repetitive DNA and nonrepetitive sequences. About 62% of the human genome is comprised of intergenic DNA most of which (46%) is repetitive DNA while the nonrepetitive fraction represents only about 16% of the genome (from Figure 7.13 in Brown 2007). These numbers may be modified in the future, as a study by De Koning et al. (2011) suggested that the current estimate of 46% repetitive DNA in the human genome might represent an underestimate owing to the possibility that current methods for finding repetitive DNA may lack sensitivity for detection. In other words, the accumulation of mutations over evolutionary time makes the identification of repetitive DNA such as transposable elements (see below) difficult or impossible. If true, then the amount of nonrepetitive DNA in the human genome would be even less than 16%.

The complete human genome sequence not only showed that genic regions comprise a minuscule fraction of the genome, but it also revealed that genes are dispersed throughout the genome much like oases in deserts (Venter et al. 2001). These *gene deserts*, which were defined as genomic regions of at least 500 kb in size that are devoid of genes, were found to occupy ~20% of the genome (Venter et al. 2001). Thus, a significant fraction of the intergenic DNA component of genomes is comprised of these gene deserts. In contrast to the parts of the genome containing RNA- and protein-coding sequences, gene deserts exhibit low G + C content, higher numbers of variable nucleotide sites, and decreased levels of sequence conservation (Ovcharenko et al. 2005).

Comparative genome analyses between species such as human versus mouse and human versus chicken show that intergenic DNA exhibits little sequence conservation suggesting that this genomic component is far less constrained by natural selection than the genic regions for maintaining essential biological functions. However, studies of these vast genomic landscapes are revealing the existence of biologically functional intergenic DNA and more such discoveries will undoubtedly be made in the future. It is also highly probable that vast amounts of intergenic DNA will always be considered nonfunctional. This is because careful consideration of the basic biology of certain genomic sequences (e.g., transposable elements and pseudogenes) leads one to the inescapable conclusion that most if not all of these sequences must be currently without a biological function (Graur et al. 2013). We will further explore the topic of functional versus nonfunctional DNA and its relevance to phylogenomics in Chapter 3.

Repetitive DNA—This class of DNA is comprised of two main groups: *tandemly repeated DNA* and *genome-wide repeats* (Brown 2007). Tandem repeats are found adjacent to each other or in clusters of repeats in close proximity to each other on the same chromosomes, whereas genome-wide repeats are dispersed randomly throughout genomes.

Different types of simple tandem repeats have been classified according to the overall length of the repeat sequence: (1) long stretches of *satellite* DNA, which span hundreds of kb and are largely confined to centromeric, telomeric, and subtelomeric regions; (2) *minisatellites*, which extend up to 20 kb along a chromosome and can occur throughout the genome; and (3) *microsatellites*, which are up to around 150 bp long and are also scattered across eukaryotic genomes (Scribner and Pearce 2000; Brown 2007). Most such repeats occur within intergenic regions and thus they are generally free to mutate without any positive or negative consequence to the individual organism (Scribner and Pearce 2000). However, some microsatellites occur within protein-coding regions and can therefore lead to diseases (e.g., CAG repeats excessive in number within the *huntingtin* gene cause Huntington's Disease; Ashley and Warren 1995).

Genome-wide repeats, which are also called *transposons, transposable elements, mobile genetic elements*, and "*jumping genes*" consist of gene families with members dispersed throughout the genome. The process of one transposable element moving from one location in a genome to another is called *transposition*. Barbara McClintock discovered transposons in the 1940s during her research on maize genetics (McClintock 1950) and was subsequently awarded a Nobel Prize for this work (Ravindran 2012).

Transposable elements are largely inserted into random locations within genomes and thus they are common in intergenic regions; they can also be inserted within functional genes and other transposons. When transposons insert themselves into functionally important regions such as regulatory regions or in the reading frame of proteins they can alter gene expression patterns or severely disrupt gene function and cause diseases (Watson et al. 2014). Some transposable elements simply represent a single element that "jumps" around the genome to different chromosomal locations, whereas other types of transposons copy themselves and thereby proliferate in number within a genome.

Interestingly, much of the size variation among eukaryotic genomes is not due to the variable numbers of genes or other important functional elements of the genome, but is instead explained by varying amounts of genome-wide repeats (Brown 2007). For example, among vertebrates genome-wide repeats only account for about 3%–13% of the reptilian genome (including birds), whereas they represent 35%–50% of the mammalian genome (Janes et al. 2010). An even more dramatic range is observed in plants, as genome-wide repeat content in the carnivorous bladderwort plant (*Utricularia gibba*) genome is only 3% while for maize it approaches 85% (Lee and Kim 2014)!

Nonrepetitive intergenic DNA—When the genic and repetitive DNA fractions of the genome are accounted for, the remaining portion is referred to as *nonrepetitive intergenic DNA*. As previously mentioned, approximately ~16% (510 Mb) of the human genome is comprised of nonrepetitive intergenic DNA (Brown 2007). Like repetitive DNA, a major hallmark of nonrepetitive DNA is its overall lack of evolutionary conservatism. However, researchers are finding exceptional elements within these genomic regions that are evidently of vital importance to the organism. For example, not long after the human genome sequence was published, researchers discovered a whole new class of noncoding and nonrepetitive DNA that is biologically important. These "ultraconserved" elements, which have had their sequences essentially "frozen in time" (i.e., without modification) for hundreds of millions of years, apparently function as long-range regulatory elements for neighboring developmental genes (Bejerano et al. 2004; Katzman et al. 2007). Other interesting findings concerning the genomic structure and function involving noncoding/nonrepetitive DNA have been discovered as we will see in Chapter 3.

Gene deserts—Although gene deserts in general show low levels of sequence conservation, they exhibit variability in terms of their composition and evolution. This observation prompted Ovcharenko et al. (2005) to characterize two types of gene deserts, which are called *stable gene deserts* and *variable gene deserts*.

Stable gene deserts have >2% of their sequences conserved, are enriched for long-range regulatory elements, are depauperate in transposable elements, and usually exist as single syntenic blocks maintained since the time of divergence between mammalian and avian lineages (Ovcharenko et al. 2005). The preservation of these large syntenic blocks over evolutionary time is intriguing especially given the overall lack of sequence conservatism exhibited by these sequences. However, research findings from different studies are now painting a picture that suggests these syntenic blocks—which include the regulatory elements and flanking sequences—must have their structure maintained in order to preserve biological functions involving the regulation of developmental processes. First, these regulatory elements are highly conserved sequences maintained by purifying natural selection (Katzman et al. 2007). Secondly, duplications of these regions are apparently not tolerated (Bejerano et al. 2004). Thirdly, the regions flanking these regulatory elements are refractory to transposon insertions (Simons et al. 2006). The human genome contains ~3 million transposable elements, which corresponds to one transposon every ~500 bp on average. Given this observation, it is surprising that 860 transposon-free regions (TFRs), which are defined as transposon free segments that are at least 10 kb long (longest is 81 kb), have been found (Simons et al. 2006). These TFRs are strongly associated with the distribution of highly conserved regulatory elements in stable gene deserts. Although some transposable elements are found in stable gene deserts, their syntenic positions have been maintained since the human and mouse lineages diverged from each other even though the actual transposons are lineage-specific (i.e., independently acquired; Simons et al. 2006).

In contrast, variable gene deserts have <2% of their sequences conserved are depauperate in regulatory elements, are enriched in transposable elements, and do not have syntenic blocks (Ovcharenko et al. 2005). Thus stable gene deserts appear to represent critically important genomic structures that are not able to tolerate transposon insertions and duplications, whereas variable gene deserts appear to have few if any selective constraints and thus might represent true genomic junkyards (Ovcharenko et al. 2005).

2.2 DNA SEQUENCE EVOLUTION

2.2.1 Patterns and Processes of Base Substitutions

Mutation occurs via different processes and at different levels in genomes: chromosomes can break into smaller pieces or they can fuse with each other, pieces of chromosomes can be inverted or translocated, mobile genetic elements mutate chromosomes by inserting themselves into random genomic locations, crossing over during meiosis alters DNA sequences, and, at the smallest scale, single base substitutions, insertions, and deletions change DNA sequences. It is essential for phylogenomics researchers to attain a solid understanding about patterns and processes of DNA mutations—particularly germline mutations—because this information is used to obtain improved estimates of gene tree topologies, evolutionary distances, and divergence times among sequences (Wakeley 1996; Yang and Yoder 1999; Graur and Li 2000).

There are two different ways DNA sequences can mutate at the level of a DNA site or short string of sites: (1) *site substitutions* or "point mutations" in which one nucleotide base is switched to another type and (2) *insertions or deletions* or "indels," which are caused when one or more consecutive nucleotide sites are added or deleted by mistake during the DNA replication process. Although both types of mutations represent potentially useful "information" in a sequence dataset, historically, the vast majority of molecular phylogenetic studies have analyzed sequence variation due to site substitutions. The primary reasons for this is that DNA sequences contain many more point mutations than indels and because it is more straightforward to model the base substitution process than modeling indel evolution.

There are a total of six distinct types of nucleotide substitutions, which are divided up into two classes: *transitions* and *transversions* (Figure 2.2). Two of these substitution types represent transitions, which occur when one purine changes into another purine (A ↔ G) or when a pyrimidine changes to another pyrimidine (C ↔ T), while the remaining four types of substitutions classed as transversions occur when a purine changes into a pyrimidine or vice versa (Figure 2.2).

2.2.1.1 Transition Bias

What are the patterns and processes of base substitution observed in DNA? This is a complicated subject with research still ongoing to better understand this aspect of molecular evolution. Nonetheless, we can still briefly explore this subject in order to obtain some understanding about the nature of base substitutions. First, let's consider the question: *Is DNA substitution a random process with respect to the directionality of change* (i.e., *each substitution*

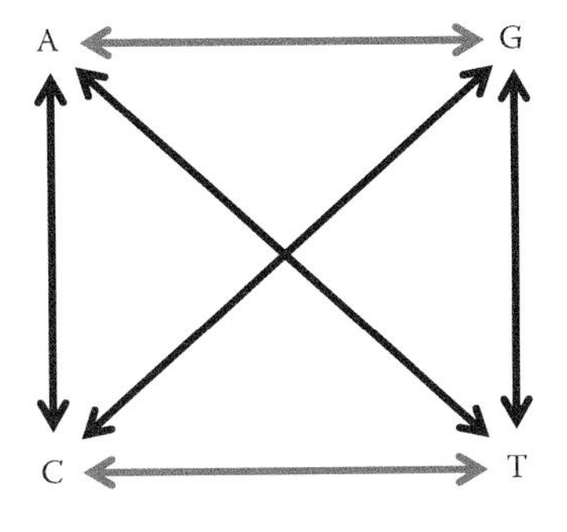

Figure 2.2. Diagram showing all six types of base substitutions. *Transitions* (blue arrows) are A ↔ G or C ↔ T changes, whereas *transversions* (black arrows) are A ↔ C, G ↔ T, A ↔ T, or G ↔ C substitutions.

is equally likely)? Before we can address this question, we need to define a relevant quantity, which is called *transition bias* or the transition:transversion rate ratio (Wakeley 1996). This important ratio is simply the estimated overall transition rate relative to the overall transversion rate and it is commonly abbreviated as *TI/TV rate ratio* (Wakeley 1996). If substitutions randomly occur, then we would expect to see a *TI/TV* rate ratio of 1:2 in real data (Graur and Li 2000). Although we do believe that DNA substitutions occur randomly with respect to the evolutionary fitness of the individual, an abundance of evidence and theory (as we will see later in this section and in the following Section 2.2.1.2) indicate that these mutations are not random when we are talking about the direction of change (Graur and Li 2000).

Brown et al. (1982) examined primate mitochondrial DNA and discovered that transitions vastly outnumbered transversions. These authors dismissed the hypothesis that natural selection was the sole explanatory factor to explain their observations because the *TI/TV* rate ratio bias was observed in tRNAs and in protein-coding genes each of which has different selection pressures. Instead, they attributed the high number of transitions to the mutation process itself. The "mutation process" Brown et al. (1982) referred to is the one that produces *spontaneous substitutions* or point mutations. Note that these substitutions only become mutations after they become fixed changes in the genome that are passed to the next generation (Graur and Li 2000). Substitutions can become so fixed if they either escape the cellular repair network or if they have no consequence to an individual's fitness (i.e., selectively neutral mutation). A later study by Tamura and Nei (1993) also found a preponderance of transitions relative to transversions in the mitochondrial control region in humans and chimpanzees, which represented a significant finding because control region sites are thought to experience little or no selection and hence the observed *TI/TV* ratio may reflect the rates of spontaneous substitutions in mitochondrial DNA (Graur and Li 2000).

High transition rates have also been found in nuclear DNA but the situation is more complicated and not fully understood. First, there is emerging evidence that rates of spontaneous substitution leading to transitions and transversions not only vary within genomes (Graur and Li 2000) but also among species (Yang and Yoder 1999; Keller et al. 2007). At the within genome level, research has shown that some bases are more mutable than others (Graur and Li 2000). For example, parts of the human genome that are rich in G + C bases (e.g., pseudogenes) display a *TI/TV* ratio of around five, which is 2.5-fold higher than the estimate observed for surrounding intergenic regions, which tend to be G + C poor regions (see Table 2.1 in Zhang and Gerstein 2003). In G + C rich sequences of mammalian genomes, a high percentage of CG dinucleotides (not to be confused with base pairs) or *CpG* sites ("CpG" represents—C—phosphate—G—on the same strand of DNA) experience a high rate of mutation into TpG dinucleotides, which occurs because the cytosines in CG dinucleotides are targets for methylation; methylated cytosines, in turn, induce C:G → T:A transitions (Petrov and Hartl 1999; Graur and Li 2000). Genomic regions that no longer contain many CpG sites such as "old" pseudogenes that have already undergone transition mutations at original CpG sites (Casane et al. 1997; Zhang et al. 2002) and intergenic DNA with no function (i.e., low G + C content; Graur and Li 2000) show *TI/TV* rate ratios, which likely reflect the rates of spontaneous substitutions due to DNA replication errors.

Studies to date suggest that the spontaneous substitution rate may not be fixed across species. For example, as was mentioned the genome-wide *TI/TV* rate ratio for humans was estimated to be about 2:1, a figure was that based on the analyses of CpG discounted pseudogenes (Zhang and Gerstein 2003). Costa et al. (2016) obtained a similar estimate for intergenic regions in the genomes of hominoids. However, in nonvertebrates the *TI/TV* rate ratio may vary. Petrov and Hartl (1999) observed a transition bias in nonfunctional parts of the *Drosophila* genome that is comparable to the mammalian ratio of two, whereas a later study on grasshoppers by Keller et al. (2007) observed a lower *TI/TV* rate ratio. These results for insect genomes suggest that the *TI/TV* rate ratio may vary across species more than is currently known.

Thus, spontaneous substitutions can originate during DNA replication or via nonreplicative processes (Graur and Li 2000; Smeds et al. 2016). Which process accounts for more germline point mutations? It has been believed that errors in DNA replication account for most spontaneous substitutions (Graur and Li 2000; Watson et al. 2014). Recent evidence from birds corroborates this hypothesis, as Smeds et al. (2016) found that only ~13% of detected germline point mutations in an avian pedigree were explained by CpG site transitions.

2.2.1.2 *Transition Bias and DNA Replication Errors*

The extreme fidelity of DNA replication is one of the remarkable wonders of nature. Nonetheless, we know that single base errors do occasionally occur during replication, which results in commonly observed transition biases. *What causes these single base misincorporations and why are they generally nonrandom with respect to their directionality?* The molecular mechanism(s) that give rise to spontaneous substitutions have not been conclusively determined. However, the long-lived *rare tautomer hypothesis*, which was conceived by Watson and Crick (1953a), is currently the favored explanation for this mutation process (Graur and Li 2000; Harris et al. 2003; Wang et al. 2011; but see Echols and Goodman 1991). Although it remains to be seen if this hypothesis is correct, at the present time it nonetheless offers a simple and elegant explanation for why transition bias exists in genomes. Before we examine this hypothesis in detail, we will briefly review aspects of DNA replication fidelity.

The fidelity of DNA replication is largely the result of so-called "Watson–Crick geometry," which characterizes the geometrical configuration of the correct base pairings (i.e., when adenine base is paired with thymine or when guanine is paired with cytosine) during DNA synthesis (Watson and Crick 1953b; Kunkel and Bebenek 2000; Harris et al. 2003; Watson et al. 2014). The correct geometry is only achieved when a purine (adenine or guanine) is paired with a pyrimidine (cytosine or thymine)—thereby creating the correct spacing between the opposing bases for hydrogen bonding—and a sufficient number of hydrogen bonds are formed (i.e., two for the A:T pair and three for the G:C pair; Watson and Crick 1953b). However, the purine–pyrimidine pairing requirement alone is insufficient to explain the specificity of base pairing. Thus, it is likely the hydrogen bonding between the two opposed bases that best accounts for this phenomenon (Watson et al. 2014).

Watson and Crick (1953a) suggested that spontaneous base substitutions could arise when rare or "disfavored" minor tautomers of each natural base occasionally become incorporated into growing DNA chains during replication. The four bases that comprise DNA exist in two tautomeric states: the *major* (amino and keto) tautomeric states represent the normal or preferred base configurations, whereas the *minor* (imino and enol) states are rarely formed (Watson et al. 2014). These alternative tautomeric bases differ

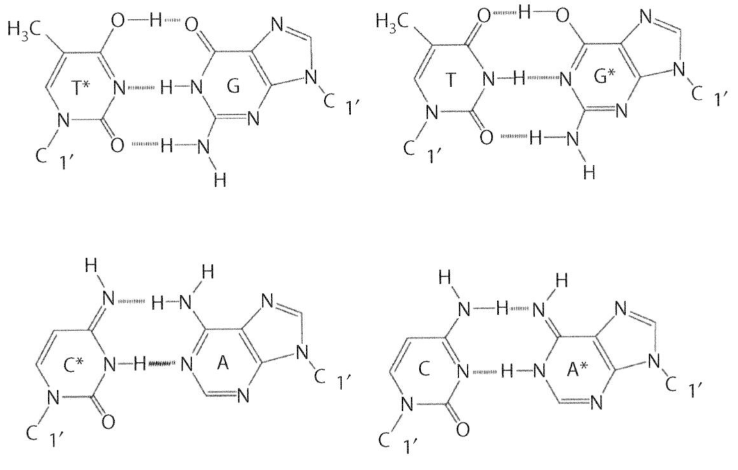

Figure 2.3. Hydrogen-bonding arrangements in pseudo-Watson–Crick base pairs. This figure shows how the minor tautomers (flagged by asterisks) of the natural bases form hydrogen bonds with complementary natural bases. Top: T*:G and T:G* pairs show the enol:keto and keto:enol pairings, respectively. Bottom: C*:A and C:A* pairs depict the imino:amino and amino:imino pairings, respectively. Hydrogen bonds are shown as hatched lines between the functional groups of opposing bases. (Reprinted from Harris, V. H. et al. 2003. *J Mol Biol* 326:1389–1401. With permission.)

from each other owing to changes in the attached functional groups, which in turn, alters their hydrogen-bonding properties (Figure 2.3). This means that the four minor tautomers will not form the usual G:C and A:T or Watson–Crick base pairs, but instead they will preferentially form G:T and C:A or "pseudo Watson–Crick" base pairs (Figure 2.3; Harris et al. 2003). Examination of the hydrogen-bonding capabilities in all possible base pairings involving the four minor base tautomers (not shown) makes clear why the pseudo Watson–Crick pairings shown in Figure 2.3 are most likely to occur during DNA replication. In other words, the pairings that satisfy the purine–pyrimidine pairing rule and maximum number of hydrogen bonds will be the stereochemically preferred pairings during DNA replication. Because the geometrical configurations of pseudo Watson–Crick pairings (Figure 2.3) mimic the traditional Watson–Crick pairings, DNA polymerase cannot discriminate the correct versus incorrect base pairs, respectively, thereby leading to occasional misincorporations during DNA replication (Harris et al. 2003). An important consequence of the rare tautomer hypothesis is that spontaneous mutations arising from incorporations of minor tautomers during DNA synthesis will result in the establishment of *transition*-type mutations in the genome.

2.2.1.3 Saturation of DNA Sites

If we were to assume that the rate of base substitution has remained constant through time, then we might naively expect that the total number of substitutions observed in two aligned homologous sequences will be linearly related to the amount of time since their most recent common ancestor or divergence time. This would be true for sequences with recent divergence times, but not for sequences with older divergence times (Upholt 1977; Brown et al. 1979, 1982). Why is this so? For sequences with older divergence times, the true total amount of evolutionary divergence will be underestimated because there will be some DNA sites that experienced multiple substitutions; hence in those situations counts of *observed* substitutions will be blind to previous unobserved substitutions (Upholt 1977). This phenomenon is known as *saturation* (Brown et al. 1982; Swofford et al. 1996; Arbogast et al. 2002) or *multiple hits* (Yang 2006) and it is of great importance to phylogenomics.

Let's take a closer look at how saturation occurs and how it can complicate efforts to estimate DNA sequence divergences. Consider a single 50 bp long gene sequence, which represents a portion of the open reading frame (ORF) for a protein-coding gene. If we sequence this locus in two different species and compare the sequences site by site, we can see that 15 of the sites are variable (Figure 2.4). Variable sites are also called *segregating* or *polymorphic sites*, whereas the sites showing no variation are labeled as *nonsegregating* or *monomorphic sites*. While looking at these two sequences in Figure 2.4, we would like to know how much evolutionary divergence has actually taken place between them. In other words, we would like to ask the question: *How many mutations have occurred between these sequences since the time of their most recent common ancestor?*

The simplest method for estimating this divergence between a pair of sequences involves calculating the *pairwise* or *p-distance*, which is expressed either as the number of substitutions/site or by percent sequence divergence (Swofford et al. 1996). The *p*-distance is equal to the total number of segregating sites divided by the total number of sites. From Figure 2.4, we can compute the *p*-distance as being $15/50 = 0.3$ substitutions/site (or 30% sequence divergence). However, remember that this *p*-distance ignores any unobserved

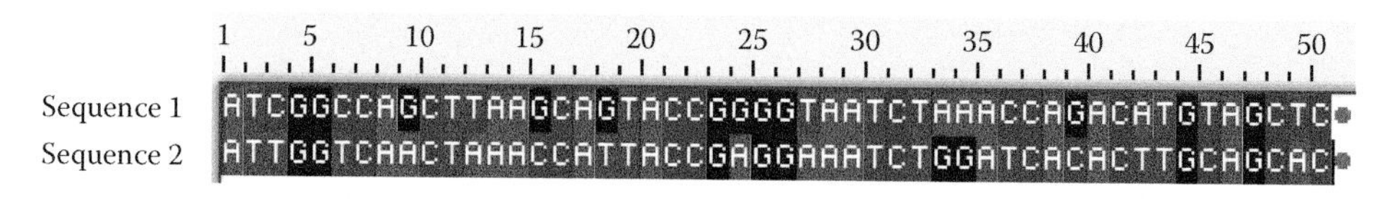

Figure 2.4. Alignment of two protein-coding DNA sequences spanning 50 sites obtained from two different species. Sequences are shown in a 5′ → 3′ orientation and site #1 at the 5′ end corresponds to the start of the reading frame (i.e., is 1st codon position). Fifteen of the sites are variable and 35 are invariable. The four bases are color-coded for easier visualization of sequence similarities and differences. Alignment was constructed using the sequence alignment software *Se-Al*. (From Rambaut, A. 2007. *Se-Al, version 2.0 a11*. Edinburgh: The University of Edinburgh.)

substitutions that could be hidden in these sequences. How bad can saturation effects be on estimates of genetic divergence?

Brown et al. (1979, 1982) provided empirical evidence for saturation effects from their studies of mitochondrial DNA evolution in primates (Figure 2.5). Notice that during the initial period of divergence—up to about 10 million years, the relationship is roughly linear (Figure 2.5). This initial phase of sequence divergence is characterized by having each new substitution occurring mostly at monomorphic sites, which generates new segregating sites. Thus, during this linear phase the estimated p-distance approximates the true number of mutations that occurred between two recently diverged sequences. However, beyond 10 million years (i.e., >20% sequence divergence), the rate of mutation accumulation begins to slow down (Figure 2.5). This is expected to occur because as time progresses there will be fewer and fewer available nonsegregating sites that can be subject to mutation for the first time and some polymorphic sites may revert to being monomorphic again. Consequently, the probability for multiple mutations within sites increases. By the time the sequences have exceeded 30% divergence, the curve in Figure 2.5 is flat, which implies that some factor (e.g., natural selection) is constraining the sequences from diverging further. If the sequences are completely unconstrained to evolve, then they can potentially continue to diverge from each other until they reach a maximum divergence of 75%, which means they are effectively random with respect to each other (Felsenstein 2004). Thus, Brown et al. (1979, 1982) results provided clear evidence that p-distances will underestimate the true distances for older divergences, which is in accord with expectation. The curve in Figure 2.5 is commonly referred to as a *saturation plot* and, as we will soon see, such plots vary depending on which sites are being considered.

Brown et al. (1982) observed another interesting aspect of DNA sequence evolution that concerns the effects of saturation on *TI/TV* rate ratios. These authors noticed that *TI/TV* estimates declined as the evolutionary distances between sequences increased (i.e., "time-dependency" of *TI/TV*). They explained this phenomenon by suggesting that, owing to multiple substitutions within sites, the number of observed transitions departs more and more from the true number with increased divergence times while the number of observed transversions remains close to the true number (i.e., do not reach point of saturation). Thus it is primarily an estimation problem caused by saturation and not due to any actual time-variable substitution rates.

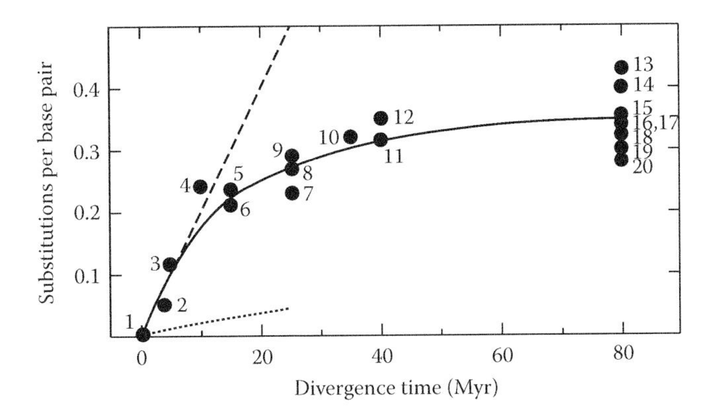

Figure 2.5. DNA sequence saturation illustrated by mammalian mtDNA divergences. This graph shows the relationship between mitochondrial p-distances (ordinate) generated for various mammal species versus the estimated divergence times in millions of years (Myr) between each pair. Divergence time estimates are based on fossil and protein data. The following p-distances are shown: (1) mean difference among humans (only intraspecific comparison on figure); (2) goat versus sheep; (3) human versus chimpanzee; (4) baboon versus rhesus; (5) guenon versus baboon; (6) guenon versus rhesus; (7) human versus guenon; (8) human versus rhesus; (9) human versus baboon; (10) rat versus mouse; (11) hamster versus mouse; (12) hamster versus rat; and (13–20), rodent versus primate species pairs. The broken line along the initial part of the curve represents the inferred mtDNA substitution rate. For comparative purposes, the substitution rate for single-copy nuclear DNA, which was obtained from an outside data source, is shown on the figure as a dotted line. (Reprinted from Brown, W. M., M. George, and A. C. Wilson. 1979. *Proc Natl Acad Sci USA* 76:1967–1971. With permission.)

2.2.1.4 Among-Site Substitution Rate Variation

Variation in nucleotide substitution rates *among* genomic sites represents another aspect of DNA sequence evolution that is of concern to phylogenomics. This is because ignorance of this phenomenon by researchers can lead to underestimates of the actual number of substitutions that had occurred (Golding 1983; Wakeley 1994; Swofford et al. 1996; Yang 1996). Yang (P. 370; 1996) described the phenomenon as follows: "*The existence of among-site rate variation means that most evolutionary changes occur at only a few sites, while many other sites never experience any substitutions.*"

An early empirical illustration of this comes from the study of Brown et al. (1982). When these authors partitioned the mitochondrial coding sequences by codon position and then made separate saturation plots for the 1st, 2nd, and 3rd codon positions, they observed three different saturation curves (Figure 2.6). The curve showing divergence among 3rd codon position bases exhibits the most rapid rate of divergence, 1st codon position bases show the second fastest rate, and the 2nd codon position shows the slowest rate of change (Figure 2.6). These curves make a lot of sense when you consider the evolutionary constraints against mutations in certain codon positions. We know that mutations at third codon positions are tolerated far more because they often do not result in amino acid replacements (because of the "wobble" phenomenon). Such mutations are referred to as *synonymous* or *silent substitutions*. If a mutation does cause an amino acid replacement, then it is called a *nonsynonymous* substitution. Nonsynonymous substitutions are much less tolerated because they can alter the protein structure in ways that result in less or no functionality and hence the mutation will likely be weeded out of the population via purifying natural selection.

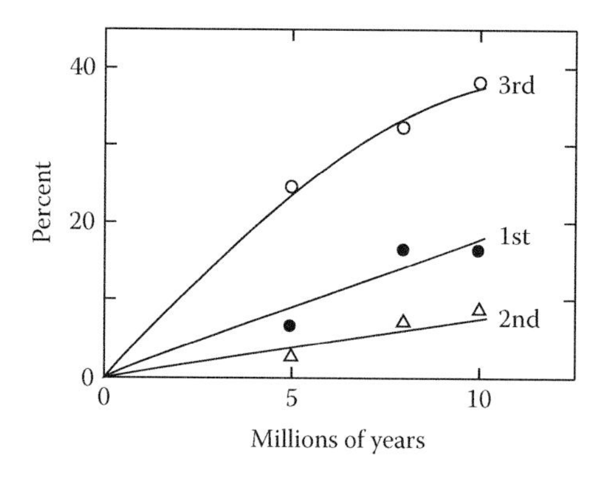

Figure 2.6. Codon bias in primate mtDNA protein-coding sequences. The graph shows observed percent sequence divergences (p-distances expressed as % values) for 1st, 2nd, and 3rd codon positions as a function of divergence time. Sequence divergences were calculated from an 896 bp sequence containing two coding regions (URF 4 and 5) from five primates and divergence times were obtained from an independent data source. (Reprinted from Brown, W. M. et al. 1982. *J Mol Evol* 18:225–239. With permission.)

The saturation plots for codon positions in Brown et al. (1982) provide a nice example of sites showing varying levels of *evolutionary conservatism*. Any sites in the genome that are under selection are expected to show such conservatism, while sites that are not maintained by natural election (directly or indirectly via linkage) are expected to show little among-site rate variation (Yang 1996). Evidence supporting the latter idea comes from the study of Costa et al. (2016) who found that nearly all of the 292 genealogically independent and presumably neutral loci in the hominoid genomes were best fit to substitution models having one substitution rate among sites.

Why is among-site rate variation a concern for us? It is a huge concern for at least two reasons. First, distances that are estimated from sequences containing sites with varying substitution rates can yield badly underestimated distances (Golding 1983; Yang 1996). Secondly, *TI/TV* rate ratios—essential parameters in many nucleotide substitution models—can also be underestimated (Wakeley 1994, 1996). Moreover, the larger the actual distances, the larger will be the underestimates (Yang 1996). These adverse effects can have severe consequences for estimates of divergence times, *TI/TV* rate ratios, and gene tree inferences (Swofford et al. 1996; Yang 1996).

2.2.2 Tandemly Repeated DNA Sequences

This class of repetitive DNA consists of chromosomal elements that have undergone duplications via *unequal crossing over* and *DNA replication errors* (Brown 2007). Tandem repeats have the characteristic of being spatially proximal to each other along a chromosome. Repeat units can be as small as a single base or long segments of DNA.

Unequal crossing over can occur between two homologous chromosomes misaligned to each other during metaphase I of meiosis. The recombinant products of unequal crossing over include a chromosome containing a duplicated region while the other homolog has this region deleted. The generation of new copies on a chromosome via unequal crossing over is one of several molecular processes that can give rise to *gene duplications*. When clusters of the same genes are produced via unequal crossing over, this results in the generation of a *gene family*.

There are several different consequences of such duplications when they involve functional genes. First, the new copies may simply contribute more of the same gene products, which may be deleterious, benign, or advantageous to the individual organism. Secondly, one or more of the new copies may evolve a novel function (Chang and Duda 2012). A third possibility is that the gene or one of its upstream regulatory elements might suffer a point mutation that renders the gene nonfunctional. Such genes that are inactivated by mutations are called *conventional pseudogenes* (Brown 2007). A conventional pseudogene can also result if the duplicated gene is missing part of its sequence (i.e., a gene fragment) as a result of unequal crossing over (Watson et al. 2014).

Another molecular mechanism by which tandem repeats can form is via "slippage" of DNA polymerase enzymes during replication (Levinson and Gutman 1987). Polymerase slippage is a phenomenon whereby the polymerase accidently inserts or deletes short stretches of DNA, which often correspond to the repeated motif already found in a sequence. For example, consider a stretch of sequence that has a string of 10 consecutive GC repeats (i.e., 5′-GCGCGCGCGCGCGCGCGCGC-3′ or "$(GC)_{10}$" repeat). Thus, after a DNA replication error, the above sequence could become a $(GC)_{11}$ sequence. There are many different repeats motifs found in genomes such as GC, AC, GT, CAG, AGCT, etc. Although DNA polymerase slippage may be the primary mechanism to initiate the formation of repeated nucleotides or motifs, once a short tandem repeat sequence is made, the conditions will exist for unequal crossing to dramatically increase the sizes of these repeat sequences (Levinson and Gutman 1987).

Microsatellite loci have the highest mutation rates for any genomic sequences, which is why such loci tend to have high allelic diversity in populations (Scribner and Pearce 2000). This attribute has thus made them the marker of choice for identifying individuals in criminal forensic studies and behavioral ecology studies (e.g., parentage studies). They are also excellent markers in some types of population genetic studies such as analyses of population structure (Waples and Gaggiotti 2006). However, microsatellites are poor markers for reconstructing gene trees because of significant problems with homoplasy; that is, it may be difficult to determine whether two identical allele sequences are related by common ancestry or through convergence (Scribner and Pearce 2000).

2.2.3 Transposable Elements

The transposition of DNA elements in genomes is a form of *site-specific* or *nonhomologous* recombination (Watson et al. 2014). There are several different transposons each with a different mechanism of transposition. Some transposons are *nonreplicative* meaning that the transposable element is enzymatically excised from its chromosomal location and then reinserted into a different genomic location. This form of transposition is called *cut and paste transposition* (Watson et al. 2014). Another form of transposition is *copy and paste transposition* or *replicative transposition* (Watson et al. 2014). Figure 2.7 shows examples of DNA transposons that use the "cut and paste" or "copy and paste" modes of transposition. Other types of transposons involve an mRNA intermediate stage to transpose. These *RNA transposons*, which are also called *retrotransposons* or *retroelements*, only use the replicative form of transposition. There are two main groups of RNA transposons: (1) *LTR retrotransposons* for "*long terminal repeat retrotransposons*" and (2) *non-LTR retrotransposons* (Watson et al. 2014).

Figure 2.8 shows an example of a DNA transposon, an LTR retrotransposon, and a non-LTR retrotransposon. Let's now take a look at the structure of these three elements. DNA transposons (Figure 2.8a) have several critical entities required for their transposition. First, DNA transposons contain the protein-coding gene for *transposase*, the enzyme that performs the transposition reaction. In addition to the *transposase* gene, the element has two terminal inverted repeats. These terminal repeats, which are also called recombination sites, are essential to transposition because they contain the recognition sequences for *transposase* (Figure 2.8a). The inverted repeats of the element will connect the element to the host chromosome once transposition is completed.

The structure of an LTR retroelement (Figure 2.8b) consists of two LTR sequences that flank a region coding for the enzymes *integrase* and *reverse*

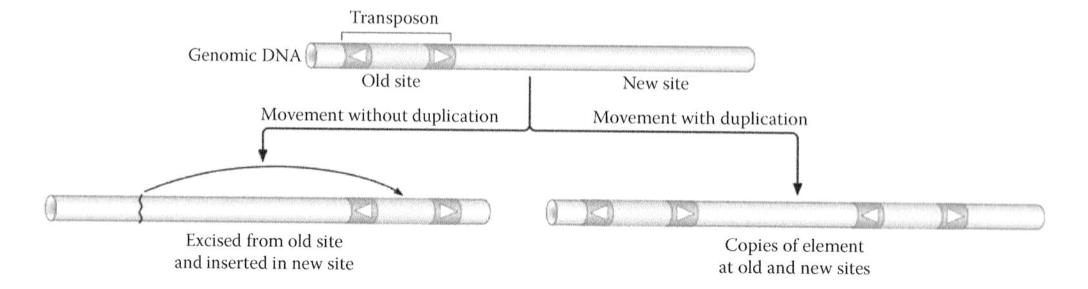

Figure 2.7. "Cut and paste" versus "copy and paste" transposition. Left side of figure shows cut and paste or nonreplicative transposition in which a transposon is excised from its host DNA and reinserted into the host genome at another location. Right side shows copy and paste or replicative transposition whereby a copy of a transposon is re-inserted into the host genome at a different location while the original copy remains in place. (Watson, James D.; Baker, Tania A.; Bell, Stephen P.; Gann, Alexander; Levine, Michael; Losick, Richard, *MOLECULAR BIOLOGY OF THE GENE*, 7th Ed., ©2014. Reprinted by permission of Pearson Education, Inc., New York, New York.)

transcriptase. The retrotransposon *integrase* enzyme functions much like *transposase*. Single inverted terminal repeats are found within each LTR and a promoter sequence for RNA polymerase is found within the upstream inverted terminal repeat (Figure 2.8b). LTR retrotransposons are sometimes called "virus-like retrotransposons" because they are likely derived from retroviruses (Watson et al. 2014). Thus, if a retrovirus invades a cell and successfully integrates its transposon genes into the host genome, then cellular RNA polymerases can produce many copies of mRNAs

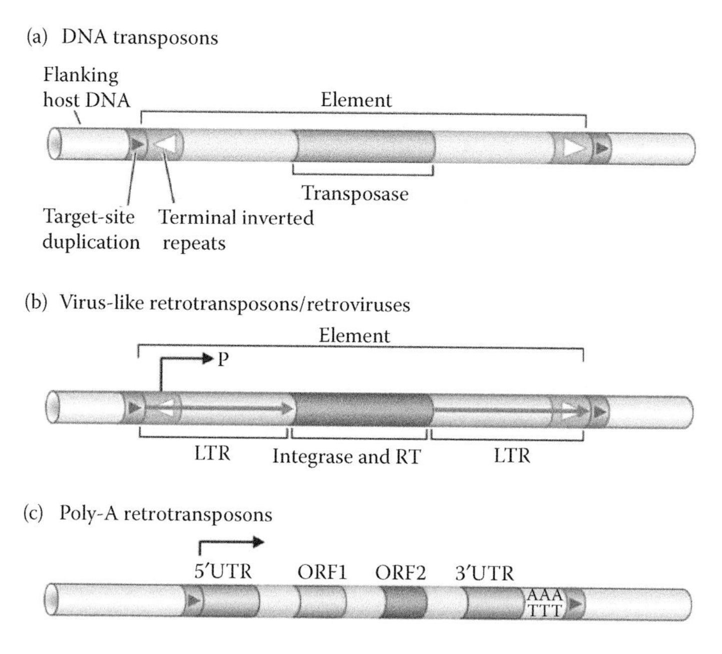

Figure 2.8. Structure of three different types of transposable elements. (a) DNA transposon elements include a *transposase* gene, which is flanked by terminal inverted repeat sequences (green and white arrows). The repeat sequences contain the recombination sites needed for transposition. (b) Virus-like retrotransposon elements contain coding regions for the enzymes *integrase* and *reverse transcriptase* and are flanked by two LTR sequences. (c) Poly-A retrotransposons or non-LTR retrotransposons elements contain coding sequences for an RNA-binding enzyme (ORF1) and an enzyme with both *reverse transcriptase* and *endonuclease* activities (ORF2). The 5′ and 3′ ends of the element are UTR sequences. (Watson, James D.; Baker, Tania A.; Bell, Stephen P.; Gann, Alexander; Levine, Michael; Losick, Richard, *MOLECULAR BIOLOGY OF THE GENE*, 7th Ed., ©2014. Reprinted by permission of Pearson Education, Inc., New York, New York.)

from the retroviral element, which are then converted into *cDNAs* or "copyDNA" before they are integrated into the host genome (Watson et al. 2014). Such host genome copies of a retrovirus are known as *endogenous retroviruses* or "ERVs" (Brown 2007). LTR retrotransposons are not only found in the genomes of plants, invertebrates, and vertebrates (Brown 2007), but this class of transposon accounts for much of the variation in genome sizes observed in plants (Ambrožová et al. 2010; Lee and Kim 2014).

Non-LTR retrotransposons (Figure 2.8c) do not have inverted terminal repeats, but they do have a coding region separated into two ORFs that code for the enzymes *reverse transcriptase* and an *integrase*-like endonuclease that performs the actual transposition. Also, fully functional LTR retrotransposons contain a promoter sequence upstream of the coding region (in a 5′ UTR or untranslated region). A 3′ UTR is located downstream of the two ORFs and the element has a poly-A string of bases on the nontemplate strand (Figure 2.8c). Although some of the details pertaining to the mechanism of transposition for LTR and non-LTR retrotransposons differ, both involve an mRNA intermediate that is produced by cellular RNA polymerases and both are reverse transcribed into a cDNA by *reverse transcriptase*. When transposition is completed, both cDNA elements reside in the host chromosome.

Non-LTR retrotransposons are sometimes called "poly-A retrotransposons" because they have a poly-A tail at the 3′ of the nontemplate DNA strand of the element (Watson et al. 2014). One family of non-LTR repeats, which are common in vertebrate genomes, consists of the *long interspersed nuclear elements* (LINEs; Watson et al. 2014). LINEs are a type of *autonomous* non-LTR retrotransposon because they encode proteins needed for reverse transcription and reintegration of LINE copies into the host genome (Watson et al. 2014). In contrast, *nonautonomous* non-LTR retrotransposons are unable to promote their own transposition because their elements only contain a promoter and poly-A tail (i.e., both ORFs are missing). An example of such a nonautonomous non-LTR retrotransposons are the short interspersed nuclear elements (SINEs), which are abundant in many vertebrate genomes. Nonetheless, SINEs can be transposed if they can borrow the key transposition proteins from LINE elements. Thus, if cellular RNA polymerases generate large numbers of mRNA transcripts from integrated LINEs, then non-LTR retroelements can, like the LTR retrotransposons, spread like wildfire in genomes. For example, ~34% of the human genome is comprised of non-LTR retrotransposons (Brown 2007). Another example is illustrated by CR1 LINE elements (for "Chicken Repeat 1"), which have approximately 200,000 copies inserted into the chicken genome (Kaiser et al. 2007). CR1 elements are common in birds and reptiles (Shedlock 2006; Shedlock et al. 2007; Janes et al. 2010; Kordis 2010).

Another interesting aspect about the biology of transposable elements is that there are generally no known mechanisms for maintaining their function indefinitely in genomes (Graur et al. 2013). Thus, mutations are apparently free to accumulate in transposable elements because such mutations generally do not affect the host organism. If a transposable element affects the host individual in a good or bad way with respect to the fitness of the host individual, it will be because of an unlikely insertion event into a sensitive part (e.g., regulatory region) of the host genome. But if the transposable element is inserted into an unimportant (i.e., nonfunctional) part of the genome, then the transposon can persist indefinitely without harming the host organism but it will thereafter be vulnerable to mutations, which would likely impair its ability to transpose itself again (Graur and Li 2000; Watson et al. 2014). Thus, each replicative transposon typically has a window of evolutionary time in which it can proliferate throughout genomes before mutations render each copy nonfunctional. Such nonfunctional transposons may persist in the genome as relics that only evolve via mutation and genetic drift.

There are a number of ways that transposable elements can be inactivated. DNA transposons can be disabled if point mutations (i.e., single base mutations) occur within the recombination sites (inverted terminal repeats) or *transposase* coding region. Likewise, point mutations in the recombination sites, promoter sequence, or coding region for the *integrase* and *reverse transcriptase* enzymes can render LTR retrotransposons nonfunctional. A functional non-LTR retrotransposon can similarly be inactivated through the occurrence of mutations in its promoter, either of its two ORFs, or in its poly-A tail. However, even if such point mutations are absent from the integrated elements, the transposons can effectively be rendered functionless due to the production of defective retrotransposons. For example, when non-LTR retroelements become transcribed into RNA intermediates, subsequent reverse transcription of these mRNAs often yields double-stranded cDNA products that are truncated at their 5′ ends. These truncated elements are missing a segment of sequence that contains part of the promoter or promoter plus portions of coding regions and thus they are unable to be transcribed (Watson et al. 2014). These truncations apparently occur when the reverse transcriptase, which starts at the 3′ of the mRNA, discontinues synthesis before a complete cDNA product is made (i.e., has a complete promoter and both ORFs). The CR1 elements provide a good example of this truncated product phenomenon. The length of a functional CR1 element is 4.5 kb, yet most of the ~200,000 copies present in the chicken genome are shorter than 400 bp owing to their badly truncated 5′ ends (Kaiser et al. 2007). As a consequence of these truncations these defective elements are considered "dead-on-arrival" when they are inserted into the host genome (Haas et al. 1997; Petrov and Hartl 1999).

2.2.4 Processed Pseudogenes

Although *reverse transcriptase* enzymes encoded in LINEs are highly specific for their own retrotransposon mRNA transcripts, occasionally these enzymes accidentally synthesize cDNAs from cellular mRNAs (i.e., mRNAs transcribed from nontransposon genes in the host genome; Watson et al. 2014). Upon insertion into the genome, these *processed pseudogenes* are not functional because they lack a promoter, introns, and other sites depending on the severity of the 5′ end truncation (Graur and Li 2000; Brown 2007; Watson et al. 2014). Thus, following insertion, these genes evolve differently than their functional cousins, as they are free to accumulate mutations on any site without consequence to the organism (i.e., codon bias no longer exists). In other words, free from selective constraint all sites in a pseudogene are expected to have the same substitution rate.

2.2.5 Mitochondrial Pseudogenes ("Numts")

As early as the 1960s researchers obtained evidence that mitochondrial-like sequences were residing within eukaryotic nuclear genomes (Bensasson et al. 2001). However, the existence of such nuclear copies of mitochondrial DNA would not be confirmed until the 1980s when researchers obtained DNA sequence-based evidence from fungi (van den Boogaart et al. 1982; Farrelly and Butow 1983; Wright and Cummings 1983), invertebrates (Gellissen et al. 1983; Jacobs et al. 1983), and vertebrates (Hadler et al. 1983; Tsuzuki et al. 1983; Wakasugi et al. 1985). Ellis (1982) showed that organellar DNA transfer to the nuclear genome also occurred with chloroplast DNA. An even more dramatic finding was the discovery by Lopez et al. (1994) that a 7.9 kb long mitochondrial sequence had been inserted into the nuclear genome of domesticated cats. Since then a large number of studies have documented the existence of **nu**clear **mit**ochondrial **s**egments, called "numts" (Lopez et al. 1994), in various eukaryotic nuclear genomes (Bensasson et al. 2001). Such numt sequences range in size from tens of bases long up to the size of an entire mitochondrial genome (Zhang and Hewitt 1996; Bensasson et al. 2001; Henze and Martin 2001).

2.2.5.1 Numt Abundance in Eukaryotic Genomes

Preliminary surveys of genomic data from representative eukaryote species suggested that numts are widespread among species (see reviews in Zhang and Hewitt 1996; Bensasson et al. 2001; Richly and Leister 2004). A later more comprehensive analysis (Hazkani-Covo et al. 2010) of 85 fully sequenced eukaryotic genomes, which included 20 fungi, 11 protists, 7 plants, and 47 animals, corroborated these early findings. However, significant variation in the amount of numts per genome also exists among species. For example 8/85 examined genomes apparently did not contain numts while three other genomes had >500 kb of numt sequences (Hazkani-Covo et al. 2010). The highest numt content among animals and plants was found in the possum (*Monodelphis domestica*) and rice (*Oryza sativa Indica*) genomes, which had 2,000 kb and 800 kb of numt sequences, respectively (Hazkani-Covo et al. 2010). Fungi and protists, on the other hand, appear to have lower numt content than animals and plants (Hazkani-Covo et al. 2010). In general, there is a strong positive correlation between genome size and the total number of numts (Bensasson et al. 2001; Hazkani-Covo et al. 2010).

The total number of numts per genome reflects the actions of three processes. First, numt loci that originated as a result of transfer of mitochondrial DNA to that genomic location can be considered *primary integrations*, whereas copies of *numts* that arose because of intragenomic duplication events may be referred to as *secondary integrations* (Tourmen et al. 2002). Because *numts* are not known to have a self-replicating mechanism (like true replicative transposons), they apparently increase their numbers using mechanisms that generate tandemly repeated or segmental duplications (Bensasson et al. 2003; Hazkani-Covo et al. 2010). As an example, the Lopez et al. (1994) 7.9 kb *numt* in the cat genome represents a single primary integration event, whereas its 38–76 tandemly repeated copies represent secondary integrations. Lastly, the deletion of genomic segments containing *numts* represents a third mechanism that can influence the total *numt* count (Hazkani-Covo et al. 2010).

2.2.5.2 Mechanisms of Primary Numt Integration

While it had been clear to researchers for a long time that transfer of mitochondrial DNA to the nuclear genome occurred in many eukaryotes, the mechanism(s) responsible for this transfer process were not immediately obvious. Two hypothetical mechanisms that could explain this phenomenon included: direct transfer of mitochondrial DNA into the nucleus where it could be integrated into the nuclear genome and, secondly, reverse transcription of mitochondrial mRNA into cDNA in the cytoplasm or nucleus prior to being integrated into the nuclear genome (Blanchard and Schmidt 1996; Bensasson et al. 2001). Early workers favored the latter hypothesis owing to two observations: that *numts* were frequently observed in close proximity to retroelements (Blanchard and Schmidt 1996); and secondly, some plant nuclear genomes contained *numts* that were evidently derived from edited mRNA transcripts (e.g., lack introns; Henze and Martin 2001). The study by Blanchard and Schmidt (1996) found some evidence showing that both mechanisms occur (Bensasson et al. 2001). However, Blanchard and Schmidt (1996) argued that the DNA-based transfer process better explained the existence of most *numts* across eukaryotes for the following reasons. First, they pointed out that the association of retroelements with *numts* could simply be coincidental because retroelements are ubiquitous in eukaryotic genomes. Secondly, they noted that many *numts* merely consisted of gene fragments or nontranscribed DNA, which likely did not originate from RNA. Lastly, they found that *numt* integration sites (i.e., the two junctions flanking each *numt*) were not consistent with transposon recombination sites, but instead resembled the short direct repeats associated with the process of *nonhomologous end-joining* (Blanchard and Schmidt 1996). Nonhomologous end-joining is the primary process that most, if not all, eukaryotic cells use to mend double-stranded chromosomal breaks (Watson et al. 2014). To repair a double-stranded break, the cell uses mitochondrial DNA fragments (and perhaps other available DNA) as "filler DNA" to rejoin two chromosomal pieces. Because DNA is inserted into a chromosome—in this case mitochondrial DNA, this is a mutagenic process (Watson et al. 2014). Thus, if *numts* are introduced into functional parts of the genome, then this process can lead to disease. However, this rarely occurs because the vast majority of *numts* are inserted into intergenic regions and thus they represent neutral mutations (Hazkani-Covo et al. 2010). Accordingly, the benefits of repairing double-stranded breaks using nonhomologous end-joining vastly outweigh the risks (Watson et al. 2014).

Another argument in favor of the direct DNA-transfer hypothesis comes from many independent discoveries of long segments of mitochondrial DNA—some of which include introns and nontranscribed regions—inside nuclear genomes (Henze and Martin 2001; Hazkani-Covo et al. 2010). For example, in addition to the 7.9 kb mitochondrial sequence found in the domestic cat genome (Lopez et al. 1994), a 12.5 kb *numt* was observed in the genomes of various species of *Panthera* (Kim et al. 2006; Antunes et al. 2007) and a 14.6 kb fragment—nearly the entire mitochondrial genome—was found in the human genome (Mourier et al. 2001). An even more startling example was the discovery of a 620 kb long segment of mitochondrial DNA—the *entire* mitochondrial genome plus some internally duplicated regions, which were found in the *Arabidopsis thaliana* nuclear genome (Lin et al. 1999; Henze and Martin 2001; Stupar et al. 2001). The RNA-intermediate hypothesis cannot account for these observations (Henze and Martin 2001). In later years, additional studies would instead support the alternative hypothesis—that primary integrations of *numts* are the result of direct transfers of bulk mitochondrial DNA into the nucleus where they are incorporated into double-stranded breaks via nonhomologous end-joining (Ricchetti et al. 1999; Yu and Gabriel 1999; Bensasson et al. 2001; Hazkani-Covo and Covo 2008; Hazkani-Covo et al. 2010).

The finding that nonhomologous end-joining appears to be the primary mechanism for new *numt* integration also provides an explanation for why larger genomes tend to have larger numbers of *numts* than do smaller genomes. Larger genomes will have more opportunities for *numt* integration simply because in larger genomes there will be more chromosomal breaks in need of repair; in other words, the limiting factor for primary integration of *numts* is the number of required repairs of chromosomal breaks (Hazkani-Covo et al. 2010).

2.2.5.3 Differences between Numts and Mitochondrial DNA

The process of primary integration of *numts* into eukaryotic genomes appears to be continuous over time (Mourier et al. 2001; Tourmen et al. 2002; Bensasson et al. 2003; Hazkani-Covo et al. 2003; Antunes et al. 2007). Even within a genome, such as the human genome, *numts* show widely varying divergence times from their mitochondrial genome progenitors as well as among homologous secondary integration copies (Bensasson et al. 2003). Recently inserted *numts* exhibit little or no sequence divergences compared to their mitochondrial counterparts, whereas ancient *numts* are much less recognizable in genomes owing to the many base substitutions, indels, inversions, and deletions they have suffered (Jacobs and Grimes 1986; Tourmen et al. 2002). *In silico* searches of complete genome sequences for *numts* can produce varying estimates of the total number of *numts* depending on the type and stringency of computerized searches (Bensasson et al. 2003; Hazkani-Covo et al. 2003, 2010; Antunes et al. 2007). This not only suggests that estimates of the "total" number of *numts* per genome are likely underestimates (Tourmen et al. 2002; Antunes et al. 2007), but that *numts* are, unlike their mitochondrial counterparts, without selective constraint and thus are free to decay via mutation (Arctander 1995; Sorenson and Quinn 1998).

Several other lines of evidence exist suggesting that *numts* (at least in animals) are free from selective constraint and therefore evolve like pseudogenes. First, many *numts* are only gene fragments or noncoding mitochondrial DNA (Blanchard and Schmidt 1996) thus they likely do not have a function. Secondly, the genetic code for mitochondrial DNA is different from that for nuclear DNA, which implies that mRNAs generated from *numts* would yield defective protein products (Gellissen and Michaelis 1987; Perna and Kocher 1996). Lastly, *numts* often have indels, which creates frameshift mutations and premature stop codons—both of which would result in nonfunctional proteins (Jacobs and Grimes 1986; Smith et al. 1992; Arctander 1995; Bensasson et al. 2001).

Let's now take a closer look at how the mutational regimes differ between *numts* and their mitochondrial counterparts. Mitochondrial and chloroplast genomes undergo low levels of cytosine methylation at CpG sites but it is extensive in the nuclear genomes of higher plants (Huang et al. 2005 and references therein). In a study of the *Arabidopsis thaliana* and *Oryza sativa* genomes by Huang et al. (2005), they found that cytosines at CpG sites in *numts* were rapidly methylated and subsequently underwent C to T transitions (or G to A transitions on the opposite strand). Moreover, they noted that transitions at CpG sites were 5–10 times higher than rates at other *numt* sites and therefore they suggested that this represented a major force in the mutational decay of *numt* loci. It is unclear how extensive CpG methylation is in animals, but it has been found in some groups and thus similar elevated substitution rates at CpG sites are also observed (Petrov and Hartl 1999; Bensasson et al. 2001; Hazkani-Covo et al. 2010). A second difference concerns the overall substitution rates: because *numt* sites are embedded in nuclear DNA, their base substitution rates are expected to be 10-fold slower than for mitochondrial DNA sites (Brown et al. 1979). Also, the *TI/TV* rate ratio for *numt* sites is expected to conform to the level observed in nuclear DNA (i.e., *TI/TV* rate ratio ~ 2) rather than the level inferred for mitochondrial DNA (i.e., *TI/TV* rate ratio ~ 15; Zhang and Hewitt 1996; Graur and Li 2000). Thus, *numt* sites require more time to reach saturation than comparable mitochondrial sites (Zischler et al. 1998; Bensasson et al. 2001). A fourth difference is that while mitochondrial genes are under selective constraints, particularly at certain codon positions (Brown et al. 1982), *numts* on the other hand are expected to have the same substitution rate across sites (Arctander 1995; Yang 1996; Zhang and Hewitt 1996; Bensasson et al. 2001, 2003). Lastly, in contrast to most mitochondrial sites, *numts* can experience insertions and deletions without the individual suffering adverse effects (Bensasson et al. 2001; Tourmen et al. 2002).

REFERENCES

Ambrožová, K., T. Mandáková, P. Bureš et al. 2010. Diverse retrotransposon families and an AT-rich satellite DNA revealed in giant genomes of *Fritillaria* lilies. *Ann Bot*. doi: 10.1093/aob/mcq235.

Antunes, A., J. Pontius, M. J. Ramos, S. J. O'Brien, and W. E. Johnson. 2007. Mitochondrial introgressions into the nuclear genome of the domestic cat. *J Hered* 98:414–420.

Arbogast, B. S., S. V. Edwards, J. Wakeley, P. Beerli, and J. B. Slowinski. 2002. Estimating divergence times from molecular data on phylogenetic and population genetic timescales. *Annu Rev Ecol Syst* 33:707–740.

Arctander, P. 1995. Comparison of a mitochondrial gene and a corresponding nuclear pseudogene. *Proc R Soc Lond B: Biol Sci* 262:13–19.

Ashley Jr, C. T. and S. T. Warren. 1995. Trinucleotide repeat expansion and human disease. *Annu Rev Genet* 29:703–728.

Ballard, J. W. O. and D. M. Rand. 2005. The population biology of mitochondrial DNA and its phylogenetic implications. *Annu Rev Ecol Evol Syst* 36:621–642.

Bejerano, G., M. Pheasant, I. Makunin et al. 2004. Ultraconserved elements in the human genome. *Science* 304:1321–1325.

Bensasson, D., M. W. Feldman, and D. A. Petrov. 2003. Rates of DNA duplication and mitochondrial DNA insertion in the human genome. *J Mol Evol* 57:343–354.

Bensasson, D., D. X. Zhang, D. L. Hartl, and G. M. Hewitt. 2001. Mitochondrial pseudogenes: Evolution's misplaced witnesses. *Trends Ecol Evol* 16:314–321.

Birol, I., A. Raymond, S. D. Jackman et al. 2013. Assembling the 20 Gb white spruce (*Picea glauca*) genome from whole-genome shotgun sequencing data. *Bioinformatics*. doi: 10.1093/bioinformatics/btt178.

Blanchard, J. L. and G. W. Schmidt. 1996. Mitochondrial DNA migration events in yeast and humans: Integration by a common end-joining mechanism and alternative perspectives on nucleotide substitution patterns. *Mol Biol Evol* 13:537–548.

Boore, J. L. 1999. Animal mitochondrial genomes. *Nucleic Acids Res* 27:1767–1780.

Brown, T. A. 2007. *Genomes* 3. New York: Garland Science/Taylor & Francis.

Brown, W. M., M. George, and A. C. Wilson. 1979. Rapid evolution of animal mitochondrial DNA. *Proc Natl Acad Sci USA* 76:1967–1971.

Brown, W. M., E. M. Prager, A. Wang, and A. C. Wilson. 1982. Mitochondrial DNA sequences of primates: Tempo and mode of evolution. *J Mol Evol* 18:225–239.

Casane, D., S. Boissinot, B. J. Chang, L. C. Shimmin, and W.-H. Li. 1997. Mutation pattern variation among regions of the primate genome. *J Mol Evol* 45:216–226.

Chang, D. and T. F. Duda. 2012. Extensive and continuous duplication facilitates rapid evolution and diversification of gene families. *Mol Biol Evol*. doi: 10.1093/molbev/mss068.

Costa, I. R., F. Prosdocimi, and W. B. Jennings. 2016. In silico phylogenomics using complete genomes: A case study on the evolution of hominoids. *Genome Res* 26:1257–1267.

De Koning, A. P., W. Gu, T. A. Castoe, M. A. Batzer, and D. D. Pollock. 2011. Repetitive elements may comprise over two-thirds of the human genome. *PLoS Genet* 7:e1002384.

Echols, H. and M. F. Goodman. 1991. Fidelity mechanisms in DNA replication. *Annu Rev Biochem* 60:477–511.

Ellis, J. 1982. Promiscuous DNA–chloroplast genes inside plant mitochondria. *Nature* 299:678–679.

Farrelly, F. and R. A. Butow. 1983. Rearranged mitochondrial genes in the yeast nuclear genome. *Nature* 301:296–301.

Felsenstein, J. 2004. *Inferring Phylogenies*. Sunderland: Sinauer.

Galtier, N., B. Nabholz, S. Glémin, and G. D. D. Hurst. 2009. Mitochondrial DNA as a marker of molecular diversity: A reappraisal. *Mol Ecol* 18:4541–4550.

Gellissen, G., J. Y. Bradfield, B. N. White, and G. R. Wyatt. 1983. Mitochondrial DNA sequences in the nuclear genome of a locust. *Nature* 301:631–634.

Gellissen, G. and G. Michaelis. 1987. Gene transfer. *Ann N Y Acad Sci* 503:391–401.

Golding, G. B. 1983. Estimates of DNA and protein sequence divergence: An examination of some assumptions. *Mol Biol Evol* 1:125–142.

Graur, D. and W.-H. Li. 2000. *Fundamentals of Molecular Evolution*, 2nd edition. Sunderland: Sinauer.

Graur, D., Y. Zheng, N. Price, R. B. Azevedo, R. A. Zufall, and E. Elhaik. 2013. On the immortality of television sets: "Function" in the human genome according to the evolution-free gospel of ENCODE. *Genome Biol Evol* 5:578–590.

Haas, N. B., J. M. Grabowski, A. B. Sivitz, and J. B. Burch. 1997. Chicken repeat 1 (CR1) elements, which define an ancient family of vertebrate non-LTR retrotransposons, contain two closely spaced open reading frames. *Gene* 197:305–309.

Hadler, H. I., B. Dimitrijevic, and R. Mahalingam. 1983. Mitochondrial DNA and nuclear DNA from normal rat liver have a common sequence. *Proc Natl Acad Sci USA* 80:6495–6499.

Harris, V. H., C. L. Smith, W. J. Cummins et al. 2003. The effect of tautomeric constant on the specificity of nucleotide incorporation during DNA replication: Support for the rare tautomer hypothesis of substitution mutagenesis. *J Mol Biol* 326:1389–1401.

Hazkani-Covo, E. and S. Covo. 2008. Numt-mediated double-strand break repair mitigates deletions during primate genome evolution. *PLoS Genet* 4:e1000237.

Hazkani-Covo, E., R. Sorek, and D. Graur. 2003. Evolutionary dynamics of large numts in the human genome: Rarity of independent insertions and abundance of post-insertion duplications. *J Mol Evol* 56:169–174.

Hazkani-Covo, E., R. M. Zeller, and W. Martin. 2010. Molecular poltergeists: Mitochondrial DNA copies (numts) in sequenced nuclear genomes. *PLoS Genet* 6:e1000834.

Henze, K. and W. Martin. 2001. How do mitochondrial genes get into the nucleus? *Trends Genet* 17:383–387.

Huang, C. Y., N. Grünheit, N. Ahmadinejad, J. N. Timmis, and W. Martin. 2005. Mutational decay and age of chloroplast and mitochondrial genomes transferred recently to angiosperm nuclear chromosomes. *Plant Physiol* 138:1723–1733.

Jacobs, H. T. and B. Grimes. 1986. Complete nucleotide sequences of the nuclear pseudogenes for cytochrome oxidase subunit I and the large mitochondrial ribosomal RNA in the sea urchin *Strongylocentrotus purpuratus*. *J Mol Biol* 187:509–527.

Jacobs, H. T., J. W. Posakony, J. W. Grula et al. 1983. Mitochondrial DNA sequences in the nuclear genome of *Strongylocentrotus purpuratus*. *J Mol Biol* 165:609–632.

Janes, D. E., C. L. Organ, M. K. Fujita, A. M. Shedlock, and S. V. Edwards. 2010. Genome evolution in Reptilia, the sister group of mammals. *Annu Rev Genomics Hum Genet* 11:239–264.

Kaiser, V. B., M. van Tuinen, and H. Ellegren. 2007. Insertion events of CR1 retrotransposable elements elucidate the phylogenetic branching order in galliform birds. *Mol Biol Evol* 24:338–347.

Katzman, S., A. D. Kern, G. Bejerano. 2007. Human genome ultraconserved elements are ultraselected. *Science* 317:915–915.

Keller, I., D. Bensasson, and R. A. Nichols. 2007. Transition–transversion bias is not universal: A counter example from grasshopper pseudogenes. *PLoS Genet* 3:e22.

Kim, J. H., A. Antunes, S. J. Luo et al. 2006. Evolutionary analysis of a large mtDNA translocation (numt) into the nuclear genome of the *Panthera* genus species. *Gene* 366:292–302.

Kordis, D. 2010. Transposable elements in reptilian and avian (sauropsida) genomes. *Cytogenet Genome Res* 127:94–111.

Kunkel, T. A. and K. Bebenek. 2000. DNA replication fidelity 1. *Annu Rev Biochem* 69:497–529.

Lee, S. I. and N. S. Kim. 2014. Transposable elements and genome size variations in plants. *Genomics Inform* 12:87–97.

Levinson, G. and G. A. Gutman. 1987. Slipped-strand mispairing: A major mechanism for DNA sequence evolution. *Mol Biol Evol* 4:203–221.

Lin, X., S. Kaul, S. Rounsley et al. 1999. Sequence and analysis of chromosome 2 of the plant *Arabidopsis thaliana*. *Nature* 402:761–768.

Lohse, M., O. Drechsel, S. Kahlau, and R. Bock. 2013. Organellar Genome DRAW—A suite of tools for generating physical maps of plastid and mitochondrial genomes and visualizing expression data sets. *Nucleic Acids Res*. doi: 10.1093/nar/gkt289.

Lopez, J. V., N. Yuhki, R. Masuda, W. Modi, and S. J. O'Brien. 1994. Numt, a recent transfer and tandem amplification of mitochondrial DNA to the nuclear genome of the domestic cat. *J Mol Evol* 39:174–190.

Margulis, L. 1970. *Origin of Eukaryotic Cells: Evidence and Research Implications for a Theory of the Origin and Evolution of Microbial, Plant, and Animal Cells on the Precambrian Earth*. New Haven: Yale University Press.

McClintock, B. 1950. The origin and behavior of mutable loci in maize. *Proc Natl Acad Sci USA* 36:344–355.

Moritz, C., T. E. Dowling, and W. M. Brown. 1987. Evolution of animal mitochondrial DNA: Relevance for population biology and systematics. *Annu Rev Ecol Syst* 1987:269–292.

Mourier, T., A. J. Hansen, E. Willerslev, and P. Arctander. 2001. The Human Genome Project reveals a continuous transfer of large mitochondrial fragments to the nucleus. *Mol Biol Evol* 18:1833–1837.

Ovcharenko, I., G. G. Loots, M. A. Nobrega, R. C. Hardison, W. Miller, and L. Stubbs. 2005. Evolution and functional classification of vertebrate gene deserts. *Genome Res* 15:137–145.

Pakendorf, B. and M. Stoneking. 2005. Mitochondrial DNA and human evolution. *Annu Rev Genomics Hum Genet* 6:165–183.

Palumbi, S. R. 1996. Chapter 7. Nucleic acids II: The polymerase chain reaction. In *Molecular Systematics*, 2nd edition, eds. D. M. Hillis, C. Moritz, and B. K. Mable, 205–247. Sunderland: Sinauer.

Perna, N. T. and T. D. Kocher. 1996. Mitochondrial DNA: Molecular fossils in the nucleus. *Curr Biol* 6:128–129.

Petrov, D. A. and D. L. Hartl. 1999. Patterns of nucleotide substitution in *Drosophila* and mammalian genomes. *Proc Natl Acad Sci USA* 96:1475–1479.

Rambaut, A. 2007. *Se-Al, version 2.0 a11*. Edinburgh: The University of Edinburgh.

Randi, E. 2000. Mitochondrial DNA. In *Molecular Methods in Ecology*, ed. A. J. Baker, 136–167. Oxford: Blackwell.

Ravindran, S. 2012. Barbara McClintock and the discovery of jumping genes. *Proc Natl Acad Sci USA* 109:20198–20199.

Ricchetti, M., C. Fairhead, and B. Dujon. 1999. Mitochondrial DNA repairs double-strand breaks in yeast chromosomes. *Nature* 402:96–100.

Richly, E. and D. Leister. 2004. NUMTs in sequenced eukaryotic genomes. *Mol Biol Evol* 21:1081–1084.

Sagan, L. 1967. On the origin of mitosing cells. *J Theor Biol* doi:10.1016/0022-5193(67)90079-3.

Scribner, K. T. and J. M. Pearce. 2000. Microsatellites: Evolutionary and methodological background and empirical applications at individual, population, and phylogenetic levels. In *Molecular Methods in Ecology*, ed. A. J. Baker, 235–273. Oxford: Blackwell.

Shedlock, A. M. 2006. Phylogenomic investigation of CR1 LINE diversity in reptiles. *Syst Biol* 55:902–911.

Shedlock, A. M., C. W. Botka, S. Zhao et al. 2007. Phylogenomics of nonavian reptiles and the structure of the ancestral amniote genome. *Proc Natl Acad Sci USA* 104:2767–2772.

Simons, C., M. Pheasant, I. V. Makunin, and J. S. Mattick. 2006. Transposon-free regions in mammalian genomes. *Genome Res* 16:164–172.

Skippington, E., T. J. Barkman, D. W. Rice, and J. D. Palmer. 2015. Miniaturized mitogenome of the parasitic plant *Viscum scurruloideum* is extremely divergent and dynamic and has lost all nad genes. *Proc Natl Acad Sci USA* 112:E3515–E3524.

Smeds, L., A. Qvarnström, and H. Ellegren. 2016. Direct estimate of the rate of germline mutation in a bird. *Genome* Res 26:1211–1218.

Smith, M. F., W. K. Thomas, and J. L. Patton. 1992. Mitochondrial DNA-like sequence in the nuclear genome of an akodontine rodent. *Mol Biol Evol* 9:204–215.

Sorenson, M. D. and T. W. Quinn. 1998. Numts: A challenge for avian systematics and population biology. *Auk* 115:214–221.

Souto, H. M., P. A. Ruschi, C. Furtado, W. B. Jennings, and F. Prosdocimi. 2014. The complete mitochondrial genome of the ruby-topaz hummingbird *Chrysolampis mosquitus* through Illumina sequencing. *Mitochondrial DNA* 27:769–770.

Stupar, R. M., J. W. Lilly, C. D. Town et al. 2001. Complex mtDNA constitutes an approximate 620-kb insertion on *Arabidopsis thaliana* chromosome 2: Implication of potential sequencing errors caused by large-unit repeats. *Proc Natl Acad Sci USA* 98:5099–5103.

Swofford, D. L., G. J. Olsen, P. J. Waddell, and D. M. Hillis. 1996. Chapter 11. Phylogenetic inference. In *Molecular Systematics*, 2nd edition, eds. D. M. Hillis, C. Moritz, and B. K. Mable, 407–514. Sunderland: Sinauer.

Tamura, K. and M. Nei. 1993. Estimation of the number of nucleotide substitutions in the control region of mitochondrial DNA in humans and chimpanzees. *Mol Biol Evol* 10:512–526.

Tourmen, Y., O. Baris, P. Dessen, C. Jacques, Y. Malthièry, and P. Reynier. 2002. Structure and chromosomal distribution of human mitochondrial pseudogenes. *Genomics* 80:71–77.

Tsuzuki, T., H. Nomiyama, C. Setoyama, S. Maeda, and K. Shimada. 1983. Presence of mitochondrial-DNA-like sequences in the human nuclear DNA. *Gene* 25:223–229.

Upholt, W. B. 1977. Estimation of DNA sequence divergence from comparison of restriction endonuclease digests. *Nucleic Acids Res* 4:1257–1266.

van den Boogaart, P., J. Samallo, and E. Agsteribbe. 1982. Similar genes for a mitochondrial ATPase subunit in the nuclear and mitochondrial genomes of *Neurospora crassa*. *Nature* 298:187–189.

Venter, J. C., M. D. Adams, E. W. Myers et al. 2001. The sequence of the human genome. *Science* 291:1304–1351.

Wakasugi, S., N. Hisayuki, F. Makoto, T. Teruhisa, and S. Kazunori. 1985. Insertion of a long KpnI family member within a mitochondrial-DNA-like sequence present in the human nuclear genome. *Gene* 36:281–288.

Wakeley, J. 1994. Substitution-rate variation among sites and the estimation of transition bias. *Mol Biol Evol* 11:436–442.

Wakeley, J. 1996. The excess of transitions among nucleotide substitutions: New methods of estimating transition bias underscore its significance. *Trends Ecol Evol* 11:158–162.

Wang, W., H. W. Hellinga, and L. S. Beese. 2011. Structural evidence for the rare tautomer hypothesis of spontaneous mutagenesis. *Proc Natl Acad Sci USA* 108:17644–17648.

Waples, R. S. and O. Gaggiotti. 2006. INVITED REVIEW: What is a population? An empirical evaluation of some genetic methods for identifying the number of gene pools and their degree of connectivity. *Mol Ecol* 15:1419–1439.

Watson, J. D., T. A. Baker, S. P. Bell, A. Gann, M. Levine, and R. Losick. 2014. *Molecular Biology of the Gene*, 7th edition. New York: Pearson Education, Inc.

Watson, J. D. and F. H. C. Crick. 1953a. Genetical implications of the structure of deoxyribonucleic acid. *Nature* 171:964–967.

Watson, J. D. and F. H. C. Crick. 1953b. Molecular structure of nucleic acids. *Nature* 171:737–738.

Wilson, A. C., R. L. Cann, S. M. Carr et al. 1985. Mitochondrial DNA and two perspectives on evolutionary genetics. *Biol J Linn Soc Lond* 26:375–400.

Wright, R. M. and D. J. Cummings. 1983. Integration of mitochondrial gene sequences within the nuclear genome during senescence in a fungus. *Nature* 302:86–88.

Yang, Z. 1996. Among-site rate variation and its impact on phylogenetic analyses. *Trends Ecol Evol* 11:367–372.

Yang, Z. 2006. *Computational Molecular Evolution* (Vol. 21). Oxford: Oxford University Press.

Yang, Z. and A. D. Yoder. 1999. Estimation of the transition/transversion rate bias and species sampling. *J Mol Evol* 48:274–283.

Yu, X. and A. Gabriel. 1999. Patching broken chromosomes with extranuclear cellular DNA. *Mol Cell* 4:873–881.

Zhang, D. X. and G. M. Hewitt. 1996. Nuclear integrations: Challenges for mitochondrial DNA markers. *Trends Ecol Evol* 11:247–251.

Zhang, Z. and M. Gerstein. 2003. Patterns of nucleotide substitution, insertion and deletion in the human genome inferred from pseudogenes. *Nucleic Acids Res* 31:5338–5348.

Zhang, Z., P. Harrison, and M. Gerstein. 2002. Identification and analysis of over 2000 ribosomal protein pseudogenes in the human genome. *Genome Res* 12:1466–1482.

Zischler, H., H. Geisert, and J. Castresana. 1998. A hominoid-specific nuclear insertion of the mitochondrial D-loop: Implications for reconstructing ancestral mitochondrial sequences. *Mol Biol Evol* 15:463–469.

CHAPTER THREE

Properties of DNA Sequence Loci: Part II

In Chapter 2, we reviewed the composition of eukaryotic genomes and aspects of molecular evolution in order to lay the groundwork for our understanding of the important properties of DNA sequence loci. Some locus properties may be beneficial to one type of phylogenomic study or problematic to another. For example, to the worker who is interested in studying the evolutionary diversification of a gene family such as the mitochondrial pseudogenes of a particular genome, the property of being a duplicated or "multicopy" gene will be of intrinsic interest. In contrast, to the worker who is interested in using mitochondrial genes as evolutionary markers for purposes of tracking organismal diversification or for identifying species (i.e., DNA barcoding), the existence of mitochondrial pseudogenes is a major source of concern and therefore special precautionary steps must be taken to avoid potential problems.

We will now continue our discussion of loci properties by focusing on several commonly made assumptions in phylogenomic analyses particularly those pertaining to tree of life studies. These assumptions include that each locus (1) exists as a single-copy in the genome; (2) has not been under the direct or indirect influence of natural selection (i.e., is selectively neutral); (3) has a gene tree that is independent of the gene trees of other sampled loci; (4) has not experienced a recombination event since the most recent common ancestor of the sampled sequences (i.e., all sites have the same gene tree); (5) has a constant substitution rate among lineages (i.e., molecular clock); and (6) has been chosen randomly from the genome (i.e., no ascertainment bias). Following this discussion, we will briefly revisit the conceptual meanings of important terms such as locus, gene, allele, gene copy, and haplotype before we review the actual types of loci used in phylogenomic studies.

3.1 SIX ASSUMPTIONS ABOUT DNA SEQUENCE LOCI IN PHYLOGENOMIC STUDIES

3.1.1 Assumption 1: Loci Are Single-Copy in the Genome

In Chapter 2, we saw that certain types of loci have been duplicated one or more times and thus exist as multiple copies in genomes. These include tandemly repeated loci, transposable elements, *numts*, and perhaps others. While the existence of these gene families is of great interest to researchers interested in molecular evolution, these multicopy genes can be a nightmare for the organismal biologist. This is because gene trees inferred from different copies of a homologous gene reveal the history of gene duplication events, but they obfuscate the phylogenetic history of the study organisms even if the gene tree is correct (Moritz and Hillis 1996). Thus, the single-copy locus assumption is fundamental to phylogenomics studies of the tree of life (Moritz and Hillis 1996; Delsuc et al. 2005; Philippe et al. 2005; Philippe and Blanchette 2007).

The term *homology* refers to common ancestry and thus DNA sequences displaying similar sequences are presumed to be homologous (Moritz and Hillis 1996). However, because homology must be inferred it is important to realize that similarity does not equate with homology (Moritz and Hillis 1996 and references therein). Nonetheless, in practice researchers typically use software to align sequences based on the site-by-site similarities or what is referred to as *positional homology* (Moritz and Hillis 1996). Obtaining a satisfactory alignment is usually trivial when aligning either recently diverged sequences or protein-coding sequences, whereas it can be much more challenging or impossible if numerous insertion and deletion mutations have occurred such as in the loop structures of ribosomal RNA genes or in introns between anciently diverged organisms (Moritz and Hillis 1996).

For molecular studies, it is necessary to distinguish several different homology concepts owing to the nature of how molecular sequences can evolve particularly with respect to the processes of gene duplication and horizontal gene transfer (HGT). Fitch (1970) first brought this topic to the attention of molecular phylogeneticists when he pointed out that two proteins (or their underlying DNA sequences) can be similar either because (A) they descended with divergence from a gene in a common ancestor via speciation (viz. Darwin's descent with modification) or (B) because the two sequences descended with convergence from two separate ancestors via gene duplication (Fitch 1970). To distinguish these forms of "homology," Fitch (1970) coined the term *orthology* to characterize similar sequences in scenario (A) and *paralogy* to describe similar sequences following scenario (B). Thus, if a researcher uses a particular genomic locus that has not been duplicated, then all sequences obtained from other individuals (or species) will likely represent orthologous copies. On the other hand, if a researcher interested in organismal phylogeny obtains a mixed sample of "homologous" sequences that includes orthologous and paralogous sequences, then the inferred gene tree for these sequences will almost certainly provide nonsensical or misleading results.

As an example of how paralogous DNA sequences can cause problems in organismal phylogenomic studies let's look at the well-studied problem of mitochondrial pseudogenes. *Numts* are especially worrisome when PCR is used to obtain the target mitochondrial sequences because there are at least two adverse scenarios possible. In the first scenario, the target mitochondrial gene and *numt* are coamplified in PCR (Zhang and Hewitt 1996; Sorenson and Quinn 1998; Bensasson et al. 2001). Unless the two amplification products have the identical sequence, which could occur if the *numt* represents a recent primary integration, then it is likely that the resulting Sanger sequence will be of poor quality if not ruined (we will see why in Chapter 6). In the second scenario, a *numt* is preferentially amplified over the target mitochondrial gene (Smith et al. 1992; Collura and Stewart 1995), which can lead to datasets consisting of all sequences from one *numt* locus, mixtures of *numt* paralogues, or mixtures of *numts* and the target mitochondrial locus.

Numt-contaminated datasets can cause a variety of problems ranging from obtaining poor quality DNA sequences to spurious interpretations of phylogenomic results. Specifically, results based on *numt*-contaminated datasets can mislead researchers interested in using the gene tree as a proxy of the species tree (Zhang and Hewitt 1996; Bensasson et al. 2001). *Numts* can also be problematic in studies of mitochondrial diseases in which discovery of "novel" mutations in *numts* might be falsely linked to diseases (Hazkani-Covo et al. 2010). Also because substitution rates for *numt* sites are 10-fold slower than their mitochondrial counterparts (Brown et al. 1979), the accidental analysis of *numt* sequences could lead to grossly underestimated evolutionary distances. This is especially a concern for DNA barcoding studies because genetic distances are often used to identify cryptic species and thus spurious distance estimates may lead to underestimates of biodiversity. *Numts* can also confound phylogeographic analyses because their effective population sizes are four times larger than mitochondrial genes (Zhang and Hewitt 1996; Zink and Barrowclough 2008). Sorenson and Quinn (1998) reviewed the *numt* phenomenon in birds and suggested a number of laboratory strategies for preventing the unintended sequencing of *numts* as well as methods for identifying putative *numts* in existing DNA sequence datasets.

Not all types of paralogous loci will likely cause problems in tree of life studies. Some tandemly repeated loci such as nuclear rDNA arrays, because of their proximity to each other along the chromosome, usually evolve via *concerted evolution* (Zimmer et al. 1980; Hillis and Dixon 1991; Moritz and Hillis 1996). Thus, as the tandemly arrayed loci evolve together, the conceptual distinction between the original orthologous versus adjacent paralogous copies is largely erased (Moritz and Hillis 1996). For example, Hillis and Davis (1986) analyzed variation within tandem arrays of nuclear ribosomal DNA genes in a 50 million year old group of ranid frogs in order to examine the evolution of the genes and to infer a molecular phylogeny.

Lastly, there is yet another form of homology that we need to define for cases in which genes are transferred horizontally from one species' genome to another. This is frequently referred to as HGT or lateral gene transfer. The term *xenology* was coined to describe genes that are transferred from the genome of one species to the genome of another species (Gray and Fitch 1983). HGT has played a significant role in the evolution of prokaryotic and eukaryotic genomes. For further discussion and perspectives about the various homology concepts, the reader should consult the article by Mindell and Meyer (2001).

3.1.2 Assumption 2: Loci Are Selectively Neutral

Neutral coalescent theory-based approaches to estimating phylogeographic and species tree parameters typically require that each locus has not been directly or indirectly influenced by natural selection (e.g., Hudson and Coyne 2002). In other words, neutral sites evolve only via mutation and genetic drift (e.g., Fu and Li 1993; Wakeley 2009). This assumption is often stated as loci are assumed to be *selectively neutral*. By harvesting large numbers of selectively neutral loci from genomes, researchers can take advantage of coalescent-based methods for inferring the evolutionary history of populations and species. However, this discussion about selectively neutral DNA presumes that genomes contain functionless DNA, an idea that has been controversial for a long time.

3.1.2.1 Does "Junk DNA" Exist?

During the 1960s, two different but related lines of thought were aired among geneticists, which set into motion a long-standing debate of importance to the field of molecular evolution. First, researchers postulated the existence of functionless or "junk" DNA (Graur et al. 2015 and references therein). Ohno (1972) later defined junk DNA as DNA on which natural selection does not operate. In other words, junk DNA does not presently have any advantageous or deleterious effect on the fitness of its carrier. Secondly, the introduction of the neutral theory of molecular evolution by Kimura (1968) and King and Jukes (1969) represented the other important development (see Gillespie 2004 for review; Eisen 2012). The empirical results of the first genetic electrophoresis study by Lewontin and Hubby (1966) revealed much higher levels of genetic (protein) variation in natural *Drosophila pseudoobscura* populations than would have been expected due to prevailing thoughts at that time about molecular evolution—namely, that most mutations were harmful and therefore would be promptly eliminated from populations via purifying natural selection (a.k.a. the "Darwinian" evolution viewpoint; Yang 2006; Eisen 2012). But the existence of such high levels of genetic variability in *Drosophila* populations demanded an explanation and the thoughts about "neutral evolution" in the papers by Kimura and by King and Jukes helped accommodate these surprising results (Yang 2006; Eisen 2012). This opposing viewpoint was encapsulated in the title of the King and Jukes (1969) paper: *Non-Darwinian Evolution*.

What exactly is junk DNA and does it really exist? Recall from Chapter 2 that intergenic DNA is largely comprised of DNA with no known function. Comparative genome studies provide evidence supporting the hypothesis that most intergenic DNA is without biological function. Meader et al. (2010) and Ponting and Hardison (2011) suggested that about 85% of the human genome is comprised of nonfunctional DNA. The results of another study suggest that an even higher fraction—95% of the human genome—is comprised of nonfunctional DNA (Lindblad-Toh et al. 2011). However, this idea that any genome could contain a large amount of functionless DNA has remained controversial and the human genome, in particular, has been the primary battleground (ENCODE Project Consortium 2012; Hurtley 2012; Pennisi 2012; Doolittle 2013; Graur et al. 2013; Palazzo and Gregory 2014; Graur et al. 2015; Palazzo and Lee 2015).

For many researchers there is no longer any question as to whether junk DNA exists. Recent papers (Doolittle 2013; Graur et al. 2013; Palazzo and Gregory 2014; Palazzo and Lee 2015) have provided strong arguments in favor of junk DNA's existence. The junk DNA controversy has arisen because researchers conflated different meanings of the word "function." For example, ENCODE researchers studying the human genome used data indicating the existence of a biochemical process (e.g., transcription) as de facto evidence for those genomic sequences being "functional" without consideration if those sequences may have been maintained by natural selection (Eisen 2012; Graur et al. 2013). This does not mean that *some* of the currently classified nonfunctional DNA in genomes won't eventually be shown to have an essential function maintained by purifying selection (Eisen 2012; Graur et al. 2013). However, at least in vertebrate genomes, there are vast tracts of intergenic DNA that are devoid of any evolutionary conservatism (Ovcharenko et al. 2005).

Nobel Prize winner Sydney Brenner offered perspective by way of a metaphor as to why we should be more comfortable with the concepts of functional and nonfunctional DNA. According to Brenner (1998), in the broadest sense, nonfunctional DNA is simply *rubbish DNA*. Within this category, Brenner described two subcategories defined on the basis of their evolutionary effects. The first category is *junk DNA*, which is DNA that does not help or harm the organism but could be useful in the future (viz. Ohno 1972). Brenner used the metaphor of rubbish in someone's garage—if the rubbish doesn't cause any problems for the homeowner, then such junk can persist in that space indefinitely. The second category is *garbage DNA*, which constitutes DNA that adversely affects the individual's fitness. In this case, natural selection would sooner or later eliminate such DNA from the genome much in the way that someone discards hazardous garbage stored in their garage (e.g., toxic chemicals).

Graur et al. (2015) proposed an evolutionary classification scheme for DNA elements that is based in part on Brenner's outline. In this scheme, genomic elements are first classified as being "functional" or "rubbish" depending on whether natural selection has maintained those elements or not. Thus, functional DNA includes DNA that has been selected for a particular function, whereas rubbish DNA has not been selected for a function. Functional DNA, in turn, is comprised of two subcategories called "literal DNA" and "indifferent DNA." Literal DNA is defined as DNA whose actual nucleotide sequence has largely been maintained by selection, while indifferent DNA is defined as DNA that is also under the effects of selection but only for its presence in the genome. Rubbish DNA is also comprised of two subcategories: "junk DNA" is neutral with respect to the fitness of the organism, whereas "garbage DNA" diminishes the fitness of the individual.

Thus, until demonstrated otherwise, it is safe to assume that functionless DNA does exist in genomes. As we will see in Chapter 8, this junk DNA component of genomes represents a vast and largely untapped source for DNA sequence loci that are especially well suited to some types of phylogenomic analyses (e.g., coalescent-based studies of species trees).

3.1.2.2 The Neutrality Assumption and the Indirect Effects of Natural Selection

Some genomic sites such as those within introns can tolerate base substitutions and indels, which probably do not affect the fitness of the individual. Therefore, in a sense, they are selectively neutral mutations. Although sites not maintained by selection may appear to be "neutral" sites, this does not necessarily mean that all of these sites will meet the neutrality assumption in phylogenomic studies. The assumption of selective neutrality requires each locus to be free not only of direct natural selection but also from the indirect effects as well. In other words, a site (or locus) that meets this assumption will have a genealogy (gene tree) that has not been distorted by any form of natural selection. However, purifying or positive selection acting on functionally important sites can influence the genealogical histories of linked nonfunctional sites; when this occurs, these gene trees will not reflect the gene tree shapes predicted by neutral theory (Kaplan et al. 1989; Charlesworth et al. 1993, 1995; Charlesworth 2012).

Earlier in this chapter we saw that a large fraction of some genomes (e.g., human) is comprised of nonfunctional DNA. Does any of this nonfunctional DNA consist of sites that are also free of the *indirect* effects of selection and thus represent selectively neutral DNA? Lee and Edwards (2008) compared the nucleotide diversities (π), which is a measure of genetic diversity, of 29 anonymous loci versus six intron loci and one mitochondrial (ND2) gene. Anonymous loci are from random genomic locations and thus many of them are expected to be located far from functional genomic elements, whereas introns contain some functional sites and are located adjacent to exons. In Figure 3.1, we see that the genetic diversities of anonymous loci tend to be higher than for the introns or ND2 coding region. These results are compelling because it suggests that genomes may contain many loci that are effectively free of all

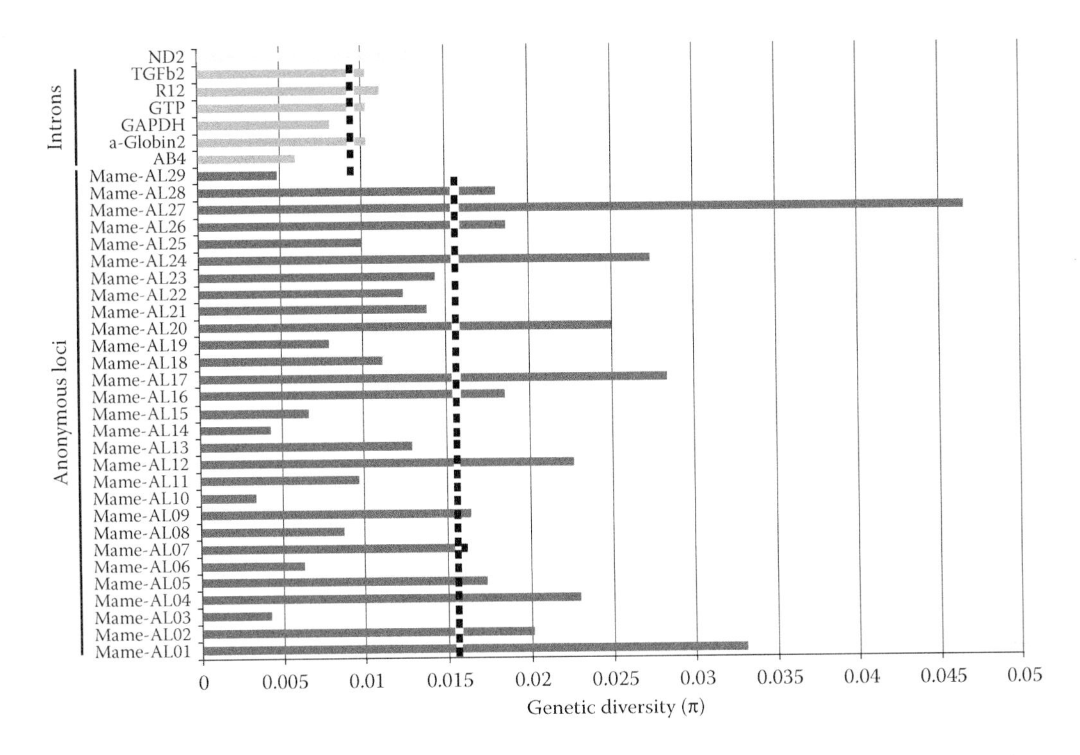

Figure 3.1. Comparison of genetic diversities among different types of DNA sequence loci (Lee and Edwards 2008; fig. 2). The legend reads, "Genetic diversity (π) across 29 nuclear markers, six introns, and ND2 region. Dotted vertical bars represent the mean genetic diversity of introns and nuclear anonymous markers." (Reprinted from Lee, J. Y. and S. V. Edwards. 2008. *Evolution* 62:3117–3134. With permission.)

forms of natural selection and hence best meet the assumption of neutrality. Why are these anonymous loci exhibiting higher levels of genetic diversity than the introns and ND2 gene?

A substantial body of theory has built up in the last few decades, mainly based on studies of the *Drosophila* genome, which shows how genomic sites maintained by selection can reduce or increase the genetic diversity (relative to neutral theory expectations) of linked sites not directly under selection (Felsenstein 1974; Maynard-Smith and Haigh 1974; Thomson 1977; Kaplan et al. 1989; Begun and Aquadro 1992; Charlesworth et al. 1993, 1995; Wiehe and Stephan 1993; Hudson and Kaplan 1995; Innan and Stephan 2003; Stephan 2010; Charlesworth 2012). Although work in this area was underway in the 1970s, later empirical findings by researchers greatly stimulated further elaboration of theoretical and empirical work in the 1990s (Stephan 2010; Charlesworth 2012). One of the key empirical discoveries was that genome-wide variation in π (a measure of genetic diversity) within the *Drosophila melanogaster* genome was correlated with local recombination rates (Begun and Aquadro 1992). This genetic pattern, which is illustrated in Figure 3.2, shows that the genetic diversities of 15 loci in the *D. melanogaster* genome are strongly associated with local recombination rates (see also Figure 1 in Begun and Aquadro 1992 and the discussion by Sella et al. 2009). As most functional genomic loci are presumably under purifying selection, we would predict that selection targeting functional sites should reduce the genetic diversity of other nearby sites simply by virtue of their close linkage. In view of this, the empirical results in Figure 3.2 suggests that recombination has the ability to dampen the effects of indirect selection on other sites and that this effect increases with higher recombination rates. As we will soon see, physical distances and effective population sizes also play important roles in diminishing the effects of indirect selection on linked nonfunctional loci. Similar empirical findings have been documented in other organisms (see Innan and Stephan 2003 for review). We will review the leading explanations for this pattern and discuss the implications for phylogenomic loci.

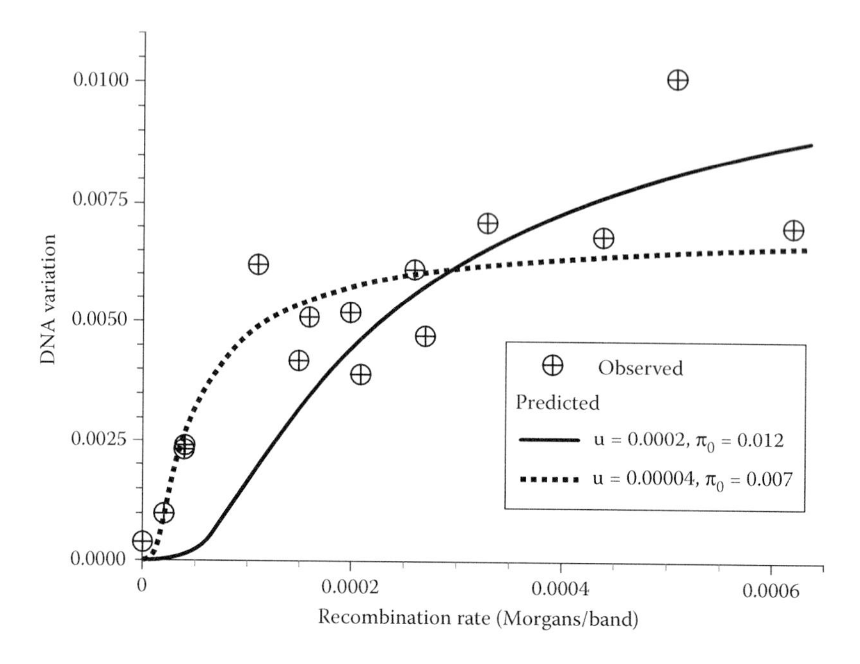

Figure 3.2. DNA variation as a function of local recombination rates in *D. melanogaster* (Hudson and Kaplan 1995; fig. 2). The legend reads, "Observed and predicted levels of DNA variation as a function of local recombination rates. The observed levels of DNA variation are estimates of theta for 15 loci on the third chromosome of *D. melanogaster* obtained from E. Kindahl and C. Aquadro (personal communication). These loci are, from left to right on the figure, *Lsply, Pc, Antp, Gld, Ubx, tra, fz, Mlc2, ry, Sod, Tl, Rh3, Est6, E(spl)*, and *Hsp26*. The local recombination rates for these loci were also provided by E. Kindahl and C. Aquadro. The predicted levels of variation are obtained with (9) with the parameter values indicated on the figure." (Reprinted from Hudson, R. R. and N. L. Kaplan. 1995. *Genetics* 141:1605–1617. With permission.)

The phenomenon of *genetic hitch-hiking* is a form of indirect or correlated selection in which a beneficial mutation quickly spreads throughout the population, which results in a reduction of the genetic diversity at that locus as well as at closely linked sites (Maynard-Smith and Haigh 1974; Thomson 1977; Kaplan et al. 1989; Stephan et al. 1992; Wiehe and Stephan 1993). This phenomenon is also called *selective sweeps* (e.g., Wakeley 2009). Another form of indirect selection called **background selection** occurs when purifying selection acts directly against deleterious mutations, which has the effect of diminishing the genetic diversity of that locus and closely linked sites (Charlesworth et al. 1993, 1995; Hudson and Kaplan 1995; Charlesworth 2012). Felsenstein (1974) had previously described this phenomenon as the *Hill-Robertson Effect* (Charlesworth 2012). *Balancing selection* can also influence the genetic diversity of linked sites. However, in contrast to hitch-hiking, this mode of selection maintains a greater amount of allelic diversity at closely linked sites than predicted by neutral theory (Kaplan et al. 1989). The MHC genes in vertebrates are a classic example of genes influenced by balancing selection and they can exhibit extraordinary levels of genetic diversity (e.g., 200 alleles; Hess and Edwards 2002; Balakrishnan et al. 2010).

The amount of influence that a selected site will have on nonfunctional sites located on the same chromosome is determined by the strength of the linkage relationship. This linkage can be thought of as a correlation in which sites that are tightly linked together have genealogies that are highly correlated with each other. For example, sites within a short locus are often linked to the extent that they all have identical gene trees. This is actually a common phylogenomic assumption, which we will discuss in detail later in this chapter. In contrast, two sites that are weakly linked together such as those separated by large physical distances along the chromosome (on the order of hundreds or thousands of kb) could have uncorrelated genealogies depending on other factors (see below). In the context of indirect selection, if a site that is under selection is tightly linked to a nonfunctional site, the effect will then be stronger because the two sites will remain associated for longer amounts of time (Santiago and Caballero 1998). On the other hand, if the rates of local recombination and outcrossing are high, then the effect declines rapidly with physical distance (Charlesworth et al. 1993; Hudson and Kaplan 1995).

Physical distance alone is an inadequate measure for characterizing the linkage relationship between two loci when we are concerned about the property of genealogical independence. First, it does not consider the fact that local recombination rates may vary across the genome (e.g., Reich et al. 2001; McVean et al. 2004; Hartl and Jones 2006). Secondly, a physical distance measure does not account for the effective population size, which may also vary across the genome (e.g., Hudson and Kaplan 1995). Physical distance, local recombination rates, and effective population sizes all play a role in determining the strength of association between two loci and their gene trees (Kaplan et al. 1989; Charlesworth et al. 1993).

The *recombinational distance* or Cm (= $2N_ecm$) combines these factors—effective population size (N_e), per generation recombination rate (c), and physical distance in bp between a selected site and a distant site not directly under selection (m)—into a single measure that can gauge the degree of genealogical independence between two genomic sites (Kaplan et al. 1989). How large does the recombinational distance need to be in order for a nonfunctional site to escape the effects of selection? Kaplan et al. (1989) defined M as the maximum recombinational distance in which a selected site can influence a distant site that is not directly under selection. In other words, when the physical distance between sites is ≥M/C bp, then the two sites will be independent of each other (Kaplan et al. 1989).

Unfortunately, the recombinational distance is more of conceptual than practical value for us at this time. For example, from this quantity we can see that the indirect effects of selection on a distant nonfunctional site (or locus) can be weakened with large population sizes, high rates of crossing over, or large physical distances (Kaplan et al. 1989; Charlesworth et al. 1993). The recombinational "distance" reminds us that population size also plays a large role (Hudson and Kaplan 1995). If a linkage map is available for a species, then a rough distance measure consisting of map units might be used *in lieu* of M (or M/C). For example, Hudson and Kaplan (1995) suggested that sites separated by "a few map units" in the *D. melanogaster* genome would likely not influence each other, at least via background selection. Costa et al. (2016) discuss a genealogically-based approach for specifying threshold distances between known functional elements and distantly located nonfunctional elements in order to identify neutral loci candidates.

3.1.3 Assumption 3: Sampled Loci Have Independent Gene Trees

A common assumption of phylogeographic and species tree analyses that use multilocus coalescent methods holds that all sampled loci (or individual sites) have *genealogically independent histories* or, more simply, *independent gene trees* (Jennings 2016). Such genealogically independent loci are often referred to as "independent loci" or "unlinked loci." The property of independence among sampled loci is important from a statistical point of view because it means that individual loci and hence gene trees inferred from them—can be considered *true replicates* that depict the ancestry of a genome (Edwards and Beerli 2000; Arbogast et al. 2002; Wakeley 2009). Thus, by increasing the numbers of independent loci in a sample, a researcher can obtain more accurate and precise parameter estimates than from datasets with fewer loci (Pluzhnikov and Donnelly 1996; Jennings and Edwards 2005; Felsenstein 2006; Lee and Edwards 2008; Smith et al. 2013; Costa et al. 2016). We will now consider a hypothetical example for a common type of phylogeographic analysis that shows the benefits of many independent loci.

Let's say you want to test the hypothesis that two closely related species diverged from each other during the Pleistocene (i.e., between 0.01 and 2 million years ago). You plan to use two different datasets, which are analyzed separately: one dataset has 12 mitochondrial loci while the other has 12 independent nuclear genes. Let's further assume the resulting two divergence estimates are the following: mtDNA divergence is 3 million years ago, whereas the nuclear estimate is 1.5 million years ago. Given the discrepancy between the two divergence time estimates, which one would you prefer? You should prefer the nuclear dataset results for two reasons. First, a sample of independent nuclear loci permits you to use multilocus coalescent-based methods, which more accurately estimate population divergence times than single locus estimates (Edwards and Beerli 2000). Recall from Chapter 2 that the mitochondrial genome effectively does not recombine therefore all of its sites can be represented by a single gene tree (e.g., Wilson et al. 1985). Secondly, the nuclear gene estimate will be bracketed by confidence intervals, whereas such confidence intervals cannot be made for mitochondrial genes because they collectively represent a sample size of one gene. Large numbers (dozens or more) of independent loci can produce more accurate and precise parameter estimates than datasets consisting of fewer independent loci. See the studies by Jennings and Edwards (2005), Lee and Edwards (2008), Smith et al. (2013), and Costa et al. (2016) for empirical examples.

3.1.3.1 How Many Independent Loci Exist in Eukaryotic Genomes?

Until recent years, there were few published phylogenomic studies with statistically meaningful numbers of presumably independent and neutral DNA sequence loci (e.g., Chen and Li 2001; Jennings and Edwards 2005; Lee and Edwards 2008). However, thanks to the advent of NGS, researchers are now able to assemble datasets with orders of magnitude more loci for nonmodel organisms than was possible before. For example, NGS-based studies of birds (Faircloth et al. 2012; Jarvis et al. 2014; Prum et al. 2015) obtained hundreds to thousands of presumably independent loci for use in multilocus coalescent analyses in order to infer species trees. Complete genome *in silico* approaches are also generating large datasets consisting of hundreds of independent loci (McCormack et al. 2012; Costa et al. 2016). These enormous datasets show that the genomics era is already making a huge impact on phylogenomics. However, as it is becoming easier and less expensive to obtain large numbers of loci, Costa et al. (2016) asked a new question of importance to phylogenomic studies: *How many genealogically independent loci exist in eukaryotic genomes?*

Let's consider this question using our own genome. Given the vastness of the 3.2-Gb human genome, there would seemingly be an unlimited supply of independent loci. Felsenstein (P. 466, 2004) points out that there may be a million different gene trees that describe the ancestry of the human genome. Let's pause for a moment to think about this figure. While this is a realistic estimate for all gene trees that characterize the evolution of our genome, it is important to realize that this number represents all the different gene trees regardless of whether they are genealogically independent of each other or not. Thus, the term "different" used in this context simply means that the topologies of the trees change as different sites or blocks of linked sites are considered along the lengths of chromosomes. Accordingly, "different gene trees" is not necessarily the same as "genealogically independent gene trees." This conceptual distinction is important. We will soon see that the number of *independent* gene trees in the human genome is likely a miniscule fraction of a million.

Hudson and Coyne (2002) derived an equation for estimating the number of independent loci or what they termed *independent genealogical units* "IGUs" in a genome, which are defined as the number of loci in a genome whose passage to monophyly is nearly independent of that for all other sampled loci. These workers pointed out that under a neutral model the value of r^2, which is the expected degree of statistical association between two loci, will be $\sim 1/4N_ec$ when $4N_ec$ is large (Ohta and Kimura 1971). Thus, if statistical independence between two loci is achieved when $r^2 = 0.001$, then $4N_ec = 1/0.001 = 1{,}000$. Therefore, the approximate number of IGUs is expected to be:

$$\text{IGUs} = \frac{4N_ec}{1{,}000} \qquad (3.1)$$

where N_e is the effective population size and *c* is the per generation recombination rate between two loci (Hudson and Coyne 2002; Costa et al. 2016 provided this equation in the generic format shown here). Using this equation, Hudson and Coyne (2002) estimated that the common fruit fly (*D. melanogaster*) genome has approximately 11,500 IGUs. How many IGUs are in our genome? Costa et al. (2016) performed the following calculations in order to estimate the number of IGUs in the human genome. The linkage map for humans is 3,614 centimorgans or "cM" (Kong et al. 2002) and estimates for the effective population size range from 7,500 (Tenesa et al. 2007) to 10,000 (Takahata 1993). Using this information and Equation 3.1 the estimated minimum and maximum number of IGUs in the human genome are

$$\text{IGUs}_{\text{min}} = \frac{(4 \times 7{,}500 \times 3{,}614\ \text{cM} \times 0.01\ \text{cross-overs per generation per cM})}{1{,}000}$$

$$\text{IGUs}_{\text{min}} = 1{,}084$$

and,

$$IGUs_{maximum} = \frac{(4 \times 10{,}000 \times 3{,}614 \text{ cM} \times 0.01 \text{ cross overs per generation per cM})}{1{,}000}$$

$$IGUs_{maximum} = 1{,}446$$

These results suggest that the human genome only contains ~1,000–1,400 independent loci, which is perhaps shocking on two levels. First, it implies that the number of independent gene trees comprises a tiny fraction (~1/1,000) of the total number of gene trees that characterize the ancestry of the human genome. Second, the number of IGUs in the genome of *D. melanogaster* is ten times higher than the number estimated for the human genome, even though the genome size of the former is 18 times smaller than the genome of the later (Brown 2007). However, the difference in the numbers of IGUs between fruit fly and human genomes is explained by the far larger effective population size of *D. melanogaster* ($N_e = 10^6$; Kreitman 1983).

To gain additional perspective on the numbers of IGUs per genome we will now examine several other animal species for which adequate data exist. Table 3.1 shows the relevant population size and genome data for several animal lineages including fruit fly, human, hominoid, zebra finch, grass finch, and tiger salamander. The numbers of IGUs vary substantially largely depending on the assumed N_e values: for tiger salamanders, the number of IGUs is only ~210 if a low N_e is assumed, whereas this number jumps to 21,000 when a larger population size is used to calculate the IGU number (Table 3.1). Species with extremely large N_e such as the zebra finch may have tens of thousands of IGUs as suggested by these calculations. The long-term average N_e for humans is much smaller than the long-term average N_e for the hominoids, which explains why the numbers of IGUs for the former is an order of magnitude lower than the latter (Costa et al. 2016; Table 3.1).

While some of the IGU estimates per genome in Table 3.1 are quite large, it is important to realize that these estimates are for *all* IGUs. This means that IGUs represented by multiple copy and nonneutral loci are included as well (Costa et al. 2016). Because multilocus coalescent-based analyses require that all sampled loci meet the assumptions of being lone orthologous copies, neutral, and independent, the number of *single-copy and neutral IGUs* will likely be far smaller than the maximum number of IGUs (Costa et al. 2016). In other words, when the IGUs that are part of multicopy gene families and are under selection or linked to sites that are under selection are excluded from consideration, the number of single-copy and neutral IGUs will likely be far lower than these maximum numbers. When Costa et al. (2016) used bioinformatics software to find the maximum number of single-copy and presumably neutral IGUs in the hominoid genome, they found only 292 such loci, which is about ~2% of the estimated maximum number of IGUs in the genome (Table 3.1).

These preliminary findings suggest that the maximum number of single-copy and neutral IGUs in animal genomes must be in the hundreds to low thousands and not in the tens of thousands, which raises an important implication. Enabled by NGS-related technologies, researchers are now able to mine genomes for hundreds to thousands of DNA sequence loci. Thus, if a phylogenomics study involves the use of multilocus coalescent methods, then the maximum number of single-copy and neutral IGUs should not be exceeded otherwise the researcher will falsely inflate the number of loci and thereby commit an error known as pseudoreplication (Costa et al. 2016). Among other problems, pseudoreplicating loci will likely result in the researcher obtaining confidence intervals around historical demographic parameter estimates that are incorrect.

3.1.3.2 Criteria for Delimiting Loci with Independent Gene Trees

Because NGS is allowing researchers to harvest unprecedented numbers of loci from genomes—even numbering into the thousands, a need exists for methods that can identify loci that meet the independent loci assumption. In practice, researchers have used two different criteria to delimit independent loci in phylogenomic studies. One criterion, first used by O'Neill et al. (2013), identifies independent loci if they undergo independent assortment, whereas a second criterion developed more recently by Costa et al. (2016) is based on the decoupling of two loci due to long-term effective recombination (Jennings 2016). How do these criteria differ from each other and what are their efficacies?

TABLE 3.1
Estimated numbers of IGUs in various animal genomes

	Fruit fly	Human (low–high)	Hominoid	Zebra finch (low–high)	Grass finch (low–high)	Tiger salamander (low–high)
Population size (N_e)	10^6	7,500–10,000	10^5	1.3×10^6	5.2×10^5	10^3–10^5
Genome map length (cM)	287	3,614	3,614	1,068–1,341	1,068–1,341	5,251
Genome size (bp)	1.8×10^8	3.2×10^9	3.2×10^9	1.2×10^9	1.2×10^9	3.5×10^8
Number of IGUs per genome	11,500	1,100–1,400	14,000	56,000–70,000	25,000–31,000	210–21,000
Number of bp per IGU	15,700	2.3×10^6–2.9×10^6	230,000	17,000–21,000	39,000–48,000	1.7×10^4–1.7×10^6

SOURCE: Data and results obtained: Fruit fly (Hudson, R. R. and J. A. Coyne. 2002. *Evolution* 56:1557–1565.); Human and Hominoid (Costa, I. R. et al. 2016. *Genome Res* 26:1257–1267.); and Tiger salamander (Jennings, W. B. 2016. bioRxiv doi: http://dx.doi.org/10.1101/066332). Data for Zebra and Grass finches were obtained from the following sources: Zebra finch population size (Balakrishnan, C. N. and S. V. Edwards. 2009. *Genetics* 181:645–660.); Zebra finch map lengths (Stapley, J. et al. 2008. *Genetics* 179:651–667; Backström, N. et al. 2010. *Genome Res* 20:485–495.); Zebra finch genome size (Warren, W. C. et al. 2010. *Nature* 464:757–762.); and Grass finch ancestral population size (Jennings, W. B. and S. V. Edwards. 2005. *Evolution* 59:2033–2047.).

NOTE: The term "Hominoid" is used to describe a representative hominoid genome for purposes of estimating the number of IGUs among extant hominoid species (see Costa et al. 2016). The genome size and map length of the grass finch genome is assumed to be the same as for the zebra finch.

If we only focused on loci found on different chromosomes, then these criteria are equivalent because such loci will necessarily have independent gene trees (Wakeley 2009). However, when we consider loci found on the same chromosomes, this becomes an issue of importance for us as we will see below. Though both criteria can identify loci with independent gene trees, one of these criteria is far too stringent—its use can drastically limit the number of loci used in a study.

Let's first examine the methodology developed by Costa et al. (2016). Recall from Equation 3.1 that N_e strongly influences the total number of IGUs in a genome. Thus, if one assumes that N_e and recombination rate are constant, then, in practice, a physical distance in bp (or kb) might be used to delineate IGUs in studies. Because such a minimum physical distance or *distance threshold* between IGUs found on the same chromosome incorporates N_e and recombination rate, it is a direct means for delimiting loci with putatively independent genealogies. For example, from Table 3.1 we see that *D. melanogaster* has an estimated 11,500 IGUs in its genome. Given the 180 Mb genome of *D. melanogaster*, we would therefore expect to see, on average, one IGU every ~15,700 bp along its chromosomes. Thus, sampling a locus every 15,700 bp will yield a dataset consisting of the maximum number of IGUs.

Now let's consider a criterion used to delineate independently assorting loci. From classical genetics, we know that loci separated by 50 cM are effectively unlinked from each other; that is, they are expected to undergo independent assortment the same as if they were on different chromosomes (Hartl and Jones 2006). Applying this criterion to *D. melanogaster*, we see from Table 3.1 that the genome for this species has a map length of 287 cM and thus we could only expect to obtain a maximum of six or seven IGUs. The fruit fly genome has seven chromosomes and so this works out to be ~1 IGU per chromosome! Comparisons between these two criteria using any of the species in Table 3.1 produces comparable results—in all cases the estimated number of IGUs using Equation 3.1 is far higher than when the independent assortment (i.e., 50 cM threshold) is used. Thus, while each criterion does its job in identifying IGUs, the independent assortment criterion is much too stringent and would therefore unfairly restrict the number of IGUs that could be used in a multilocus coalescent-based analysis (Jennings 2016). This assessment regarding the conservative nature of an independent assortment criterion to identify genealogically independent loci agrees with Felsenstein (P. 484, 2004), who noted that the distance between two loci with totally different trees is actually very short, even ≪30 cM. In light of these results, it is recommended that researchers not use an "independent assortment" criterion to delimit independent loci in phylogenomic studies (Jennings 2016).

3.1.4 Assumption 4: No Historical Recombination within Loci

Another common assumption in phylogenomic studies holds that, for each locus, there has been *no* recombination within any of the sampled DNA sequences since the time of their most recent common ancestor. This assumption is typically required by multilocus coalescent analyses in phylogeographic (Yang 2002; Rannala and Yang 2003; Edwards et al. 2005; Jennings and Edwards 2005) and species tree studies (Edwards 2009; Lanier and Knowles 2012). This assumption is important because these types of studies consider each gene tree as an independent replicate and thus incorrectly reconstructed gene trees due to past intralocus recombination event(s) is expected to adversely impact these types of studies (Hare 2001; Edwards 2009; Lanier and Knowles 2012). Thus, a locus that has experienced no past recombination since the most recent common ancestor of the sampled sequences is represented by one genealogical history (Wakeley and Hey 1997). For example, as we saw in Chapter 2, all sites in the mitochondrial genome are effectively linked together such that all sites represent one super-locus or super-gene. However, as we will see in Section 3.1.4.1, recombination via crossing over will alter the properties of DNA sequence loci such that multiple distinct gene trees are needed to account for the histories of different sites in a particular recombined "locus."

3.1.4.1 Intralocus Recombination and Gene Trees

It only takes one recombination event within a DNA sequence locus to result in a situation in which two topologically different gene trees are needed to account for the genealogical histories of all sites on the same locus. This is important because many types of phylogenomic analyses assume a one-to-one relationship between loci and gene trees (i.e., one gene tree represents each locus defined by the researcher). To illustrate this phenomenon, we will borrow the example by Felsenstein (2004), which shows what happens to a hypothetical 204 base long locus after a one intralocus recombination event occurs (Figure 3.3). If the breakpoint due to a recombination event occurred between sites 138 and 139, then *two* gene trees having slightly different topologies are needed to show the genealogical histories of all sites in the locus: one tree is needed to show the history of sites 1–138, while a second similar, but topologically different, tree is needed to account for the history of sites 139–204 (Figure 3.3). Thus, in the aforementioned example the locus consists of two *nonrecombined blocks* of sites: the first block represents sites 1–138 and the second block represents sites 138–204 linked together. In this case, if a researcher unknowingly attempts to infer a gene tree using all the sites from this locus, then the resulting tree can, at best, only reflect one of the two possible tree topologies, which would not capture the complete genealogical histories of all sites.

3.1.4.2 What Is the Optimal Locus Length?

When designing new loci (Chapter 8) the researcher often has to choose the approximate length, in base pairs, of each locus. Given that most loci in nuclear genomes are subject to recombination via crossing over, an important question to consider is *What is the optimal locus length?* Unfortunately, there is no simple answer to this question because a tradeoff exists. On the one hand, longer sequences (all else equal) will tend to produce more resolved and robust gene tree reconstructions because larger numbers of variable sites will contribute to the gene tree estimate. However, a longer locus will also increase the chances of including recombined sites (Felsenstein 2004), which would render such a locus in violation of the no intralocus recombination assumption. While shorter loci will have fewer or no recombination break points within them, these shorter loci will also tend to have fewer informative sites and therefore they will be more susceptible to phylogenetic reconstruction errors. Historically, many of the developed nDNA loci have been less than 1 kb long, an upper bound mostly imposed by the costs and limits of Sanger sequencing. However, new loci development methods are available that enable researchers to design loci of far longer lengths if desired (Chapter 8). Thus, if you are designing your own DNA sequence loci, then how long should they be? Or, if you are using loci developed by others, should you worry that your locus contains more than one nonrecombined block of sites?

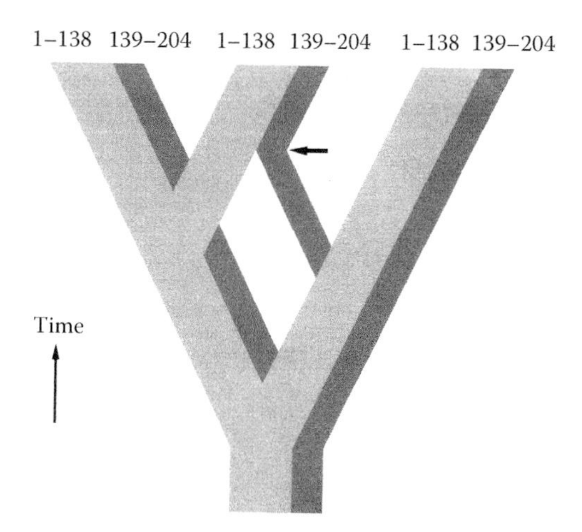

Figure 3.3. Effect of a single recombination event in a locus that is 204 sites long (Felsenstein 2004, fig. 26.8). The legend reads, "A coalescent tree with a single recombination event. The recombination is between sites 138 and 139 at the point indicated by the horizontal arrow. The result is a pair of related coalescent trees." (Reprinted from Felsenstein, J. 2004. *Inferring Phylogenies*. Sunderland: Sinauer. With permission.)

Felsenstein (2004) reviewed the theory for understanding the evolutionary factors that determine the physical distance (in base pairs) along a chromosome that is comprised of sites with the same gene tree before a recombination break point is encountered between two adjacent sites. As we saw earlier, the human recombination map length is 3,614 cM (Table 3.1), which means, on average, about 36 cross-overs (or "Morgans") occur in each human genome. Given that the human genome consists of 3.2×10^9 bases (Table 3.1), the expected number of bases between each recombination site is expected to be around 10^8 bases (Robertson and Hill 1983; Felsenstein 2004). However, the true frequency of recombination break points must be evaluated in light of the demographic history of the sample of individual genomes under consideration (Robertson and Hill 1983; Felsenstein 2004). If we consider the genomic sequences of two individuals sampled randomly from the population, then the average number of recombination events that occurred between them since their last common ancestor is equal to $4N_er$, where r is the recombination fraction between two points in the genome (Felsenstein 2004). In order to find the long-term average number of bases separating recombination break points (i.e., average nonrecombined block length), we must first set $4N_er = 1$. The parameter r can then be estimated if a value for N_e is supplied in this formula. Felsenstein (2004) calculated this break-point interval (hereafter "BPI") using the human genome: setting N_e to 10,000 (Takahata 1993), we see that $r = 1/4N_e = 1/40,000 = 2.5 \times 10^{-5}$. This estimate for the long-term average recombination frequency is multiplied by the average number of bases per cross-over per genome (10^8 bases/cross-over), which give us an average nonrecombined block length or BPI of 2,500 bases (Felsenstein 2004). This suggests that, on average, one can expect 2,500 bases to separate each recombination break-point in the human genome (Felsenstein 2004). However, this BPI depends heavily on population size and thus if the human population size is increased an order of magnitude (100,000), then the average interval length is expected to be only 250 bases long (Felsenstein 2004), which is comparable to the original estimate of ~120 bases by Robertson and Hill (1983) who used an estimate of $N_e = 200,000$ in their calculation.

What are the BPIs in the genomes of other species? We can calculate the expected BPI in other species with the following equation based on the work of Robertson and Hill (1983) and Felsenstein (2004):

$$\mathrm{BPI} = \frac{(1/4N_e)\times(\text{genome bp})}{((0.01\ \text{cross-overs/cM})\times\text{genome cM})} \qquad (3.2)$$

Using data for population size, genome size, and genome map length for the zebra finch in Table 3.1 and Equation 3.2 we can calculate the expected BPI in the genome of this species:

$$\begin{aligned}\mathrm{BPI} &= (1/(4\times(1.3\times10^6)))\times(1.2\times10^9)/\\&\quad(0.01\times1,341)\\&= 17\ \text{bp}\end{aligned}$$

Zebra finches have very large population sizes and thus their average nonrecombined block has an expected interval length of only 17 bases long, which is comparable to the estimated ~16 base interval in *D. melanogaster* (calculation not shown). Australian grass finches likely have a similar genome size and recombination map to the zebra finch, but they have smaller population sizes (Table 3.1) and thus their BPI is expected to be larger at ~54 bases (calculation not shown). If we assume that most animals have population sizes between humans and fruit flies, then the average length of a nonrecombined block of sites is expected to range from a maximum of ~2,500 bases down to a minimum of ~16 bases. This is an alarming result because gene trees are routinely inferred from individual loci that span hundreds or more bases. Lanier and Knowles (2012) suggested that the no recombination assumption is commonly violated in species tree studies and the calculations above shows us why this must be so. However, as Felsenstein (2004) pointed out, recombination rates vary across genomes with some regions having hotspots and others coldspots (e.g., McVean et al. 2004). We can use this fact to our advantage in constructing phylogenomic datasets consisting of nonrecombined loci, at least for recently diverged species.

One remedy is to identify presumably nonrecombined blocks of sites within a sequence dataset and then retain only the largest nonrecombined block for downstream gene tree analyses (Jennings and Edwards 2005). Hudson and Kaplan (1985) developed a method called the "four-gamete test," which can identify the *minimum* number of past recombination events within a DNA sequence locus. To see how this method works, we will examine a simple example that is illustrated in Figure 3.4. In this example, we have a locus that is 52 bases long and is represented by four DNA sequence alleles. In reality, loci to be tested will usually have hundreds or more contiguous sites. Nonetheless, for illustrative purposes we can still evaluate this short locus using the four-gamete test. The first step is to identify all the segregating (i.e., polymorphic or variable) sites, which in our locus includes sites 4, 20, and 49 (Figure 3.4a). In the next step, we perform pairwise comparisons between these polymorphic sites and then count the number of "gametes" or unique 2-base combinations: doing this shows us that sites 4 and 20, considered together, yield four unique base combinations, while the other pair of sites (20 and 49) only have three unique gametes each (Figure 3.4b). In the final step, we count one recombination event for every pair of sites showing all four possible gametes and then sum

(a)

	1 4 20 49 52
Allele 1	AAAAGACCTACGCAGTTAACGGACGGTTAGGGTCGAGATCATCAAGATGATA
Allele 2	AAAGGACCTACGCAGTTAACGGACGGTTAGGGTCGAGATCATCAAGATAATA
Allele 3	AAAAGACCTACGCAGTTAATGGACGGTTAGGGTCGAGATCATCAAGATAATA
Allele 4	AAAGGACCTACGCAGTTAATGGACGGTTAGGGTCGAGATCATCAAGATAATA

(b)

Sites 4–20	Sites 20–49
A–C	C–G
G–C	C–A
A–T	T–A
G–T	T–A
4 "gametes"	3 "gametes"

Figure 3.4. The four-gamete test of Hudson and Kaplan (1985). (a) The test is applied to a minimum of four orthologous DNA sequences. In this example, the sequences are 52 base pairs (sites) long with three segregating sites (i.e., 4, 20, and 49). The test only considers the segregating sites (black letters) and not the monomorphic sites (gray) in a sample. (b) Two pair-wise comparisons between consecutive segregating sites are needed to complete a test for the minimum number of historical recombination events in this sample (R_m). The comparison between sites 4 and 20 reveal four different "gametes," which is indicative of a past recombination event between these sites, whereas the comparison involving sites 20 and 49 show only three unique gamete types, which suggests no recombination occurred between these sites. Thus, the estimated minimum number of historical recombination events in these sites is $R_m = 1$.

them to obtain a value for the R_m or the minimum number of past recombination events. Our results yield an $R_m = 1$, which means that we have identified one historical recombination event within our locus. This result tells us that in the past there was at least one recombination event that occurred somewhere between sites 4 and 20 in our locus. We cannot know exactly where the breakpoint was because these intervening sites are all invariable. However, we can still use this information to modify our locus so that it conforms to the no intralocus recombination assumption. The procedure is simply to truncate the locus down to the largest nonrecombined block of sites, which in our example is defined by sites 20–52. The other sites from the smaller block, which includes the recombined region, are discarded. Although this test is easy enough to perform by hand on small datasets (e.g., with the minimum four sequences or a low number of polymorphic sites), it is better to search sequences for evidence of historical recombination events using a computer program such as *DNAsp* (Rozas et al. 2003), *IMgc* (Woerner et al. 2007), or *RDP3* (Martin et al. 2010).

There are several reasons why this statistic represents an estimate of the *minimum* number of past recombination events. First, back mutation can convert a polymorphic site to an invariable site and thus possibly render a past recombination event invisible to the test. Secondly, a double-recombination event in the same intervening region between two consecutive polymorphic sites can also "erase" the recombination history in this genomic segment. Lastly, although the minimum number of sequences required for the four-gamete test is four, it is always possible that using a larger number of sequences will reveal new polymorphic sites and hence this could conceivably lead to the discovery of additional past recombination events. Thus, the four-gamete test will most likely underestimate the true number of past recombination events. However, as Lanier and Knowles (2012) point out, this is not expected to be a problem because not all recombination events alter the topology of a gene tree. Therefore, this test should only be used for purposes of determining whether a given locus dataset shows evidence of past recombination. If the test is applied to a dataset and *Rm* is estimated to be zero, then the original locus can be assumed to meet no intralocus recombination assumption, whereas if $Rm > 0$, then the largest nonrecombined block can be identified and retained for further coalescent-based analyses.

An important assumption of the four-gamete test holds that each site has only mutated once and therefore follows the "infinite sites" model of sequence evolution. According to the infinite sites model, each site can only undergo a single substitution. However, if any sites in a locus have undergone multiple substitutions, then the four-gamete test may yield false positive results and thus overestimate the number of historical recombination events in a locus. False positive results can also be caused by DNA sequencing errors, haplotype phasing errors, and sequence alignment errors (Martin et al. 2010).

One problem with this approach is that by truncating loci down to the largest presumably nonrecombined block, you solve one problem (i.e., meet the no recombination assumption) but create another. Truncating loci means eliminating variable sites, which will elevate the incidence of phylogenetic reconstruction errors (Lanier and Knowles 2012). In effect, eliminating one type of gene tree reconstruction error simultaneously causes another type of gene tree error. Multilocus coalescent methods used in phylogeographic and species tree studies often rely mainly on the individual gene tree topologies and thus topological errors whether they are due to recombination or lack of phylogenetic signal in the sequences may adversely affect results.

One approach that can be used to solve both problems takes advantage of the heterogeneity of recombination rates across genomes. Although an average expected nonrecombined block may be short (e.g., 50 bp), it will likely be possible to find longer blocks, which can be used in phylogenomic analyses. Thus, an initial dataset can be collected, which consists of loci that are fairly long (e.g., >2 kb). Recombination tests are then applied to find the longest nonrecombined blocks in each locus. The longest nonrecombined block is then retained from each locus (one block per locus) for downstream analyses. This approach would help reduce the adverse effects of recombination and gene tree reconstruction errors. It should be noted, however, that this methodology will only be useful for datasets consisting of recently diverged species (i.e., sequences that conform to the infinite sites model). However, given that most multilocus coalescent-based studies involve closely related species, this should not be a problem. Unfortunately, it may not be possible to identify recombined sites in loci consisting of more anciently diverged sequences because multiple hits or alignment errors due to indels would preclude the use of recombination tests, which assume an infinite sites model. On a more encouraging note, the simulation study of Lanier and Knowles (2012) suggests that robust species tree inferences can be obtained even if the intralocus assumption is violated. Future studies should perform sensitivity analyses whereby multilocus coalescent-based analyses are conducted on recombination-corrected and uncorrected datasets to determine how sensitive the resulting phylogeographic parameters or species trees are to the no intralocus recombination assumption.

3.1.5 Assumption 5: Loci Evolved Like a Molecular Clock

Zuckerlandt and Pauling (1965) noticed that substitution rates for amino acids in different hemoglobin chains for human, horse, and cattle appeared to be constant. This observation of lineage rate constancy of substitutions prompted these authors to propose the *molecular evolutionary clock hypothesis*. Although this was originally proposed for protein data, the molecular clock hypothesis has since been extensively studied in DNA sequences. The clock-like manner by which substitutions seem to occur in some genomic sites is a useful property for a DNA sequence locus to have. The molecular clock is often used to estimate the timing of historical events, which can be gene divergences or population and species divergences (Swofford et al. 1996; Arbogast et al. 2002; Yang 2006). The molecular clock is also a standard assumption of coalescent-based gene tree analyses (Felsenstein 2004) and it can even be used to root gene trees (Huelsenbeck et al. 2002; Jennings et al. 2003; Felsenstein 2004; Yang 2006). If the clock is used to date a historical divergence event, then the mutation rate must be calibrated, which means that the units of substitutions must be converted into time units. These calibrations are obtained from time-dated fossils.

The so-called "strict" molecular clock assumption states that the mutation rate (μ) is constant among all sequences in a sample (i.e., among lineages). Although this condition is expected to be observed among closely related species owing to their similar physiologies, DNA repair, and generation times, the clock may gradually break down for more divergent species (Felsenstein 2004; Yang 2006). In cases where molecular data do not meet the strict clock assumption, investigators have used other clock-like methods such as *relaxed molecular clocks* (Felsenstein 2004) and *local clock* approaches (Thorne and Kishino 2005; Yang 2006).

3.1.6 Assumption 6: Loci Are Free of Ascertainment Bias

Molecular systematists and population geneticists have long known that some loci are more variable than other loci. Moreover, because variable loci contain "more information" than less variable loci, it may be tempting to choose the most variable loci for a study while discarding less variable ones. However, this act of "cherry picking" the most variable loci can lead to a form of bias known as *ascertainment bias* (Kuhner et al. 2000; Nielsen 2000; Wakeley et al. 2001; Brumfield et al. 2003; Felsenstein 2004; Nielsen et al. 2004; Rosenblum and Novembre 2007). If loci or sites are chosen nonrandomly, then it is important to try and correct the ascertainment biases (e.g., Kuhner et al. 2000; Nielsen 2000; Wakeley et al. 2001; Felsenstein 2004).

3.2 DNA SEQUENCE LOCI: TERMINOLOGY AND TYPES

Now that we reviewed the composition of genomes, various aspects of molecular evolution, and the common assumptions for DNA sequence loci used in phylogenomic studies, we are finally ready to review the types of loci commonly used in studies. As was already mentioned, studies focusing on the evolution of genes or genomes will know right away which locus or loci to use, whereas studies directed at reconstructing the tree of life must choose from among the various loci types. Accordingly, much of the following discussion will be mainly relevant for those interested in tree of life studies. Later in this book, in Chapters 7 and 8, we will see how to acquire these loci. Before we start this discussion, however, we must address an important issue relevant to everyone conducting phylogenomic studies, which concerns the inconsistent use of some commonly used genetics terms.

3.2.1 On Genes, Alleles, and Related Terms

The terms *gene*, *locus*, *allele*, *gene copy*, and *haplotype* are not always used in a consistent manner, which is unfortunate because misusing these terms can lead to confusion when trying to understand phylogenomic concepts and studies. Gillespie (P. 6–10, 2004) recognized this problem and brought much needed clarity. However, this issue is important enough that it deserves to be revisited here.

First, the meaning of the word "gene" has varied depending on the background of the worker. For example, molecular biologists use the traditional definition of gene and thus only regard functional segments of DNA as being genes such as RNA-coding and protein-coding sequences (e.g., Brown 2007; Watson et al. 2014). Population geneticists, on the other hand, tend to use this term more broadly with some workers relying on the function-based definition because it relates to some organismal trait, while others consider *any* genomic segment—even one base pair—regardless of whether or not it has a known function as a gene (e.g., Felsenstein 2004; Wakeley 2009). As Gillespie (2004) pointed out, the term "locus" refers to a specific genomic location where a segment of DNA is located. Thus, to the molecular biologist and some population geneticists, a gene is a functional segment of DNA and locus refers to its genomic location. In contrast, other population geneticists may refer to any segment of DNA—regardless whether it is functional or not—as a gene or locus. Owing to the confusion surrounding the various uses of the word gene, Gillespie (2004) suggested that this term has lost its usefulness in population genetics studies. However, given the growing popularity of phylogenomic studies that use the term "gene trees," in this book we will retain the term gene and use it interchangeably with locus.

The term "allele" has long been a fundamental unit in genetics, but what exactly is an allele? If we define a gene (or locus) as a particular segment of genomic DNA, then the allele is the actual physical manifestation (DNA sequence) of the gene found in a particular individual. Another term that is sometimes used *in lieu* of allele, particularly in population genetics, is *gene copy* (e.g., Felsenstein 2004). A third term that is used synonymously with allele and gene copy is *haplotype* owing to the haploid nature of an allele or gene copy. The term haplotype has been used often in phylogeographic studies to describe alleles perhaps because of the predominant use of mitochondrial DNA during the first two decades of this discipline (e.g., Avise 2000). As you will recall from Chapter 2, any DNA sequence obtained from a mitochondrial genome is a haplotype because there is only one copy of the mitochondrial genome inherited from a parent (most often the mother). But if we look at a human autosomal gene, we cannot forget that there will be *two copies* of that gene in each individual (i.e., one from each parent). Although restating this fundamental of genetics may seem unnecessary, we will see elsewhere in this book some cases in which it is not uncommon for researchers to overlook this idea thereby leading to some confusion. Throughout this book we will consider the terms allele, gene copy, and haplotype as having identical meanings and it does not matter whether an allele is a 500 bp sequence or a single base.

Unfortunately, the term gene (or locus) has also been confused with allele (or gene copy or haplotype). For example from medicine, if a doctor says that a person is "gene positive" for Huntington's disease (HD), what the doctor is really saying is that the person has inherited the *allele* (or gene copy or haplotype) that gives rise to the disease (HD is an autosomal dominant disease). Most people inherit two normal (nondisease) alleles from their parents, but those that are so-called gene positive inherit a copy of the defective allele. In an example from population genetics, it is not uncommon to see an author talking about genes in a phylogenetic study but in reality the author is describing gene *copies* (or alleles or haplotypes) for a gene. Needless to say, this can become confusing if a study includes numerous genes (or loci) each of which is represented by multiple haplotype sequences. Thus, it is essential for phylogenomics researchers to use these terms in a consistent manner.

3.2.2 Commonly Used DNA Sequence Loci in Phylogenomic Studies

3.2.2.1 Mitochondrial DNA Loci

Ever since the late 1970s and continuing to the present day, mitochondrial DNA loci have been the most extensively used genetic markers in molecular phylogenetic studies. Despite the rapid growth of genomics over the past decade, mitochondrial loci are still tremendously popular evolutionary markers owing to their many desirable properties. As we saw in Chapter 2, these properties include their haploid nature, negligible recombination, high substitution rate relative to nuclear sites, and low effective population size compared to most nuclear loci. Regarding this last property, the relatively smaller effective population sizes of mitochondrial DNA implies that, on a locus per locus basis, mitochondrial loci will better track the species tree than will autosomal loci (Moore 1995; Zink and Barrowclough 2008). Note, however, that despite claims of mitochondrial DNA being "selectively neutral," the earlier discussion in this chapter about the indirect effects of natural selection (e.g., background selection) show why this is very unlikely to be true (Wilson et al. 1985; Ballard and Rand 2005; Galtier et al. 2009). Another contributing factor to the popularity of mitochondrial DNA, especially during the early years of molecular systematics, was the introduction of the first universal PCR primers by Kocher et al. (1989), which allowed researchers to obtain mitochondrial sequence data from a wide variety of metazoan species. Not only have mitochondrial loci played a huge role in molecular systematics, but they also were largely responsible for launching phylogeography (Avise 2000) and DNA barcoding (Hebert et al. 2003).

Although nuclear loci are rapidly gaining momentum in phylogenomics studies, it should not be forgotten that mitochondrial DNA will always provide an independent evolutionary perspective of historical events and thus it can complement the perspective offered by nuclear loci (e.g., Prychitko and Moore 1997; Reilly et al. 2012). The following summary of mitochondrial loci is not comprehensive. For reviews and perspectives on the utility of mitochondrial DNA in different types of phylogenomic studies the reader should consult the works by Wilson et al. (1985), Moritz et al. (1987), Moore (1995), Palumbi (1996), Avise (2000), Hebert et al. (2003), Ballard and Rand (2005), Zink and Barrowclough (2008), and Galtier et al. (2009).

Mitochondrial protein-coding loci—Mitochondrial protein-coding loci have been perhaps the most important class of mitochondrial DNA. As we saw in Chapter 2, the existence of codon bias in protein-coding genes means that different codon positions will exhibit different levels of variability. This aspect of their evolution is of practical importance because it can be useful for reconstructing both recent and old divergences in a gene tree. Another nice feature of protein-coding sequences, at least compared to most noncoding sequences, is that multiple sequence alignments with protein-coding sequences are usually simple to perform even among highly diverged sequences. The mitochondrial *cytochrome oxidase I* gene or "COI" has been a particularly important mitochondrial gene due to its central role in the DNA barcoding program (Hebert et al. 2003).

Mitochondrial ribosomal RNA loci—The mitochondrial genomes of animals contain two ribosomal RNA (rRNA) genes (Mindell and Honeycutt 1990; Palumbi 1996). One of the gene sequences codes for the 12S rRNA or "small subunit" while the other codes for the 16S rRNA or "large subunit" (Palumbi 1996). In vertebrates the 12S rDNA exists as a ~950 bp long sequence, whereas the 16S rDNA is ~1,600 bp long. In contrast to nuclear rDNA sequences, the 12S and 16S sequences in the mitochondrial genome are contiguous (do not contain any spacer DNA; Mindell and Honeycutt 1990). In order to obtain 12S and 16S sequences, researchers typically acquire these gene sequences from ~300 to 800 bp loci, which yield partially overlapping sequences that are later joined into continuous sequences or "contigs" during data analysis. Palumbi (1996) provides lists of widely used 12S and 16S PCR primers—most of which perform well on diverse metazoan species—as well as information pertaining to their physical locations and orientations in the mitochondrial genome.

Both the 12S and 16S rDNA segments in the mitochondrial genome contain stretches of bases that are extremely conserved (i.e., "stem" regions) and stretches that are highly variable (i.e., "loop" regions; Palumbi 1996). Thus, there is great heterogeneity in the evolutionary rates among sites. The among-site variation within these genes makes the process of multiple sequence alignment easy or difficult depending on which stretches of sites are being aligned. The highly conserved blocks are simple to align while other sections are difficult if not impossible. For stretches of sites that have questionable positional homologies, Swofford and Olsen (1990) recommend discarding them because it is safer to remove these sites rather than to infer a gene tree based on spurious alignments.

Despite these minor complications, both the 12S and 16S genes have long been favorite DNA sequence loci of molecular systematists because of their effectiveness in resolving the phylogenetic relationships within eukaryotes particularly among species, genera, and families (e.g., Mindell and Honeycutt 1990; Hillis and Dixon 1991; Palumbi 1996; Darst and Cannatella 2004). The mitochondrial 16S rRNA gene also has special properties, which makes it an attractive candidate for being a DNA barcode locus in amphibians (Vences et al. 2005).

Mitochondrial control region loci—The control region is the most variable part of the mitochondrial genome (Vigilant et al. 1989; Tamura and Nei 1993; Palumbi 1996; Randi 2000). The control region is hypervariable because, unlike the protein-coding regions, it can tolerate both base substitutions and indels of varying lengths yet still remain functional (Vigilant et al. 1989). However, this property of control region loci means they are a poor choice for studies of highly divergent species owing to difficulties with multiple sequence alignments. Also, attempts to obtain high quality control region sequences using the Sanger method can be frustrated by the existence of indels (Chapter 6) that reside either in different heteroplasmic copies of the mitochondrial genome or from artifacts in PCR due to long microsatellite repeats. Despite these drawbacks, control region DNA has been an effective evolutionary locus for studies operating at shallow phylogenetic scales especially those at the intraspecific level (Vigilant et al. 1989, 1991; Baker et al. 1993; Tamura and Nei 1993; Avise 2000; Randi 2000). Control region loci have also been employed as DNA barcode-like sequences in conservation genetics studies (Baker and Palumbi 1994). For further information about the structure and organization of the control region the reader can consult Avise (P. 27, 2000) for an illustration of the entire mammalian control region. Palumbi (1996) provides additional details about control region loci not discussed here.

3.2.2.2 Nuclear DNA Loci

The nuclear genome is a veritable treasure trove for DNA sequence loci yet it has largely remained inaccessible to researchers until recent years. This is because prior to the commencement of the genomics era, the development of novel DNA sequence loci was accomplished using the slow and technically challenging methodology of gene cloning. However, in recent years the availability of genomics resources has empowered researchers with a variety of exciting new approaches to developing phylogenomic loci and acquiring datasets consisting of hundreds or more loci. We will review these approaches in Chapters 7–9.

Nuclear loci enjoy important advantages over mitochondrial loci. Besides the sheer variety and number of loci that can be obtained from the nuclear genome, some nuclear loci have the desirable properties of being presumably neutral and genealogically independent of other sampled loci. These two properties, which are not found in mitochondrial loci, enable researchers to conduct statistically rigorous multilocus analyses in tree of life studies. However, nuclear loci are not without disadvantages. For example, earlier we saw how intralocus recombination can complicate gene tree inferences. In Chapter 6, we will see how the presence of multiple haplotypes at each diploid (or polyploid) locus can also complicate matters though NGS-based methods for data acquisition are largely immune to this problem. Despite these issues, tree of life studies are now making extensive use of large numbers of nuclear loci and this trend will increase further as access to genome-scale datasets becomes simpler, faster, and less expensive in coming years. The following is a brief introduction to the common types of nuclear loci used in phylogenomic studies. For more extensive and in depth discussions of these loci the reader should see the reviews by Palumbi (1996) and Thomson et al. (2010) as well as the references cited below.

Exon-primed-intron-crossing loci (EPICs)—Motivated by the limitations of mitochondrial DNA, researchers attempted to find alternative loci in the nuclear genome so that additional independent perspectives on the evolutionary history of populations or closely related species could be obtained. Nuclear introns represented one such important new class of DNA sequence locus. Introns are largely evolutionarily unconstrained except for their highly conserved 3′ and 5′ termini, which consist of functionally critical sites (Fedorov et al. 2002). Thus, introns display a high level of variability compared to protein-coding and RNA-coding loci, which makes them ideal for studies of populations or closely related species. With this property of introns in mind, Lessa (1992) developed PCR-based intron loci and illustrated their applicability to the study of pocket gopher (*Thomomys bottae*) populations in California. Moreover, to increase the chances of PCR success across different genetic samples, Lessa designed the two PCR primers in highly conserved exon regions flanking the intron of interest rather than nesting them within the intron itself. Palumbi and Baker (1994), who developed their own universal exon–intron loci for vertebrates, called these loci *Exon-Primed-Intron-Crossing loci* or "EPICs," a label subsequently adopted by researchers (Shaffer and Thomson 2007). Many studies have confirmed that intron loci are highly variable even at the population level (e.g., Lessa 1992; Palumbi and Baker 1994; Prychitko and Moore 1997; Sequeira et al. 2006). EPIC loci have generated high-resolution gene trees that are comparable to mitochondrial loci (Palumbi 1996; Weibel and Moore 2002).

Although EPIC loci have enjoyed success in many studies, they suffer from some problems that do not similarly affect mitochondrial loci. Aside from the multiple haplotype per individual problem already mentioned, EPIC loci are largely free to accumulate multiple base substitutions and indels, which can complicate efforts to align sequences from highly diverged species such as above the genus level (Prychitko and Moore 1997; Sequeira et al. 2006 and references therein). Indels of varying lengths are commonly observed in introns (e.g., Ohresser et al. 1997; Hassan et al. 2002) and the more evolutionarily divergent the species the more difficult it is to make an acceptable alignment. However, for intraspecific studies or interspecific studies involving closely related species, multiple sequence alignments are usually trivial (e.g., Sequeira et al. 2006).

Notwithstanding, EPIC loci still retain a number of advantages over mitochondrial loci. First, because mitochondrial loci collectively represent one independent locus (Wilson et al. 1985), such loci cannot be used to infer all details about the evolutionary history of populations or species (Edwards and Bensch 2009). Instead, many independent nuclear loci are required (e.g., Edwards and Beerli 2000). Despite their difficulties, EPICs can better capture the evolutionary history of populations or closely related species because a large number of independent EPIC loci can be harvested from genomes. Many researchers have suggested that most of the sites within introns are nearly neutral in character (Lessa 1992; Prychitko and Moore 1997; Friesen 2000; Daguin et al. 2001; Fedorov et al. 2002). However, as we saw earlier this chapter (see Figure 3.1), the genetic variability of introns is likely to be reduced due to the effects of indirect natural selection acting at nearby sites. Therefore, EPIC loci may not be appropriate for phylogenomic analyses that require the neutrality assumption.

Many EPIC loci have PCR primers that are "universal" in character, which means they can be used to acquire the same sequences from a wide range of metazoan species (e.g., Chenuil et al. 2010). We will discuss how universal primers are made in Chapter 8 but for now it suffices to know that such primers greatly simplify the acquisition of DNA sequence datasets. Fully sequenced EPIC loci contain stretches of highly conserved exon sequences that flank the entire intron sequence nested inside (Lessa 1992). If the entire sequence is analyzed, then the mixture of high and low variation sites can allow for robust reconstructions of gene trees at both shallow (e.g., intraspecific) and deeper phylogenetic levels (Li et al. 2010). Alternatively, gene trees could be inferred using only the exon sequences if the intron sequences are not alignable (e.g., Li et al. 2010) or the 3′ and 5′ exon sequences can be trimmed away so that only the intron sequences are used in phylogenetic analyses (e.g., Sequeira et al. 2006). For additional information about EPIC loci see Palumbi (1996) and Friesen (2000).

Anonymous loci—One year after the publication of the first EPIC paper by Lessa (1992), Karl and Avise (1993) proposed a new class of DNA sequence loci called *single-copy nuclear anonymous loci*, which could be applied to the study of populations. These so-called "anonymous loci" were given this name because they were developed from random and unknown genomic locations (Karl and Avise 1993; Jennings and Edwards 2005).

Anonymous loci have several advantages that make them ideal for studies involving populations or closely related species. First, because most anonymous loci reside in intergenic regions, they are thought to be free from the influence of natural selection (Thomson et al. 2010; Costa et al. 2016). Anonymous loci are therefore expected to exhibit higher levels of genetic variation than loci under the direct or indirect effects of selection (see Figure 3.1 and Costa et al. 2016 for empirical evidence). Indeed, among all types of phylogenomic loci currently available, anonymous loci are the only loci that may meet the neutrality assumption (i.e., totally free of the effects of natural selection). Another advantage is that large numbers (e.g., hundreds or more) of single-copy and genealogically independent anonymous loci can be mined from genomes (Chapters 8 and 9; Costa et al. 2016). Finally, because anonymous loci are obtained without regard to their variability, they are free from ascertainment biases.

Like intron loci, anonymous loci have a number of drawbacks such the difficulty in aligning sequences from highly diverged species, susceptibility to the effects of intralocus recombination, and difficulties in haplotype resolution. Despite this, anonymous loci still represent the ideal locus type for phylogenomic studies that employ multi-locus coalescent-based analyses because the gene trees for these loci are expected to best resemble the structure of coalescent trees (see Felsenstein 2004 and Wakeley 2009 for reviews of coalescent theory).

Anonymous loci have been developed in a wide variety of vertebrates including hominoids (Chen and Li 2001; Costa et al. 2016), Australian grass finches (Jennings and Edwards 2005), eastern fence lizards (Rosenblum et al. 2007a,b), turtles (Shaffer and Thomson 2007), Australian fairy wrens (Lee and Edwards 2008), black salamanders (Reilly et al. 2012), sea snakes and geckos (Bertozzi et al. 2012), chorus frogs (Lemmon and Lemmon 2012), Atlantic Rainforest antbirds (Amaral et al. 2012), and Mojave fringe-toed lizards (Gottscho et al. 2014), among many others. As we will see in Chapters 8 and 9, the use of anonymous loci is expected to dramatically increase as more researchers take advantage of new NGS and complete genome methods for obtaining large anonymous loci datasets.

Nuclear protein-coding exon loci (NPCLs)—In contrast to introns and anonymous loci, nuclear exons represent a class of more evolutionarily conserved DNA sequence loci. Nuclear exons are also referred to as *Nuclear protein-coding exon loci* or "NPCLs" to distinguish them from organellar protein-coding loci (Thomson et al. 2010). Because of their conserved nature and general lack of indels (indels, if they occur, must be in multiples of 3-bp to preserve the correct reading frame), NPCLs sequences are usually easy to align even for some highly diverged species. NPCLs usually do not show sufficient variation among sequences at the intraspecific level or among closely related species for producing well-resolved gene trees. Thus, the variation present in NPCLs is at a level best suited to reconstructing the gene trees for highly diverged species (e.g., genus or family). For example, in a molecular phylogenetic study of Australian pygopodid lizards, a gene tree inferred from nuclear *c-mos* gene (Saint et al. 1998) sequences did not resolve all species relationships within each genus but it did recover the same generic clades that had been found in a mitochondrial gene tree and a tree based on morphological characters (Jennings et al. 2003). Thus, NPCLs will be better choices for phylogenomic studies involving speciation events that occurred tens of millions of years ago or earlier.

Until recently, the number of NPCL loci available to researchers was limited. However, once genomics resources (e.g., annotated full genome sequences) for a variety of organisms became available, researchers began using bioinformatics-based methods to design large numbers of NPCLs that could be applied to a wide array of nonmodel organisms (reviewed in Chapter 8). For example, Li et al. (2007) developed 154 presumably single-copy NPCLs that can be applied to a wide variety of fish species. In a similar study, Portik et al. (2011) generated 104 NPCL loci that can be used to obtain sequence data for squamate reptiles.

Traditionally, NPCL loci have been PCR-amplified from genomic DNA followed by Sanger sequencing. However, in recent years researchers have developed exciting new NGS-enabled *transcriptome*-based methods for acquiring hundreds or thousands of exonic sequences from multiple individuals in a single experiment. Thus, rather than directly PCR-amplifying exons from genomic DNA templates, extracted mRNA transcripts are first converted into a *cDNA library* before being sequenced using NGS methods (Chapter 7). For example, Bi et al. (2012) used this approach to obtain DNA sequence data from over 10,000 exonic loci for *Tamias* chipmunks.

Nuclear ribosomal RNA loci—Nuclear ribosomal RNA genes exhibit a number of characteristics that differentiates them from mitochondrial rRNA genes. First, these genes are members of an "array," which consists of three rDNA coding segments (18S, 5.8S, and 28S), two internal transcribed spacer elements (ITS-1 and ITS-2), an external transcribed spacer (ETS), and a non-transcribed spacer (NTS) flanking each end of the array. The structure of this array, which is highly conserved in eukaryotes, consists of the following ordering of elements starting at the 3′ end of the rDNA coding strand: 3′—NTS, ETS, 18S, ITS-1, 5.8S, ITS-2, 28S, NTS—5′ (Mindell and Honeycutt 1990; Hillis and Dixon 1991). In contrast to mitochondrial rRNA genes, which occur as single copies in the mitochondrial genome, nuclear rDNA arrays have been tandemly duplicated hundreds or thousands of times depending on species (Mindell and Honeycutt 1990; Palumbi 1996). An NTS element separates each array copy within a cluster of arrays on a chromosome (Mindell and Honeycutt 1990; Palumbi 1996). Because these array copies have been subjected to the homogenizing effects of concerted evolution, DNA sequences obtained from any array copy can still be effective at resolving the phylogenetic relationships among organismal lineages (Hillis and Dixon 1991; Moritz and Hillis 1996). Another difference between nuclear and mitochondrial rDNA genes is that the former are more highly conserved than the latter and thus they complement each other in terms of the phylogenetic time depths that each can resolve; specifically, nuclear rDNAs are best for pre-Cenozoic divergences (>65 million years ago) while mitochondrial rDNAs are better for studying the evolutionary diversification of groups within the Cenozoic (Hillis and Davis 1986; Mindell and Honeycutt 1990; Hillis and Dixon 1991). Some regions of rDNA genes are so highly conserved that they have revealed the basic domains of life (Woese and Fox 1977; Woese et al. 1990) and resolved the evolutionary relationships among phyla (Hillis and Dixon 1991). Another desirable property of nuclear rDNA arrays is that the different elements exhibit variable conservation of sites and thus researchers can target particular elements for sequencing depending on the presumed levels of evolutionary divergence among the study organisms in a given study (Hillis and Dixon 1991). The rDNA genes exhibit the highest levels of conservation followed by the transcribed spacer elements and then the NTS elements (Hillis and Dixon 1991; Palumbi 1996). Moreover, there is also substantial among site variability within and among the rDNA genes (Hillis and Dixon 1991). Lists of PCR primers and guidance for sequencing entire rRNA arrays can be found in Hillis and Dixon (1991) and Palumbi (1996). For reviews of rDNA structure, evolution, and phylogenetic utility the reader should consult the reviews by Mindell and Honeycutt (1990) and Hillis and Dixon (1991).

Anchored loci—The development of *ultraconserved elements-loci* or "UCE-loci" (Faircloth et al. 2012) and *anchored enrichment-loci* or "AE" loci (Lemmon et al. 2012; Lemmon and Lemmon 2013) represents two exciting and important phylogenomic innovations that arose in recent years. Although the methods used to design UCE-anchored and AE loci involve different approaches (see Faircloth et al. 2012 and Lemmon et al. 2012 for descriptions of protocols), these loci share many evolutionary properties and occupy similar phylogenomic loci niches. Thus, for simplicity we will generically refer to them as *anchored loci*. The significance of the term "anchor" will be made clearer below.

Unlike the previously discussed loci, which are commonly sequenced using the traditional PCR-Sanger sequencing routine, anchored loci are sequenced using NGS-based methods to which they are perfectly suited. Individual sets of UCE-anchored and AE loci consist of hundreds or thousands of single-copy and presumably independent DNA sequence loci found throughout the genomes of various eukaryotic groups. Studies using anchored loci have typically generated datasets consisting of sequences for hundreds to thousands of loci from dozens (e.g., Jarvis et al. 2014) to hundreds (e.g., Prum et al. 2015) of individuals. Phylogenomic studies based on these enormous datasets are resolving species trees at both shallow (i.e., from hundreds of thousands to around ten million years ago) and deep (i.e., tens to hundreds of millions of years ago) levels of evolutionary divergence (e.g., Faircloth et al. 2012; Crawford et al. 2012; Faircloth et al. 2013; McCormack et al. 2013; Jarvis et al. 2014; Prum et al. 2015). Thus, anchored loci studies are already dramatically increasing and refining our knowledge of the tree of life. However, what exactly are UCEs and AEs? We will first consider UCEs and UCE-like elements because they have attracted considerable interest of genome researchers in recent years.

The term "ultraconserved element" was originally coined by Bejerano et al. (2004) to describe the hundreds of mostly noncoding genomic elements common to the human, mouse, and rat genomes, which are perfectly conserved (100% identical) for at least 200 bp. Nearly all of these UCEs also exist in the chicken and dog genomes, albeit they are slightly less conserved than the human-rodent UCEs, and some are found in fish (Bejerano et al. 2004). Moreover, thousands more UCEs that are perfectly conserved for more than 100 bp are found throughout mammals (Bejerano et al. 2004). Human UCEs are most often found to be overlapping exons of RNA processing genes or they exist in intergenic regions adjacent to regulatory and developmental genes (Bejerano et al. 2004). Another notable feature of human-rodent UCEs is that the vast majority of them exist as single-copy elements (Bejerano et al. 2004; Derti et al. 2006; Faircloth et al. 2012). Several other classes of highly conserved genomic elements displaying levels of sequence conservation below that of UCEs have since been discovered in comparative genome studies involving various eukaryotic species.

Siepel et al. (2005) defined a class of conserved genomic elements called "highly conserved elements" or "HCEs," which exhibit slightly less sequence conservation and tend to be longer than human-rodent UCEs. These workers performed a comparative genome analysis in an effort to locate all HCEs within the genomes from representative groups of eukaryotes (i.e., vertebrates, insects, worms, and yeasts). The resulting group-specific sets of HCEs, which they found, ranged in size from five to thousands of bp with an average size of 100–120 bp (Siepel et al. 2005). Approximately 42% of HCEs in vertebrates overlapped known exons, whereas >93% of HCEs in insect, worm, and yeast genomes were found to overlap exons. The most HCEs were found in 3′ UTRs especially those for regulatory genes (Siepel et al. 2005). Many of the HCEs associated with exons and 3′ UTRs seem to be enriched for RNA secondary structure and thus might represent unidentified coding regions (Siepel et al. 2005). Consistent with other studies, these authors noticed that stable gene deserts were well populated with HCEs and therefore this subclass of HCEs may represent noncoding elements that act as long-range regulators of developmental genes (Ovcharenko et al. 2005).

The fact that these sequences have been so highly conserved for tens or hundreds of millions of years is highly suggestive that they all are functional elements (Siepel et al. 2005). Indeed, Katzman et al. (2007) obtained evidence showing that human-rodent ultraconserved elements are under intense purifying selection. Findings from other studies also suggest that at least some of these HCEs act as long-range regulators of developmental genes or perform other gene regulatory functions (Nobrega et al. 2003; Woolfe et al. 2004; Bejerano et al. 2006; Pennacchio et al. 2006; Simons et al. 2006; Stephen et al. 2008). However, the function of most conserved elements remains unknown. These and other comparative genome studies have shown that eukaryotic genomes are well populated with these ultra- or HCEs, which, in turn, can be developed into a plethora of phylogenomic loci for tree of life studies.

The first sets of UCE- and AE-anchored loci were developed in animals, as the former was designed to obtain large multilocus datasets from amniotes (Faircloth et al. 2012) while the latter was made for acquiring similar datasets from all vertebrates (Lemmon et al. 2012). In contrast to the original set of UCE loci, which were comprised of noncoding elements, the earliest set of AE loci primarily resided in coding regions of the genome (see Figure 2 in Lemmon et al. 2012; Lemmon and Lemmon 2013; Eytan et al. 2015). Despite any differences in the types of genomic sites found in UCE- and AE-anchored loci, both types of loci share the general and crucial properties of exhibiting extreme levels of DNA sequence conservation across wide phylogenetic distances (i.e., intraphyla) and offering potentially hundreds to thousands of genome-wide loci for phylogenomic studies.

If the sites within these elements are so highly conserved (i.e., >80% identical), then how can they be used to generate empirical gene trees? After all, sequences displaying little or no variation cannot be used to accurately infer gene trees. The answer is that the sites within the UCEs/AEs are themselves not used to reconstruct gene trees. Instead, the less-conserved flanking regions are used in phylogenomic analyses. Indeed, the level of variability increases with distance away from the core regions of UCE- and AE-anchored loci (see Figure 3 in Faircloth et al. 2012 and Figure 2 in Lemmon et al. 2012). The primary role of these HCEs is to serve as *anchors* to which oligonucleotide probes are hybridized during the data acquisition process (Chapter 7). Briefly, for each UCE- or AE-anchored locus, sequences are obtained from different individual genomes that include both the probe region (anchor location) as well as some amount of sequence from the flanking regions. Thus, the sites closest to the anchor location will be most conserved and thus can be used to reconstruct the deep nodes of a species tree while the more distant and variable sites are useful for inferring the relationships among recently diverged species (Faircloth et al. 2012; Lemmon et al. 2012). The anchors, therefore, represent the key elements that allow researchers to acquire orthologous sequences from lineages of organisms that diverged from each other hundreds of millions of years ago. Moreover, despite the fact that the anchor portion of these sequences may contribute little or no information to gene tree inferences, they are also useful for facilitating multiple sequence alignments among highly diverged species (Faircloth et al. 2012; Lemmon et al. 2012). What types of sites are in these flanking regions? Owing to the generally low or nonexistent levels of sequence conservation in UCE flanking regions, these sites are presumably nonfunctional and therefore more closely resemble anonymous loci. In contrast, the flanking regions of AE-anchored loci represent a mixture of coding, intron, and other sequences (see Figure 2 in Lemmon et al. 2012).

Many recent phylogenomic studies based on anchored loci used coalescent methods to infer species trees. One potential problem with this practice is that it is not yet clear how violating the neutrality assumption of these coalescent methods will affect the evolutionary inferences obtained from the datasets. Because all anchored loci include stretches of highly conserved noncoding or coding sites, which are very likely under purifying selection (Siepel et al. 2005; Katzman et al. 2007), the flanking sites will be subjected to indirect selection. As we saw earlier in this chapter, such selection is expected to diminish the genetic diversity of linked nonfunctional sites. This means that the gene trees representing the flanking regions of anchored loci will not only reflect the historical processes of mutation and genetic drift, but also that of direct or indirect natural selection. Accordingly, a set of anchored loci will yield a distribution of presumably independent gene trees that do not match coalescent expectations. McCormack et al. (2012) pointed out that selection on UCE-anchored loci may not adversely affect species tree inferences because selection would increase the rate of lineage sorting and thus the inferred gene trees would better match the topology of the true species tree. In support of this idea, Faircloth et al. (P. 721, 2012) noted that the distribution of inferred gene tree topologies based on a set of UCE-anchored loci for hominoids (human, chimpanzee, and gorilla) was similar to well-established expectations. Given this result, these authors concluded that UCE-anchored loci follow coalescent processes. Thus, despite anchored loci not being strictly neutral, they may still have the capability to generate the true species tree topology in coalescent-based analyses. This encouraging result should be verified by more studies especially those involving simulations.

The use of anchored loci for estimating historical demographic parameters (e.g., effective population size and historical gene flow), as practiced in some studies (e.g., Smith et al. 2013), may be more problematic. This is because the loss of genetic diversity due to the effects of indirect selection is expected to lead to underestimates of effective population sizes, which are key parameters in phylogeography studies (McVicker et al. 2009; Costa et al. 2016). Anonymous loci on the other hand, are likely not impacted by direct or indirect selection and hence distributions of gene trees based on these loci should better reflect coalescent expectations. Until we know more about the effects of violating the neutrality assumption in multilocus coalescent analyses, it seems prudent to exercise some caution when interpreting phylogeographic results based on anchored UCE or AE loci (Costa et al. 2016).

REFERENCES

Amaral, F. R., S. V. Edwards, and C. Y. Miyaki. 2012. Eight anonymous nuclear loci for the squamate antbird (*Myrmeciza squamosa*), cross-amplifiable in other species of typical antbirds (Aves, Thamnophilidae). *Conserv Genet Resour* 4:645–647.

Arbogast, B. S., S. V. Edwards, J. Wakeley, P. Beerli, and J. B. Slowinski. 2002. Estimating divergence times from molecular data on phylogenetic and population genetic timescales. *Annu Rev Ecol Syst* 33:707–740.

Avise, J. C. 2000. *Phylogeography: The History and Formation of Species*. Cambridge: Harvard University Press.

Backström, N., W. Forstmeier, H. Schielzeth et al. 2010. The recombination landscape of the zebra finch *Taeniopygia guttata* genome. *Genome Res* 20:485–495.

Baker, C. and S. Palumbi. 1994. Which whales are hunted? A molecular genetic approach to monitoring whaling. *Science* 265:1538–1539.

Baker, C. S., A. Perry, J. L. Bannister et al. 1993. Abundant mitochondrial DNA variation and worldwide population structure in humpback whales. *Proc Natl Acad Sci USA* 90:8239–8243.

Balakrishnan, C. N. and S. V. Edwards. 2009. Nucleotide variation, linkage disequilibrium and founder-facilitated speciation in wild populations of the zebra finch (*Taeniopygia guttata*). *Genetics* 181:645–660.

Balakrishnan, C. N., R. Ekblom, M. Völker et al. 2010. Gene duplication and fragmentation in the zebra finch major histocompatibility complex. *BMC Biol* 8:29.

Ballard, J. W. O. and D. M. Rand. 2005. The population biology of mitochondrial DNA and its phylogenetic implications. *Annu Rev Ecol Evol Syst* 36:621–642.

Begun, D. J. and C. F. Aquadro. 1992. Levels of naturally occurring DNA polymorphism correlate with recombination rates in *D. melanogaster*. *Nature* 356:519–520.

Bejerano, G., C. B. Lowe, N. Ahituv et al. 2006. A distal enhancer and an ultraconserved exon are derived from a novel retroposon. *Nature* 441:87–90.

Bejerano, G., M. Pheasant, I. Makunin et al. 2004. Ultraconserved elements in the human genome. *Science* 304:1321–1325.

Bensasson, D., D. X. Zhang, D. L. Hartl, and G. M. Hewitt. 2001. Mitochondrial pseudogenes: Evolution's misplaced witnesses. *Trends Ecol Evol* 16:314–321.

Bertozzi, T., K. L. Sanders, M. J. Sistrom, and M. G. Gardner. 2012. Anonymous nuclear loci in non-model organisms: Making the most of high throughput genome surveys. *Bioinformatics* 28:1807–1810.

Bi, K., D. Vanderpool, S. Singhal, T. Linderoth, C. Moritz, and J. M. Good. 2012. Transcriptome-based exon capture enables highly cost-effective comparative genomic data collection at moderate evolutionary scales. *BMC Genomics* 13:403.

Brenner, S. 1998. Refuge of spandrels. *Curr Biol* 8:R669.

Brown, T. A. 2007. *Genomes* 3. New York: Garland Science/Taylor & Francis.

Brown, W. M., M. George, and A. C. Wilson. 1979. Rapid evolution of animal mitochondrial DNA. *Proc Natl Acad Sci USA* 76:1967–1971.

Brumfield, R. T., P. Beerli, D. A. Nickerson, and S. V. Edwards. 2003. The utility of single nucleotide polymorphisms in inferences of population history. *Trends Ecol Evol* 18:249–256.

Charlesworth, B. 2012. The effects of deleterious mutations on evolution at linked sites. *Genetics* 190:5–22.

Charlesworth, B., M. T. Morgan, and D. Charlesworth. 1993. The effect of deleterious mutations on neutral molecular variation. *Genetics* 134:1289–1303.

Charlesworth, D., B. Charlesworth, and M. T. Morgan. 1995. The pattern of neutral molecular variation under the background selection model. *Genetics* 141:1619–1632.

Chen, F. C. and W.-H. Li. 2001. Genomic divergences between humans and other hominoids and the effective population size of the common ancestor of humans and chimpanzees. *Am J Hum Genet* 68:444–456.

Chenuil, A., T. B. Hoareau, E. Egea et al. 2010. An efficient method to find potentially universal population genetic markers, applied to metazoans. *BMC Evol Biol* 10:1.

Collura, R. V. and C. B. Stewart. 1995. Insertions and duplications of mtDNA in the nuclear genomes of Old World monkeys and hominoids. *Nature* 378:485–489.

Costa, I. R., F. Prosdocimi, and W. B. Jennings. 2016. In silico phylogenomics using complete genomes: A case study on the evolution of hominoids. *Genome Res* 26:1257–1267.

Crawford, N. G., B. C. Faircloth, J. E. McCormack, R. T. Brumfield, K. Winker, and T. C. Glenn. 2012. More than 1000 ultraconserved elements provide evidence that turtles are the sister group of archosaurs. *Biol Lett* 8:783–786.

Daguin, C., F. Bonhomme, and P. Borsa. 2001. The zone of sympatry and hybridization of *Mytilus edulis* and *M. galloprovincialis*, as described by intron length polymorphism at locus mac-1. *Heredity* 86:342–354.

Darst, C. R. and D. C. Cannatella. 2004. Novel relationships among hyloid frogs inferred from 12S and 16S mitochondrial DNA sequences. *Mol Phylogenet Evol* 31:462–475.

Delsuc, F., H. Brinkmann, and H. Philippe. 2005. Phylogenomics and the reconstruction of the tree of life. *Nat Rev Genet* 6:361–375.

Derti, A., F. P. Roth, G. M. Church, and C. Wu. 2006. Mammalian ultraconserved elements are strongly depleted among segmental duplications and copy number variants. *Nat Genet* 38:1216–1220.

Doolittle, W. F. 2013. Is junk DNA bunk? A critique of ENCODE. *Proc Natl Acad Sci USA* 110:5294–5300.

Edwards, S. V. 2009. Is a new and general theory of molecular systematics emerging? *Evolution* 63:1–19.

Edwards, S. V. and P. Beerli. 2000. Perspective: Gene divergence, population divergence, and the variance in coalescence time in phylogeographic studies. *Evolution* 54:1839–1854.

Edwards, S. V. and S. Bensch. 2009. Looking forwards or looking backwards in avian phylogeography? A comment on Zink and Barrowclough 2008. *Mol Ecol* 18:2930–2933.

Edwards, S. V., W. B. Jennings, and A. M. Shedlock. 2005. Phylogenetics of modern birds in the era of genomics. *Proc R Soc Lond B Biol Sci* 272:979–992.

Eisen, M. 2012. A neutral theory of molecular function. It is NOT junk, a blog about genomes, DNA, evolution, open science, baseball and other important things. http://www.michaeleisen.org/blog/?p=1172 (accessed October 16, 2015).

ENCODE Project Consortium. 2012. An integrated encyclopedia of DNA elements in the human genome. *Nature* 489:57–74.

Eytan, R. I., B. R. Evans, A. Dornburg et al. 2015. Are 100 enough? Inferring acanthomorph teleost phylogeny using Anchored Hybrid Enrichment. *BMC Evol Biol* 15:1.

Faircloth, B. C., J. E. McCormack, N. G. Crawford, M. G. Harvey, R. T. Brumfield, and T. C. Glenn. 2012. Ultraconserved elements anchor thousands of genetic markers spanning multiple evolutionary timescales. *Syst Biol* 61:717–726.

Faircloth, B. C., L. Sorenson, F. Santini, and M. E. Alfaro. 2013. A phylogenomic perspective on the radiation of ray-finned fishes based upon targeted sequencing of ultraconserved elements (UCEs). *PLoS ONE* 8:e65923.

Fedorov, A., A. F. Merican, and W. Gilbert. 2002. Large-scale comparison of intron positions among animal, plant, and fungal genes. *Proc Natl Acad Sci USA* 99:16128–16133.

Felsenstein, J. 1974. The evolutionary advantage of recombination. *Genetics* 78:737–756.

Felsenstein, J. 2004. *Inferring Phylogenies*. Sunderland: Sinauer.

Felsenstein, J. 2006. Accuracy of coalescent likelihood estimates: Do we need more sites, more sequences, or more loci? *Mol Biol Evol* 23:691–700.

Fitch, W. M. 1970. Distinguishing homologous from analogous proteins. *Syst Biol* 19:99–113.

Friesen, V. L. 2000. Introns. In *Molecular Methods in Ecology*, ed. A. J. Baker, 274–294. Oxford: Blackwell.

Fu, Y.-X. and W.-H. Li. 1993. Statistical tests of neutrality of mutations. *Genetics* 133:693–709.

Galtier, N., B. Nabholz, S. Glémin, and G. D. D. Hurst. 2009. Mitochondrial DNA as a marker of molecular diversity: A reappraisal. *Mol Ecol* 18:4541–4550.

Gillespie, J. 2004. *Population Genetics: A Concise Guide*, 2nd edition. Baltimore: The Johns Hopkins University Press.

Gottscho, A. D., S. B. Marks, and W. B. Jennings. 2014. Speciation, population structure, and demographic history of the Mojave Fringe-toed Lizard (*Uma scoparia*), a species of conservation concern. *Ecol Evol* 4:2546–2562.

Graur, D. and W.-H. Li. 2000. *Fundamentals of Molecular Evolution*, 2nd edition. Sunderland: Sinauer.

Graur, D., Y. Zheng, and R. B. Azevedo. 2015. An evolutionary classification of genomic function. *Genome Biol Evol* 7:642–645.

Graur, D., Y. Zheng, N. Price, R. B. Azevedo, R. A. Zufall, and E. Elhaik. 2013. On the immortality of television sets: "Function" in the human genome according to the evolution-free gospel of ENCODE. *Genome Biol Evol* 5:578–590.

Gray, G. S. and W. M. Fitch. 1983. Evolution of antibiotic resistance genes: The DNA sequence of a kanamycin resistance gene from *Staphylococcus aureus*. *Mol Biol Evol* 1:57–66.

Hare, M. P. 2001. Prospects for nuclear gene phylogeography. *Trends Ecol Evol* 16:700–706.

Hartl, D. L. and E. W. Jones. 2006. *Essential Genetics: A Genomics Perspective*, 4th edition. Sudbury: Jones and Bartlett Publishers.

Hassan, M., C. Lemaire, C. Fauvelot, and F. Bonhomme. 2002. Seventeen new exon-primed intron-crossing polymerase chain reaction amplifiable introns in fish. *Mol Ecol Notes* 2:334–340.

Hazkani-Covo, E., R. M. Zeller, and W. Martin. 2010. Molecular poltergeists: Mitochondrial DNA copies (numts) in sequenced nuclear genomes. *PLoS Genet* 6:e1000834.

Hebert, P. D. N., A. Cywinska, S. L. Ball, and J. R. deWaard. 2003. Biological identifications through DNA barcodes. *Proc R Soc Lond B Biol Sci* 270:313–321.

Hess, C. M. and S. V. Edwards. 2002. The evolution of the major histocompatibility complex in birds. *Bioscience* 52:423–431.

Hillis, D. M. and S. K. Davis. 1986. Evolution of ribosomal DNA: Fifty million years of recorded history in the frog genus *Rana*. *Evolution* 40:1275–1288.

Hillis, D. M. and M. T. Dixon. 1991. Ribosomal DNA: Molecular evolution and phylogenetic inference. *Q Rev Biol* 66:411–453.

Hudson, R. R. and J. A. Coyne. 2002. Mathematical consequences of the genealogical species concept. *Evolution* 56:1557–1565.

Hudson, R. R. and N. L. Kaplan. 1985. Statistical properties of the number of recombination events in the history of a sample of DNA sequences. *Genetics* 111:147–164.

Hudson, R. R. and N. L. Kaplan. 1995. Deleterious background selection with recombination. *Genetics* 141:1605–1617.

Huelsenbeck, J. P., J. P. Bollback, and A. M. Levine. 2002. Inferring the root of a phylogenetic tree. *Syst Biol* 51:32–43.

Hurtley, S. 2012. No more junk DNA. *Science* 337:1581.

Innan, H. and W. Stephan. 2003. Distinguishing the hitchhiking and background selection models. *Genetics* 165:2307–2312.

Jarvis, E. D., S. Mirarab, A. J. Aberer et al. 2014. Whole-genome analyses resolve early branches in the tree of life of modern birds. *Science* 346:1320–1331.

Jennings, W. B. 2016. On the independent loci assumption in phylogenomic studies. *bioRxiv* doi: http://dx.doi.org/10.1101/066332.

Jennings, W. B. and S. V. Edwards. 2005. Speciational history of Australian Grass Finches (*Poephila*) inferred from thirty gene trees. *Evolution* 59:2033–2047.

Jennings, W. B., E. R. Pianka, and S. Donnellan. 2003. Systematics of the lizard family Pygopodidae with implications for the diversification of Australian temperate biotas. *Syst Biol* 52:757–780.

Kaplan, N. L., R. R. Hudson, and C. H. Langley. 1989. The "hitchhiking effect" revisited. *Genetics* 123:887–899.

Karl, S. A. and J. C. Avise. 1993. PCR-based assays of mendelian polymorphisms from anonymous single-copy nuclear DNA: Techniques and applications for population genetics. *Mol Biol Evol* 10:342–361.

Katzman, S., A. D. Kern, G. Bejerano et al. 2007. Human genome ultraconserved elements are ultraselected. *Science* 317:915–915.

Kimura, M. 1968. Evolutionary rate at the molecular level. *Nature* 217:624–626.

King, J. L. and T. H. Jukes. 1969. Non-Darwinian evolution. *Science* 164:788–798.

Kocher, T. D., W. K. Thomas, A. Meyer et al. 1989. Dynamics of mitochondrial DNA evolution in animals: Amplification and sequencing with conserved primers. *Proc Natl Acad Sci USA* 86:6196–6200.

Kong, A., D. F. Gudbjartsson, J. Sainz et al. 2002. A high-resolution recombination map of the human genome. *Nat Genet* 31:241–247.

Kreitman, M. 1983. Nucleotide polymorphism at the alcohol dehydrogenase locus of *Drosophila melanogaster*. *Nature* 304:412–417.

Kuhner, M. K., P. Beerli, J. Yamato, and J. Felsenstein. 2000. Usefulness of single nucleotide polymorphism data for estimating population parameters. *Genetics* 156:439–447.

Lanier, H. C. and L. L. Knowles. 2012. Is recombination a problem for species-tree analyses? *Syst Biol* 61:691–701.

Lee, J. Y. and S. V. Edwards. 2008. Divergence across Australia's Carpentarian barrier: Statistical phylogeography of the red-backed fairy wren (*Malurus melanocephalus*). *Evolution* 62:3117–3134.

Lemmon, A. R., S. A. Emme, and E. M. Lemmon. 2012. Anchored hybrid enrichment for massively high-throughput phylogenomics. *Syst Biol* 61:727–744.

Lemmon, A. R. and E. M. Lemmon. 2012. High-throughput identification of informative nuclear loci for shallow-scale phylogenetics and phylogeography. *Syst Biol* 61:745–761.

Lemmon, E. M. and A. R. Lemmon. 2013. High-throughput genomic data in systematics and phylogenetics. *Annu Rev Ecol Evol Syst* 44:99–121.

Lessa, E. P. 1992. Rapid surveying of DNA sequence variation in natural populations. *Mol Biol Evol* 9:323–330.

Lewontin, R. C. and J. L. Hubby. 1966. A molecular approach to the study of genic heterozygosity in natural populations. II. Amount of variation and degree of heterozygosity in natural populations of *Drosophila pseudoobscura*. *Genetics* 54:595–609.

Li, C., G. Ortí, G. Zhang, and G. Lu. 2007. A practical approach to phylogenomics: The phylogeny of ray-finned fish (Actinopterygii) as a case study. *BMC Evol Biol* 7:44.

Li, C., J.-J. M. Riethoven, and L. Ma. 2010. Exon-primed intron-crossing (EPIC) markers for non-model teleost fishes. *BMC Evol Biol* 10:90.

Lindblad-Toh, K., M. Garber, O. Zuk et al. 2011. A high-resolution map of human evolutionary constraint using 29 mammals. *Nature* 478:476–482.

Martin, D. P., P. Lemey, M. Lott, V. Moulton, D. Posada, and P. Lefeuvre. 2010. RDP3: A flexible and fast computer program for analyzing recombination. *Bioinformatics* 26:2462–2463.

Maynard Smith, J. and J. Haigh. 1974. The hitch-hiking effect of a favourable gene. *Genet Res* 23:23–35.

McCormack, J. E., B. C. Faircloth, N. G. Crawford, P. A. Gowaty, R. T. Brumfield, and T. C. Glenn. 2012. Ultraconserved elements are novel phylogenomic markers that resolve placental mammal phylogeny when combined with species-tree analysis. *Genome Res* 22:746–754.

McCormack, J. E., M. G. Harvey, B. C. Faircloth, N. G. Crawford, T. C. Glenn, and R. T. Brumfield. 2013. A phylogeny of birds based on over 1,500 loci collected by target enrichment and high-throughput sequencing. *PLoS ONE* 8:e54848.

McVean, G. A., S. R. Myers, S. Hunt, P. Deloukas, D. R. Bentley, and P. Donnelly. 2004. The fine-scale structure of recombination rate variation in the human genome. *Science* 304:581–584.

McVicker, G., D. Gordon, C. Davis, and P. Green. 2009. Widespread genomic signatures of natural selection in hominid evolution. *PLoS Genet* 5: e1000471.

Meader, S., C. P. Ponting, and G. Lunter. 2010. Massive turnover of functional sequence in human and other mammalian genomes. *Genome Res* 20:1335–1343.

Mindell, D. P. and R. L. Honeycutt. 1990. Ribosomal RNA in vertebrates: Evolution and phylogenetic applications. *Annu Rev Ecol Syst* 21:541–566.

Mindell, D. P. and A. Meyer. 2001. Homology evolving. *Trends Ecol Evol* 16:434–440.

Moore, W. S. 1995. Inferring phylogenies from mtDNA variation: Mitochondrial-gene trees versus nuclear-gene trees. *Evolution* 49:718–726.

Moritz, C., T. E. Dowling, and W. M. Brown. 1987. Evolution of animal mitochondrial DNA: Relevance for population biology and systematics. *Annu Rev Ecol Syst* 18:269–292.

Moritz, C. and D. M. Hillis. 1996. Chapter 1. Molecular systematics: Context and controversies. In *Molecular Systematics*, 2nd edition, eds. D. M. Hillis, C. Moritz, and B. K. Mable, 1–13. Sunderland: Sinauer.

Nielsen, R. 2000. Estimation of population parameters and recombination rates from single nucleotide polymorphisms. *Genetics* 154:931–942.

Nielsen, R., M. J. Hubisz, and A. G. Clark. 2004. Reconstituting the frequency spectrum of ascertained single-nucleotide polymorphism data. *Genetics* 168:2373–2382.

Nobrega, M. A., I. Ovcharenko, V. Afzal, and E. M. Rubin. 2003. Scanning human gene deserts for long-range enhancers. *Science* 302:413–413.

Ohno, S. 1972. So much "junk" DNA in our genome. *Brookhaven Symp Biol* 23:366–370.

Ohresser, M., P. Borsa, and C. Delsert. 1997. Intron-length polymorphism at the actin gene locus mac-1: A genetic marker for population studies in the marine mussels *Mytilus galloprovincialis* Lmk. and *M. edulis* L. *Mol Mar Biol Biotechnol* 6:123–130.

Ohta, T. and M. Kimura. 1971. Linkage disequilibrium between two segregating nucleotide sites under the steady flux of mutations in a finite population. *Genetics* 68:571–580.

O'Neill, E. M., R. Schwartz, C. T. Bullock et al. 2013. Parallel tagged amplicon sequencing reveals major lineages and phylogenetic structure in the North American tiger salamander (*Ambystoma tigrinum*) species complex. *Mol Ecol* 22:111–129.

Ovcharenko, I., G. G. Loots, M. A. Nobrega, R. C. Hardison, W. Miller, and L. Stubbs. 2005. Evolution and functional classification of vertebrate gene deserts. *Genome Res* 15:137–145.

Palazzo, A. F. and T. R. Gregory. 2014. The case for junk DNA. *PLoS Genet* 10:e1004351.

Palazzo, A. F. and E. S. Lee. 2015. Non-coding RNA: What is functional and what is junk? *Front Genet* 6:1–11.

Palumbi, S. R. 1996. Chapter 7. Nucleic acids II: The polymerase chain reaction. In *Molecular Systematics*, 2nd edition, eds. D. M. Hillis, C. Moritz, and B. K. Mable, 205–247. Sunderland: Sinauer.

Palumbi, S. R. and C. S. Baker. 1994. Contrasting population structure from nuclear intron sequences and mtDNA of humpback whales. *Mol Biol Evol* 11:426–435.

Pennacchio, L. A., N. Ahituv, A. M. Moses et al. 2006. In vivo enhancer analysis of human conserved non-coding sequences. *Nature* 444:499–502.

Pennisi, E. 2012. ENCODE project writes eulogy for junk DNA. *Science* 337:1159–1161.

Philippe, H. and M. Blanchette. 2007. Overview of the first phylogenomics conference. *BMC Evol Biol* 7(Suppl 1)S1.

Philippe, H., F. Delsuc, H. Brinkmann, and N. Lartillot. 2005. Phylogenomics. *Annu Rev Ecol Evol Syst* 36:541–562.

Pluzhnikov, A. and P. Donnelly. 1996. Optimal sequencing strategies for surveying molecular genetic diversity. *Genetics* 144:1247–1262.

Ponting, C. P. and R. C. Hardison. 2011. What fraction of the human genome is functional? *Genome Res* 21:1769–1776.

Portik, D. M., P. L. Wood Jr, J. L. Grismer, E. L. Stanley, and T. R. Jackman. 2011. Identification of 104 rapidly-evolving nuclear protein-coding markers for amplification across scaled reptiles using genomic resources. *Conserv Genet Resour* 4:1–10.

Prum, R. O., J. S. Berv, A. Dornburg et al. 2015. A comprehensive phylogeny of birds (Aves) using targeted next-generation DNA sequencing. *Nature* 526:569.

Prychitko, T. M. and W. S. Moore. 1997. The utility of DNA sequences of an intron from the β-fibrinogen gene in phylogenetic analysis of woodpeckers (Aves: Picidae). *Mol Phylogenet Evol* 8:193–204.

Randi, E. 2000. Mitochondrial DNA. In *Molecular Methods in Ecology*, ed. A. J. Baker, 136–167. Oxford: Blackwell.

Rannala, B. and Z. Yang. 2003. Bayes estimation of species divergence times and ancestral population sizes using DNA sequences from multiple loci. *Genetics* 164:1645–1656.

Reich, D. E., M. Cargill, S. Bolk et al. 2001. Linkage disequilibrium in the human genome. *Nature* 411:199–204.

Reilly, S. B., S. B. Marks, and W. B. Jennings. 2012. Defining evolutionary boundaries across parapatric ecomorphs of Black Salamanders (*Aneides flavipunctatus*) with conservation implications. *Mol Ecol* 21:5745–5761.

Robertson, A. and W. G. Hill. 1983. Population and quantitative genetics of many linked loci in finite populations. *Proc R Soc Lond B Biol Sci* 219:253–264.

Rosenblum, E. B., N. M. Belfiore, and C. Moritz. 2007a. Anonymous nuclear markers for the eastern fence lizard, *Sceloporus undulatus*. *Mol Ecol Notes* 7:113–116.

Rosenblum, E. B., M. J. Hickerson, and C. Moritz. 2007b. A multilocus perspective on colonization accompanied by selection and gene flow. *Evolution* 61:2971–2985.

Rosenblum, E. B. and J. Novembre. 2007. Ascertainment bias in spatially structured populations: A case study in the eastern fence lizard. *J Hered* 98:331–336.

Rozas, J., J. C. Sánchez-DelBarrio, X. Messeguer, and R. Rozas. 2003. DnaSP, DNA polymorphism analyses by the coalescent and other methods. *Bioinformatics* 19:2496–2497.

Saint, K. M., C. C. Austin, S. C. Donnellan, and M. N. Hutchinson. 1998. C-mos, a nuclear marker useful for squamate phylogenetic analysis. *Mol Phylogenet Evol* 10:259–263.

Santiago, E. and A. Caballero. 1998. Effective size and polymorphism of linked neutral loci in populations under directional selection. *Genetics* 149:2105–2117.

Sella, G., D. A. Petrov, M. Przeworski, and P. Andolfatto. 2009. Pervasive natural selection in the *Drosophila* genome. *PLoS Genet* 5:e1000495.

Sequeira, F., N. Ferrand, and D. J. Harris. 2006. Assessing the phylogenetic signal of the nuclear β-Fibrinogen intron 7 in salamandrids (Amphibia: Salamandridae). *Amphibia-Reptilia* 27:409–418.

Shaffer, H. B. and R. C. Thomson. 2007. Delimiting species in recent radiations. *Syst Biol* 56:896–906.

Siepel, A., G. Bejerano, J. S. Pedersen et al. 2005. Evolutionarily conserved elements in vertebrate, insect, worm, and yeast genomes. *Genome Res* 15:1034–1050.

Simons, C., M. Pheasant, I. V. Makunin, and J. S. Mattick. 2006. Transposon-free regions in mammalian genomes. *Genome Res* 16:164–172.

Smith, B. T., M. G. Harvey, B. C. Faircloth, T. C. Glenn, and R. T. Brumfield. 2013. Target capture and massively parallel sequencing of ultraconserved elements (UCEs) for comparative studies at shallow evolutionary time scales. *Syst Biol* doi: 10.1093/sysbio/syt061.

Smith, M. F., W. K. Thomas, and J. L. Patton. 1992. Mitochondrial DNA-like sequence in the nuclear genome of an akodontine rodent. *Mol Biol Evol* 9:204–215.

Sorenson, M. D. and T. W. Quinn. 1998. Numts: A challenge for avian systematics and population biology. *Auk* 115:214–221.

Stapley, J., T. R. Birkhead, T. Burke, and J. Slate. 2008. A linkage map of the zebra finch *Taeniopygia guttata* provides new insights into avian genome evolution. *Genetics* 179:651–667.

Stephan, W. 2010. Genetic hitchhiking versus background selection: The controversy and its implications. *Philos Trans R Soc Lond B Biol Sci* 365:1245–1253.

Stephan, W., T. H. Wiehe, and M. W. Lenz. 1992. The effect of strongly selected substitutions on neutral polymorphism: Analytical results based on diffusion theory. *Theor Popul Biol* 41:237–254.

Stephen, S., M. Pheasant, I. V. Makunin, and J. S. Mattick. 2008. Large-scale appearance of ultraconserved elements in tetrapod genomes and slowdown of the molecular clock. *Mol Biol Evol* 25:402–408.

Swofford, D. L. and G. J. Olsen. 1990. Phylogeny reconstruction. In *Molecular Systematics*, eds. D. M. Hillis and C. Moritz, 411–501. Sunderland: Sinauer.

Swofford, D. L., G. J. Olsen, P. J. Waddell, and D. M. Hillis. 1996. Chapter 11. Phylogenetic inference. In *Molecular Systematics*, 2nd edition, eds. D. M. Hillis, C. Moritz, and B. K. Mable, 407–514. Sunderland: Sinauer.

Takahata, N. 1993. Allelic genealogy and human evolution. *Mol Biol Evol* 10:2–22.

Tamura, K. and M. Nei. 1993. Estimation of the number of nucleotide substitutions in the control region of mitochondrial DNA in humans and chimpanzees. *Mol Biol Evol* 10:512–526.

Tenesa, A., P. Navarro, B. J. Hayes et al. 2007. Recent human effective population size estimated from linkage disequilibrium. *Genome Res* 17:520–526.

Thomson, G. 1977. The effect of a selected locus on linked neutral loci. *Genetics* 85:753–788.

Thomson, R. C., I. J. Wang, and J. R. Johnson. 2010. Genome-enabled development of DNA markers for ecology, evolution and conservation. *Mol Ecol* 19:2184–2195.

Thorne, J. L. and H. Kishino. 2005. Chapter 8. Estimation of divergence times from molecular sequence data. In *Statistical Methods in Molecular Evolution*, ed. R. Nielsen, 233–256. New York: Springer.

Vences, M., M. Thomas, A. Van der Meijden, Y. Chiari, and D. R. Vieites. 2005. Comparative performance of the 16S rRNA gene in DNA barcoding of amphibians. *Front Zool* 2:5.

Vigilant, L., R. Pennington, H. Harpending, T. D. Kocher, and A. C. Wilson. 1989. Mitochondrial DNA sequences in single hairs from a southern African population. *Proc Natl Acad Sci USA* 86:9350–9354.

Vigilant, L., M. Stoneking, H. Harpending, K. Hawkes, and A. C. Wilson. 1991. African populations and the evolution of human mitochondrial DNA. *Science* 253:1503–1507.

Wakeley, J. 2009. *Coalescent Theory: An Introduction* (Vol. 1). Greenwood Village: Roberts & Company Publishers.

Wakeley, J. and J. Hey. 1997. Estimating ancestral population parameters. *Genetics* 145:847–855.

Wakeley, J., R. Nielsen, S. N. Liu-Cordero, and K. Ardlie. 2001. The discovery of single-nucleotide polymorphisms—And inferences about human demographic history. *Am J Hum Genet* 69:1332–1347.

Warren, W. C., D. F. Clayton, H. Ellegren et al. 2010. The genome of a songbird. *Nature* 464:757–762.

Watson, J. D., T. A. Baker, S. P. Bell, A. Gann, M. Levine, and R. Losick. 2014. *Molecular Biology of the Gene*, 7th edition. New York: Pearson Education, Inc.

Weibel, A. C. and W. S. Moore. 2002. A test of a mitochondrial gene-based phylogeny of woodpeckers (genus *Picoides*) using an independent nuclear gene, β-fibrinogen intron 7. *Mol Phylogenet Evol* 22:247–257.

Wiehe, T. H. and W. Stephan. 1993. Analysis of a genetic hitchhiking model, and its application to DNA polymorphism data from *Drosophila melanogaster*. *Mol Biol Evol* 10:842–854.

Wilson, A. C., R. L. Cann, S. M. Carr et al. 1985. Mitochondrial DNA and two perspectives on evolutionary genetics. *Biol J Linn Soc Lond* 26:375–400.

Woerner, A. E., M. P. Cox, and M. F. Hammer. 2007. Recombination-filtered genomic datasets by information maximization. *Bioinformatics* 23:1851–1853.

Woese, C. R. and G. E. Fox. 1977. Phylogenetic structure of the prokaryotic domain: The primary kingdoms. *Proc Natl Acad Sci USA* 74:5088–5090.

Woese, C. R., O. Kandler, and M. L. Wheelis. 1990. Towards a natural system of organisms: Proposal for the domains archaea, bacteria, and eucarya. *Proc Natl Acad Sci USA* 87:4576–4579.

Woolfe, A., M. Goodson, D. K. Goode et al. 2004. Highly conserved non-coding sequences are associated with vertebrate development. *PLoS Biol* 3:e7.

Yang, Z. 2002. Likelihood and Bayes estimation of ancestral population sizes in hominoids using data from multiple loci. *Genetics* 162:1811–1823.

Yang, Z. 2006. *Computational Molecular Evolution* (Vol. 21). Oxford: Oxford University of Press.

Zhang, D. X. and G. M. Hewitt. 1996. Nuclear integrations: Challenges for mitochondrial DNA markers. *Trends Ecol Evol* 11:247–251.

Zimmer, E. A., S. L. Martin, S. M. Beverley, Y. W. Kan, and A. C. Wilson. 1980. Rapid duplication and loss of genes coding for the alpha chains of hemoglobin. *Proc Natl Acad Sci USA* 77:2158–2162.

Zink, R. M. and G. F. Barrowclough. 2008. Mitochondrial DNA under siege in avian phylogeography. *Mol Ecol* 17:2107–2121.

Zuckerkandl, E. and L. Pauling. 1965. Evolutionary divergence and convergence in proteins. In *Evolving Genes and Proteins*, eds. V. Bryson and H. J. Vogel, 97–166. New York: Academic Press.

CHAPTER FOUR

DNA Extraction

Regardless whether a researcher wants to sequence a single locus or an entire genome the usual first step toward the acquisition of phylogenomic data is to perform what are called *DNA extractions*. DNA must be isolated and purified from the cells of animals, plants, fungi, or microorganisms before it can be used in downstream applications such as PCR, NGS library preparation, and DNA sequencing. That means that all RNA (usually), proteins, salts, and other cellular components must be separated from the DNA (genomic and organellar). When the process is complete, the purified DNA is usually in an aqueous buffer solution that is compatible with downstream enzymatic-driven applications. Because the process of DNA extraction involves the use of chemicals and specialized laboratory equipment, the work must be done in a molecular laboratory and appropriate safety procedures followed. As we will see in this chapter, the time required for doing DNA extractions ranges from less than an hour to several days depending on extraction method used and the nature and number of the genetic samples. The cost per sample will also depend on the type of extraction method to be used. In general DNA extraction "kits" represent the simplest and fastest methods for extracting DNA but they also cost significantly more than methods based on "old fashioned" molecular biology.

The goals of this chapter are to provide an overview of the common methods for extracting DNA, discuss the advantages and disadvantages of each method, and consider some troubleshooting strategies. Although I discuss the basic steps involved in these methods for DNA extraction, I do not provide the actual protocols that should be followed in the laboratory. Thus, the reader should consult actual step-by-step molecular biology recipes and protocols found in the literature or in the instructions provided with commercially available DNA extraction kits. Kit manuals often contain helpful technical advice on the DNA extraction process, which can further aid the researcher in better understanding the methodology involved and hence improve the quality of extraction results, but often do not sufficiently describe the chemical processes. A discussion and detailed step-by-step protocols for some older (but still very useful) DNA extraction methods can be found in Hillis et al. (1996).

4.1 DNA EXTRACTION METHODOLOGY

Until the late 1990s, molecular biologists relied upon the *phenol–chloroform–isoamyl alcohol* or "PCI" method as it is sometimes called. Though effective, this method consisted of a laborious process involving noxious organic solvents and requiring the use of a chemical fume hood. However, the introduction of molecular biology kits, just prior to the dawn of the genomics era, represented a major innovation that greatly simplified the DNA extraction process. Moreover, these extraction kits offered an additional advantage—freedom from the chemical fume hood. Hazardous volatile chemicals (e.g., phenol) are no longer necessary for high-quality DNA extractions. Most work involving an extraction kit is done using only pipettes, a vortex mixer, and microcentrifuge.

4.1.1 Summary of the DNA Extraction Process

Although the various extraction methods differ from each other in terms of specific procedures and reagents used, they involve the same basic sequence of steps: (1) *lysis of tissues/cells and degradation of proteins*; (2) *degradation of RNA* (can be optional); (3) *isolation of DNA from proteins*; (4) *desalting the DNA*; and (5) *resuspension of DNA in storage buffer solution*.

Step 1: Lysis of Tissues/Cells and Degradation of Proteins This initial step involves the breaking down of tissues into cells, the lysis of those cells to release the DNA into a common "soup," and the enzymatic destruction of proteins. A detergent-containing buffer lyses the cells while the broad spectrum protease *Proteinase K* destroys most proteins. *Proteinase K* is used for two reasons: (1) some structural proteins must be destroyed in order to facilitate the breaking down of the tissues and (2) DNA-degrading enzymes such as nucleases, endonucleases, and exonucleases must be inactivated to prevent DNA degradation. Performing this starting procedure is simple and can be done in less than an hour. For example, vertebrate tissues are processed in the following manner: the lysis buffer + *Proteinase K* are first pipetted into a 1.5 mL microcentrifuge tube, which contains the tissue sample; the tube is then heated to around 55°C until the tissue has dissolved in the solution. The heat-incubation time required ranges from several hours to several days (or more) depending on the size and nature of the tissue sample. The step is deemed complete when a solid tissue sample is no longer visible. If the starting tissue was pigmented, then the solution will change color as pigments are extracted from the cells. These do not affect subsequent DNA purification. This step is carried out in almost identical fashion among most DNA extraction methods. The main difference is due to whether the researcher uses buffers that were "homemade" in the lab or are components of a kit.

Step 2: Degradation of RNA The enzyme *Ribonuclease A*, more commonly called *RNase A*, is another enzyme that is often added to the initial lysate with the purpose of destroying all RNA molecules. Prior to implementing this step, careful thought should be given as to whether it is absolutely necessary to destroy the RNA in a DNA extract. There are two important considerations: (1) *RNase* is a very expensive reagent and thus it may not be needed if the primary downstream goal is to use PCR and Sanger sequencing and (2) the RNA component might be needed at a later time for another application such as constructing a cDNA library. However, in many types of NGS-based studies, the RNA component is often eliminated from a DNA extract before starting the construction of a sequencing library otherwise it can confound attempts to estimate the concentration of double-stranded DNA (dsDNA) in a sample. The decision to skip the *RNase* step during the DNA extraction process is not critical as RNA can later be removed via *RNase* treatment followed by DNA precipitation.

Step 3: Isolation of DNA from Proteins The next major step during the extraction process consists of separating the DNA from the other cellular components, especially proteins. Different extraction methods achieve this task by different means. A very popular DNA extraction method known as "salting out" (or "saline method") by Miller et al. (1988), uses a saturated sodium chloride solution (5–6M) to precipitate the proteins from the solution, while the DNA remains in an aqueous supernatant following centrifugation. Ice-cold ethanol (≥95%) or isopropanol (pure) is then added to the aqueous solution and mixed. The presence of salt and alcohol causes the DNA to lose its ability to remain dissolved in water thereby causing it to precipitate. After mixing the solution, the tubes are placed into a –20°C freezer for a minimum of 2 hours (preferably overnight). Note that the longer the time spent in the freezer the greater the DNA yield, up to a point—most of the DNA will have precipitated after an overnight precipitation period and thus longer waiting times are unnecessary. However, as this is often a convenient stopping point, the tubes can be left in the freezer indefinitely until the researcher is ready to resume the extraction process with the next step. After the tubes are removed from the freezer, they should immediately be placed into a microcentrifuge where they are spun at very high speed (~14,000 rpm) for ~5–15 minutes. This causes the DNA to form a solid pellet at the bottom of the tube, which, in a good light, can sometimes be visible to the naked eye. Some commercially available kits (e.g., Wizard® Genomic DNA Purification kit, Promega) are largely based on the salting out principle.

A second method for separating the protein and DNA components, which is used by many kits, involves mixing proprietary buffers from the kits with the lysate before adding the solution to a plastic spin column containing membranes with silica, glass, or diatomaceous earth (e.g., QIAquick® PCR Purification kit, Qiagen®). DNA binds to the membranes at a low pH, and tubes are centrifuged at high speed in order to draw the liquid through the column and into a waste liquid collection tube. The bound DNA is retained while the proteins and other unwanted materials to pass through into the waste tube.

The traditional PCI method, which uses phenol to separate the proteins from the DNA, is accomplished in the following manner. First, after adding phenol to the aqueous (lysate) solution and mixing, the solution is centrifuged at high speed (~14,000 rpm), which stratifies the solution into two roughly equal layers in the microtube: a clear aqueous layer above that contains mostly DNA and salts and an "organic" layer below that contains mostly phenol, proteins, and other cellular debris. The aqueous supernatant is saved while the organic layer is discarded into appropriate waste containers inside a chemical fume hood. Remember that phenol is a very hazardous chemical and should be handled carefully following all safety precautions. This phenol addition step is repeated at least one more time before a similar step is performed using chloroform. With the retained aqueous solution, a small volume of a salt solution (e.g., 2M NaCl or 3M NaOAc pH 5) is pipetted into the solution and mixed followed by the addition of ice-cold 95% ethanol. This is the same DNA precipitation step described earlier. The tubes are placed in the –20°C freezer for at least one night before pelleting the DNA with centrifugation.

Step 4: Desalting the DNA Regardless of the extraction method used, the DNA pellet recovered via precipitation or spin columns will contain excess salts. If not removed, these salts can interfere with various downstream enzymatic procedures such as PCR. The salts therefore need to be "washed" from the DNA pellet. In this procedure, a fairly substantial volume of ethanol (e.g., ~500 μL of 70%–80% ethanol) is *gently* pipetted into each tube, which effectively dilutes away the salts. If using spin columns, the ethanol and salts are removed via a centrifugation step after the ethanol is added. Other methods simply require the user to pour the ethanol out of the microcentrifuge tubes or use the pipet to *carefully* remove the supernatant. When pouring or pipetting the ethanol from the tube, extreme caution should be exercised to avoid losing the DNA pellet, which can at any time become dislodged from the bottom of the tube—and hence lost. Touching the pellet with the pipet tip can also result in a lost pellet. DNA extraction kits usually require the user to purchase their own stock of ethanol for purposes of washing the recovered DNA. Although a single wash may be sufficient to desalt the DNA, two wash steps are usually recommended.

Step 5: Resuspension of DNA in Storage Buffer Solution The final procedures in DNA extraction are to (1) dry the DNA pellet to remove residual alcohol, which can interfere with downstream enzymatic applications and (2) resuspend or elute the DNA into an appropriate storage buffer solution for DNA. When using spin column kits, the drying step is simply carried out via a brief period of centrifugation. Afterward, an "elution" buffer is pipetted onto the membrane in the spin column, which releases the bound-DNA from the membrane. A final centrifugation step causes the DNA and buffer solution to pass into a clean collection microtube, which completes the extraction process. In the salting out or PCI protocols, drying of the pellet can be accomplished via air-drying the open and inverted tube overnight or by using a vacuum centrifuge *with no heat* for ~10 minutes depending on how much residual alcohol remains. After the pellet is dry, a DNA storage buffer is pipetted into the tube and allowed to dissolve the DNA pellet.

4.1.2 A Note about DNA Storage Buffers

As we just saw, the final step in the extraction process consists of resuspending or eluting the DNA into a "storage buffer" solution. Historically, this buffer consisted of 1× strength Tris–EDTA or "1× TE" buffer (i.e., 10 mM Tris–HCl, 1 mM EDTA, pH 8.0). The Tris component acts as a pH buffer keeping the DNA in a slightly alkaline environment, which helps to inhibit DNA-destroying DNases. The EDTA also acts as a preservative but does so by chelating metal cations such as Mg^{2+}, which are cofactors for most nucleases (Carter 2000). Although 1× TE is a good preservative for DNA, a drawback is that EDTA can interfere with DNA polymerases during PCR if too many of the Mg^{2+} cations are chelated (DNA polymerases also require these cations to function). This problem, however, can be overcome if the researcher takes the time to optimize the concentration of MgCl buffer in PCR, which would then allow the DNA polymerases to function well despite the presence of EDTA. However, this procedure adds additional optimization work to PCR. Another strategy to improve the performance of enzymatic reactions involving your extracted DNA is to use a buffer containing a lower concentration of EDTA such as a 0.5× TE. Some DNA extraction kits use 0.5× TE as a DNA elution buffer (e.g., QIAquick, Qiagen). Some researchers even use a 0.1× TE, which is also called "ultra-low TE." However, keep in mind that as the strength of TE is lowered the DNA extract becomes more suitable to PCR and more vulnerable to destruction by nucleases. This is an important tradeoff. The researcher has two options that can accommodate the short-term requirements of PCR and the long-term storage of stock DNA. First, you can use 1× TE and optimize the Mg^{2+} in PCR. Instructions that accompany commercially available PCR polymerases often provide guidance on how to optimize the amount of Mg^{2+} (via a MgCl buffer solution) in a PCR reaction. Alternatively, you can mix up some DNA in low strength TE aliquots (0.1–0.5× TE) and maintain the primary DNA extract in 1× TE strength for longer term storage. For this latter strategy, it may be preferable to initially elute (or resuspend) the DNA in a lower volume of 1× TE in order to increase the DNA concentration for the long-term storage stock. This will facilitate the making aliquots of desired concentrations for PCR (Chapter 5) or NGS library construction (Chapter 7).

Many researchers have simply used molecular biology grade water for making aliquots for PCR or storing the extracted DNA. While such samples are ideal for PCR, they are not safe for long-term storage—and not just because of the threat of nuclease destruction. Water is not a good storage medium because it will facilitate degradation of the DNA sample. Thus, the DNA should be stored in a buffer solution. A number of additional methods might prove effective for long-term storage of DNA including (1) using 1× TE and freezing the sample at −80C; (2) using 95%–100% ethanol and freezing at −20 or −80C; and (3) drying the DNA before freezing at −80C. Unfortunately, no clear consensus seems to exist yet as to which storage method works best therefore the best course of action may be to divide a sample into different aliquots and try various methods (Yates et al. 1989). In other words, this is the "don't keep all your eggs in a single basket" strategy. More research is clearly needed for this very important topic.

Chemicals such as nucleases and water are not the only agents that can degrade DNA. Ultraviolet (UV) light as well as repeated episodes of freezing and thawing of extract samples will also degrade the DNA into smaller fragments. Thus, following a DNA extraction procedure, a "working stock" aliquot of each DNA sample should be made so that the number of freeze–thaw cycles for a given DNA sample can be minimized. New aliquots can be taken from the original extracts whenever DNA is needed for a procedure. This approach also makes it easy for the researcher to dilute a 1× TE-preserved DNA sample to a 0.1–0.5× concentration thus making the sample ideal for PCR or other enzyme-driven procedures.

4.1.3 Extracting DNA from Plants, Fungi, and Invertebrates

The previously discussed DNA extraction procedures are effective for many types of tissues especially those from vertebrates (my main taxonomic group of interest). However, it is well known that tissues from plants, fungi, invertebrates, and other organisms may require a modified protocol in order to obtain purified DNA samples. Readers who are interested in learning more about these special DNA extraction procedures for these organisms should consult the materials and methods sections of recent research articles on a particular group of interest in order to determine which procedures or kits are being used. Also see Palumbi (1996) for advice and DNA extraction protocols that deal with difficult-to-extract tissues.

4.1.4 Extracting DNA from Formalin-Fixed Museum Specimens

Tissue samples obtained from formalin-fixed specimens also require specialized DNA extraction methods. This problem represents one of the greatest challenges in molecular phylogenetics and phylogenomics because millions of archived specimens in natural history museums and other research collections may contain DNA suitable for sequencing and hence are of potentially great importance to tree of life studies. For example, most herpetological and ichthyological specimens are initially fluid-preserved in a 10% buffered formalin solution for a period of time lasting days or weeks. However, because formalin is extremely toxic, these specimens are subsequently transferred to a 70% ethanol solution for long-term storage. Although this preservation technique is important for preserving specimens, once specimens are fixed in formalin it becomes exceedingly difficult or impossible to extract useable DNA from their tissues. Even if DNA is isolated from formalin-fixed tissues, depending on how long the specimen was fixed and other factors, there is no guarantee that the extracted genetic material will be amenable to DNA sequencing.

Formalin adversely affects DNA in a number of ways including (1) forming cross-linkages between DNA bases and between DNA and proteins; (2) altering the nucleotides; and (3) fragmenting the DNA (Tang 2006). If specimens are preserved in unbuffered formalin (which is acidic), then the formalin will become oxidized into formic acid. This, in turn, results in substantial DNA degradation including depurination and fragmentation (Tang 2006). Fortunately, the use of unbuffered formalin to preserve natural history specimens seems to be a long discontinued practice.

Formalin-induced cross-links between proteins and DNA cannot easily be broken and thus it is difficult or impossible to isolate DNA using conventional DNA extraction methods (Shedlock et al. 1997). In order to extract DNA from formalin-preserved specimens, tissue samples must first be subjected to laborious and time-consuming "washings" in special buffers and heat treatments in order to break down cross-linked DNA–protein complexes in an effort to free DNA (Shedlock et al. 1997; Tang 2006). However, even if some DNA is obtained this way, the yield of high molecular weight DNA is usually low and is fragmented into pieces no larger than several hundreds of base pairs (Shedlock et al. 1997). Still, some studies have successfully obtained DNA sequence data from formalin-fixed museum specimens.

Shedlock et al. (1997) described a modified DNA extraction method for obtaining DNA from formalin-fixed fishes. Included in their study were 12 deep-sea fish specimens of varying condition and with in-jar storage times ranging from several years to more than 85 years (a specimen collected in 1909). Moreover, the study only focused on obtaining template DNA suitable for amplifying a 470 bp portion of the cytochrome oxidase b (*Cyt b*) gene and a 570 bp fragment of the 16S rRNA gene from the mitochondrial genome. Of the 34 total PCR reactions attempted, 28 (82%) were successful and one of these products (*Cyt b*) was obtained from an 85-year-old specimen. However, these authors noted that most of their successful amplifications were with specimens preserved for 25 years or less. Although their modified extraction procedure is laborious and requires several days to complete, this study nonetheless produced encouraging results showing that it is possible to at least obtain PCR-amplifiable mtDNA even from specimens nearly a century old.

An unfortunate aspect of preserving specimens in formalin is that extracted DNA consists of fragments that are less than a few hundred bases long. However, a recent study by Hykin et al. (2015) suggested and showed that the short-read NGS technology used by the Illumina platforms (Chapter 7) is perfectly suited to such degraded DNA samples. These authors presented a modified DNA extraction protocol for formalin-fixed reptile tissues and tested it using the Illumina platform. Their study included *Anolis* lizard specimens with preservation ages of ~30 years and ~100 years. In addition, they subsampled different anatomical parts of each specimen (i.e., liver, leg muscle, and tail tip) and used different extraction methods (i.e., PCI and a Qiagen kit) in order to determine which tissue types and extraction methods are best suited for their protocol. Their results showed that potentially useful amounts of DNA were only obtained from the 30-year-old *Anolis* specimen and only when PCI was used. An analysis of the fragment-size distributions revealed that the main peak fragment size for the 30-year-old specimen was centered over the 200–300 bp range. Interestingly, although the 100-year-old specimen yielded a lower concentration of DNA, results indicated that the peak fragment size for this extract was in the 200–400 bp range suggesting it might be possible to obtain sequence data from very old museum specimens. However, Illumina sequencing of these two samples only produced sequence data for the 30-year-old specimen extract—including enough data to fully reconstruct its entire mitochondrial genome. Hykin et al. (2015) pointed out that further refinements of their protocol including the possible use of alternative library-making methods could lead to better results.

4.2 EVALUATING THE RESULTS OF DNA EXTRACTIONS

After the extraction process is completed the next task is to check the **quality**, **concentration**, and **purity** of the extracted DNA samples. This step will provide valuable information about the quality of your tissue samples, the quality of your DNA extraction procedures, and allow you to appropriately dilute these often highly concentrated DNA samples in ways that allow for consistent PCR or NGS sequencing library results while conserving stock DNA. Three approaches for evaluating DNA samples include: *agarose gel electrophoresis*, *UV spectrophotometry*, and *fluorometry*. As we will see, each of these methods has a particular strength and thus together they can provide critical information about DNA extracts.

4.2.1 Agarose Gel Electrophoresis

Agarose gel electrophoresis has changed little over several decades owing to its simple, inexpensive, and reliable nature. For work involving the sequencing of DNA, researchers use this method for several important purposes: (1) to assess the quality (i.e., degree of degradation) and concentration of DNA obtained from an extraction; (2) to assess the quality of PCR results (Chapter 5); and (3) to assess the quality of an NGS sequencing library and for size-selecting DNA fragments for NGS (Chapter 7).

The equipment for agarose gel electrophoresis consists of two main parts: a rectangular box containing an agarose gel slab submerged in a buffer solution and a power supply unit (Figure 4.1). When the power is switched on an electrical current passes through the gel. Because DNA is a negatively charged molecule, the current will mobilize the DNA fragments causing them to travel through the gel from the negative side of the box to the positive side. A key principle of electrophoresis is that the linear DNA will travel through the gel matrix at a rate proportional to its size with shorter DNA molecules traveling at a faster rate than longer ones. Agarose gel electrophoresis is an invaluable tool for fractionating DNA of different sizes, which has important applications for checking the quality of DNA extractions, PCR, and DNA sequencing.

The agarose gel electrophoresis procedure is comprised of a series of steps, which includes: (1) *casting the gel*, (2) *loading samples into the gel*, (3) *running the gel* (i.e., *the actual electrophoresis step*), and (4) *photographing the gel*. The entire procedure requires 2–3 hours of time depending upon actual electrophoresis settings (e.g., voltage). We will now review these steps.

Step 1: Casting the Gel The first step of agarose gel electrophoresis begins when the researcher heats a 0.7%–1% agarose solution (~50–75 mL) in a microwave oven for a minute, then lets the boiling-hot liquid cool down before pouring into a casting dish and adding the combs. The comb creates the individual sample wells in the gel. Note, that if the agarose solution is too hot when poured into the casting dish then the dish itself can be damaged through warping. On the other hand, letting the liquid cool too much prior to pouring is not good either because the agarose will then start to solidify before it is poured into the dish thus making a lumpy gel. Thus, an ideal gel is made with a "warm" agarose solution—one that is still in liquid form yet has a viscous consistency. For purposes of checking the results of a DNA extraction, a 0.7%–1% gel is ideal because less concentrated gels are too delicate to handle (i.e., they break into pieces too easily when handled) and more concentrated gels are not necessary. Using a more concentrated gel means you are only wasting agarose, which is an expensive

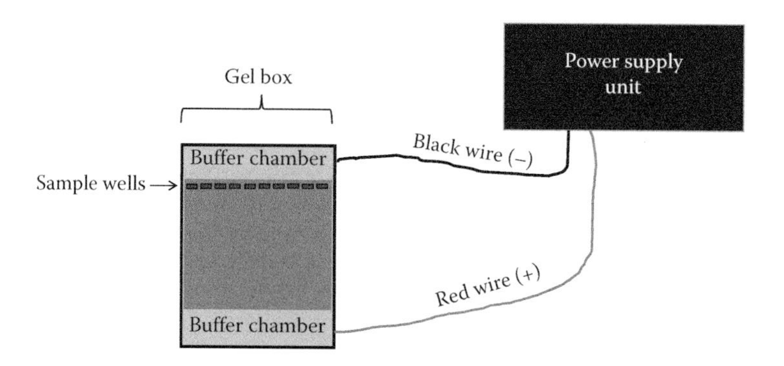

Figure 4.1. Agarose gel electrophoresis apparatus consisting of a gel box and power supply unit (top view). The gel (dark gray) sits between the two buffer chambers.

reagent. In Chapter 5, we will discuss using more concentrated gels to evaluate the results of PCR. More concentrated gels provide better resolving power for discriminating fragments of DNA by their sizes. Thus, a 2% gel is preferred for determining the sizes of PCR products, whereas a 1% gel is sufficient for evaluating the results of DNA extraction. After approximately 30 minutes the gel cools and hardens (much like fruit-flavored gelatin-type desserts) at which time it is submerged in a gel box containing a buffer, which is usually 1× TAE (Tris–Acetic Acid–EDTA) or 1× 0.5× TBE (Tris–Boric Acid–EDTA; Sambrook et al. 1989).

Step 2: Loading Samples into the Gel Next, a sample of each DNA extraction (usually ~5 μL) is mixed with a loading buffer that includes sucrose, glycerol, and/or ficoll along with one or multiple negatively charged dyes (usually ~1 μL of a 6× solution) before being pipetted into each open well in the gel. The purpose of the loading buffer (sometimes called "loading dye") is to help the DNA sink into the gel well (otherwise the DNA will float and disperse itself) and provide a visual marker in the gel to confirm that the electrophoresis process is running smoothly (i.e., colored bands can be seen migrating across the gel under white light after, say, 5–10 minutes of running). Note that these dyes are visualized in normal light and thus they do not allow for the visualization of the migrating DNA. A separate nucleic acid stain, which fluoresces under UV or colored light, must be used to actually see the DNA in a gel. In addition to the sample lanes in the gel, one well is reserved for a sample of "ladder" or (sometimes called "molecular ruler"). The ladder is a solution comprised of a set of *linear* dsDNA fragments of known size (e.g., 100 bp increments), which migrate across the gel at a rate proportional to their length. Although the ladder acts as a reference allowing you to judge the approximate sizes of DNA bands in the gel, another major function of the ladder is to provide *a positive control* for DNA in the gel. In other words, this also is a control for the electrophoresis process.

Step 3: Running the Gel Once the gel is loaded with the samples, electrical current flows through the buffer solution containing the gel, which "pulls" the negatively charged DNA through the porous interstitial spaces of the solidified agarose. Gels are typically run for 20–60 minutes depending on the voltage setting (e.g., 90 volts). When electrophoresis is complete, for safety reasons, don't forget to switch off the voltage box *and* disconnect the gel box from the power source by unplugging the red and black wires.

Step 4: Photographing the Gel In order to visualize the DNA in a gel, the investigator must first incorporate into the DNA some type of nucleic acid "stain" that fluoresces under UV light. The nucleic acid stain can be applied to the DNA in several different ways. In practice, stains have been applied to gels either before or after electrophoresis. If applied before electrophoresis, then the stain is mixed into the molten agarose prior to casting. If applied after electrophoresis, then the gel is submerged in a box containing a solution of TAE (or TBE) buffer + nucleic acid stain solution followed by a 10–30-minute waiting period, the length of which depends on the type and strength of the staining chemical. When the stain comes into contact with the DNA either during electrophoresis or while the gel is submerged in the buffer and stain box, the stain molecules bind to or become intercalated into the DNA. A major advantage to staining gels after electrophoresis is that investigators can avoid contaminating the entire electrophoresis apparatus and workspace with a potentially toxic stain; instead, a plastic dish, which can be placed inside a chemical fume hood, is used for staining gels. Some commercially available stains allow you to premix a nucleic acid stain with normal 6× loading dye. Thus, the stain comes into contact with the DNA samples at the time immediately before pipetting samples into the wells of a gel. Always consult the stain manufacturer's instructions for safe proper use.

What exactly are these nucleic acid stains? For many years, ethidium bromide (EtBr) was the primary staining reagent for DNA. However, because EtBr has long been presumed to be extremely toxic (Sambrook et al. 1989; also see Lowe 2016 for a different viewpoint), putatively safer alternative nucleic acid stains have become commercially available. Regardless whether or not these nucleic acid stains pose health hazards, common sense dictates that researchers always handle these chemicals in as safe a manner as possible. At the very least, researchers working with these chemicals should wear protective gloves and safety goggles, minimize contamination of other equipment and surfaces, perform this work in a chemical fume hood, and dispose of chemical waste in an institution-approved manner.

After staining and electrophoresis (or vice versa), the investigator places the gel on a blue light box or UV transilluminator box so that the results can be visualized and photographed for later reference. UV light will cause injuries to eyes and skin. Therefore, always wear appropriate eye and skin protection and use a shield when examining a gel on an UV transilluminator box. DNA stains such as SYBR Green or Gel Green can be visualized on a blue light box through an orange filter minimizing concerns of UV exposure. Gel photos represent invaluable records of each "experiment" and therefore should be printed, labeled, and included in laboratory notebooks. Figure 4.2 is a gel photo showing a successful set of DNA extractions. When examining a gel photo the main things to look for include the presence of any DNA and its level of degradation. If a fresh tissue sample was used in an extraction, then the resulting gel image should show two major bands of DNA in the gel: *genomic or chromosomal DNA* indicated by a band that is inside or just outside the well; *mitochondrial DNA* shown as the next discrete band (~16 kb); and some degraded DNA (mixture of nuclear and mitochondrial DNA) shown as a smear of smaller sized fragments (Figure 4.2). If the DNA was not treated with *RNase*, then you may also see a high concentration of RNA fragments, which appear as a bright band on the gel. See Sambrook et al. (1989) for discussion of agarose gel electrophoresis and protocols.

4.2.1.1 Troubleshooting

The results of electrophoresis not only provide important information about the quality and concentration of each DNA sample, but each gel image can also contain invaluable clues about the causes of any failures relating to the DNA extraction and electrophoresis processes. For example, if no DNA—not even the ladder—is visible on the gel (i.e., a "blank" gel), then the investigator can only conclude that one or more problems arose during the electrophoresis process. When this happens, nothing can be concluded about the quality of the DNA extraction process or the samples until the electrophoresis problems are solved. Various problems can arise during electrophoresis making the visualization of any DNA difficult or impossible. For example, a common culprit is a nucleic acid stain solution that has become too weak (due to overuse and/or having been degraded by light). Running a new gel with a fresh nucleic acid stain may solve the problem. Other possible causes of a failed gel include using a buffer solution contaminated by nucleases; switching the red and black wires causing the DNA to travel in a reverse direction and into the buffer chamber (using the mnemonic "run to red" will help avoid this issue); and problems with the cast gel such as having holes in the bottom of the wells causing the DNA to travel *under* the gel instead of through the gel matrix.

In other gel photos one might see normal-looking ladder bands but no other DNA elsewhere on the gel. In these cases, the investigator can conclude that a problem arose during the extraction process because existence of the ladder bands proves that the basic electrophoresis equipment, reagents, and running conditions were normal.

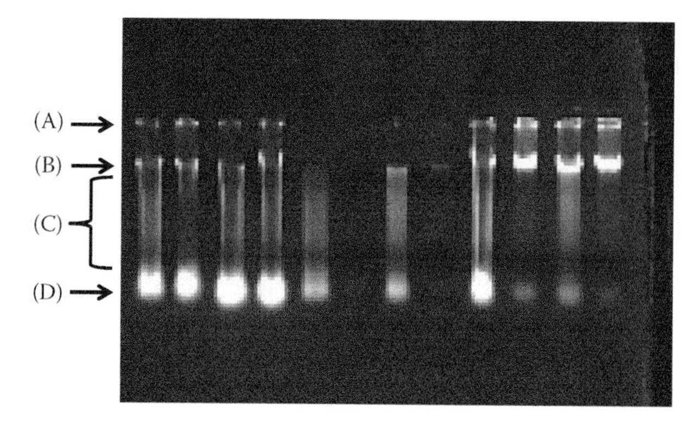

Figure 4.2. Photo of 1% agarose gel stained in ethidium bromide. DNA samples shown are from DNA extractions using the phenol–chloroform–isoamyl method. (A) Undegraded genomic DNA; (B) undegraded mitochondrial DNA; (C) degraded genomic DNA and RNA; (D) RNA. Note, loading wells are located along the top edge of the A bands. Ladder not shown.

DNA extractions can fail for a number of reasons including using defective reagents or kits, experimenter error (pipetting mistakes), and/or bad quality tissue samples.

4.2.2 UV Spectrophotometric Evaluation of DNA Samples

Another standard piece of equipment in a DNA laboratory is a tabletop device called a *UV spectrophotometer*, which is used to estimate the concentration and purity of a DNA sample. These machines operate by shining UV light through a DNA solution and then measure the light absorbance by the DNA. Since nucleic acids maximally absorb light in the UV range, the amount of light absorbed in a solution is directly proportional to the amount DNA/RNA in that sample. This property of nucleic acids enables molecular biologists to use a simple method for quantifying DNA samples and for determining their purity. There are two main limitations of UV spectrophotometry. First, UV spectrophotometry cannot discriminate RNA from DNA and thus it will only produce accurate estimates of genomic DNA concentration if the RNA component has been removed/destroyed. Thus, if a genomic DNA extract is not treated beforehand with *RNase*, then the actual concentration for the dsDNA component will be lower than the total concentration because of RNA contamination. This can be particularly important to consider when quantifying genomic DNA during NGS library construction. A second limitation is that UV spectrophotometry cannot determine whether or not a DNA sample is degraded. Thus, a sample containing only nucleotides can have the same concentration as a sample containing undegraded genomic DNA.

Spectrophotometers are simple to use. The first step is to run a "blank" before you measure your DNA sample so that a "background" absorbance value is established for your DNA solution minus the DNA. To perform the blank step, you simply run a sample of the *same* liquid your DNA is sitting in through the spectrophotometer, which will provide a "zero" concentration value. For example, if the final step of your DNA extraction kit specifies that you elute or dilute your DNA sample using a buffer provided in the kit (e.g., TE), then you should blank using the *same* buffer. If instead you performed the elution or dilution step using pure water, then you should blank using pure water. Blanking with the same solution will help ensure accuracy. The next step is to measure a sample of your DNA. If using an older model spectrophotometer, then you will likely be pipetting your sample into a quartz glass or clear plastic cuvette—the size of your small finger, whereas newer "microvolume" machines such as the NanoDrop 2000 (Thermo Scientific™) do not use a cuvette. One nice advantage of the NanoDrop is that it consumes far less sample than the older machines (1–2 μL vs. 5–10 μL of DNA). Moreover, checking the accuracy of your machine is easily done by pipetting 1 μL of a known concentration template (e.g., some 100 bp or 1 kb ladders are sold at a concentration of ~100 ng/μL).

4.2.2.1 UV Spectrophotometry to Determine Concentrations of Nucleic Acid Samples

DNA (and RNA) maximally absorbs light that is of 260 nm in wavelength. This constant is useful because the absorption of 260 nm light by dsDNA, single-stranded DNA (ssDNA), and RNA can be standardized in terms of optical density (OD) units and hence yield the following *conversion factors* (Sambrook et al. 1989):

1 OD = 50 μg/mL

1 OD = 40 μg/mL RNA and ssDNA

UV spectrophotometers typically do not perform well with highly concentrated DNA samples (e.g., >10 μg/μL). Therefore a DNA sample should be diluted 10-, 50-, or 100-fold before being assayed plus an appropriate *dilution factor* will also be needed to calculate the concentration. Using the above information we can calculate DNA concentration with the following equation:

$$\text{DNA concentration} = \text{OD} \times \text{conversion factor} \times \text{dilution factor} \qquad (4.1)$$

Here is an example concentration calculation using Equation 4.1: begin with 50 μL of extracted genomic DNA. Next, dilute 20 μL of that DNA into 980 μL of water ("1/50 dilution" or a "50× dilution factor"). Next, place the diluted sample in a cuvette and determine the OD of the sample at 260 nm. Let's say the spectrophotometric reading = 0.2 OD, then the dsDNA concentration (μg/mL) will be

DNA concentration = 0.2 × 50 μg/mL × 50

DNA concentration = 500 μg/mL

Many routine DNA methods typically involve minute volumetric measurements in microliters (μL). Accordingly, it is generally more useful to express DNA concentrations in terms of ng/μL, which is the equivalent to μg/mL. Older spectrophotometers usually measure concentration as μg/mL, whereas newer models (e.g., NanoDrop) output concentrations in ng/μL thus one does not need to manually convert OD units into concentration units. Lastly, spectrophotometers tend to not only have trouble in estimating the OD of concentrated samples, but they also tend to lose reliability with samples that are too dilute (<5 ng/μL). It is important to keep in mind that samples with very low readings—even some with a 0.0 ng/μL reading—may still contain enough DNA template to result in a successful PCR reaction—PCR is extremely sensitive as we will see in Chapter 5. Thus, these evidently DNA-free samples should not be dismissed as failed extractions and discarded. On many occasions I have observed such "zero concentration" samples yield useful PCR results and thus it is always worthwhile to try PCR anyway with such samples.

4.2.2.2 *UV Spectrophotometry to Determine the Purity of DNA Samples*

Ideally, a DNA sample will be free of all types of contaminants, which can interfere with downstream laboratory procedures such as PCR. Proteins, salt, and ethanol are the most common contaminants mixed with DNA following an extraction procedure. Fortunately, in addition to measuring concentration, a spectrophotometer can also measure the purity of a DNA sample. An index of sample purity called the "260/280 ratio" can be used to gauge the purity of a DNA extract (Sambrook et al. 1989). This ratio is simply produced after a spectrophotometer generates OD values at 260 and 280 nm wavelengths. Pure DNA has a ratio of 1.8 but values ranging from 1.3 to 2.1 indicate a relatively pure DNA sample that can generally perform well in PCR. The presence of proteins, which have a maximum absorbance at 280 nm will increase absorbance at 280 causing the 260/280 to lower in value, whereas the presence of ethanol, salts, or RNA will raise the ratio above 1.8. Applying careful techniques throughout the DNA extraction process can help minimize contamination problems. For example, using two separate 70% ethanol wash steps can help to fully desalt DNA samples as well as using extra care to ensure that ethanol has been evaporated from DNA pellets can help improve the purity of your DNA samples.

4.2.3 Fluorometric Quantitation of DNA Samples

Another common lab device used for DNA quantitation is the **fluorometer** such as the Qubit® (Thermo Scientific). The fluorometric approach has a couple advantages over microvolume UV spectrophotometric devices. First, a fluorometer can distinguish ssDNA from dsDNA (Life Technologies 2014). Thus, a fluorometer has the ability to accurately measure the concentration of dsDNA while ignoring the presence of other molecules such as RNA and lone nucleotides. Moreover, fluorometric-based concentration estimates of DNA are not affected by the presence of contaminants such as proteins, salts, and some organic compounds (Life Technologies 2014). Microvolume UV spectrophotometers measure DNA concentrations over an effective range from 2 ng/μL to 15 μg/μL, whereas the Qubit operates within the range 10 pg/μL to 1 μg/μL (Life Technologies 2014). Thus, the Qubit can effectively estimate the concentrations of very dilute DNA samples. Both the microvolume spectrophotometer and Qubit require small volumes (e.g., 1 μL) of samples to conduct a quantitation assay. Accordingly, the strength of the fluorometric approach is that it can provide the best estimates of dsDNA concentration—even with samples containing RNA, which is especially critical when quantifying DNA during NGS library construction. However, as fluorometric devices are unable to provide information about the quality (i.e., degree of DNA degradation) and purity of samples, agarose gels and microvolume UV spectrophotometers are able to provide this complementary information.

4.3 THE HIGH-THROUGHPUT WORKFLOW

Now that we have examined the basic methodology of DNA extraction, we will conclude this chapter by introducing the idea of using high-throughput procedures to further economize lab work resulting in lower labor and consumables costs per sample while dramatically increasing the numbers of sequences obtained. For research projects in which researchers need to obtain many DNA sequences from many different individuals, loci, or both, the high-throughput laboratory format can be employed to great advantage. In practice, "high-throughput" simply refers to the process of using multichannel pipettes or robotic equipment (e.g., liquid handlers) to process many samples simultaneously, which can be in groups of 8, 12, 16, 96, and even 384 samples. High-throughput methods can be easily incorporated into each stage of the basic DNA extraction-template acquisition-sequencing workflow provided the appropriate equipment upgrades are made in the laboratory. These upgrades include purchasing: 8 or 12 channel pipettes, a tabletop centrifuge that can spin 96 deep well (2 mL) or 96 well (0.2 mL) PCR/sequencing microplates, a 96 sample microplate rotor for the vacuum centrifuge, and an agarose gel electrophoresis apparatus that can accommodate 96 samples of extracted DNA or 96 PCR products. Some large molecular ecology labs and especially commercial or institutional genome institutes also have robotic equipment such as liquid handling machines, each of which can cost hundreds of thousands of dollars. However, much of the time, researchers may not have enough samples to justify using such expensive equipment; a handheld 8-channel pipette will suffice for most applications. Although there is a significant up-front cost associated with these additional equipment purchases, the savings in labor and consumables over the long-term will likely far exceed the initial investment. Another exciting aspect of this high-throughput methodology is that thanks to the outsourcing of Sanger sequencing (Chapter 6) and NGS (Chapter 7), even small laboratories that adopt some or all of these procedures can scale up their production such that their *per capita* data acquisition can be comparable to the larger molecular biology labs. Moreover, as we will see in later chapters, high-throughput practices can be integrated into PCR, Sanger sequencing, and NGS template acquisition. We will now discuss the first opportunity for incorporating high-throughput into the workflow: high-throughput DNA extractions.

4.3.1 High-Throughput DNA Extractions

There are at least four opportunities for including high-throughput procedures during the process of DNA extraction: (1) using high-throughput DNA extraction kits (or other nonkit method) to simultaneously extract DNA from 96 tissue samples; (2) using high-throughput agarose gel electrophoresis to simultaneously analyze the results of 48–96 DNA extractions; (3) using high-throughput UV spectrophotometry to analyze the results of 96 DNA extractions; and (4) preparing diluted DNA template in a high-throughput format to facilitate large-scale PCR and Sanger sequencing. We will now review each of these procedures in turn. Keep in mind that although it may not be practical or affordable to follow all of these procedures, even adopting one or two of them can greatly improve the economy and output of your lab work.

4.3.1.1 Extracting DNA from 96 Tissue Samples

If you want to extract DNA from at least 96 tissue samples at one time, then you should consider purchasing a DNA extraction kit in the 96 well format. If you have access to one of the "automatic" DNA extraction machines (e.g., autogen), then the process of extracting DNA from 96 tissue samples will be even easier and almost hands free. These large-scale DNA extraction methods typically rely on 96 deep well (2 mL) blocks throughout the procedure including the final step in which the purified DNA is eluted into a clean deep well block. One note of caution when using these methods—given the large investment in terms of genetic starting material (i.e., tissue) that is at risk in case something goes wrong—you should carefully consider beforehand the desired concentration range and volume. Of course this will depend on the amount of tissue you input into the extraction process relative to the volume of buffer used to elute the DNA in the final step. Ideal concentrations for your newly extracted DNA will be 50–1,000 ng/μL. This "concentrated stock" can be stored in the freezer until new aliquots of appropriately diluted (working stock) template DNA are needed for downstream procedures (e.g., PCR and NGS libraries). What are the ideal working concentrations for these applications? The ideal working concentration will depend on the particular procedure under consideration and thus we will wait to visit this issue when we discuss PCR (Chapter 5) and NGS (Chapter 7).

4.3.1.2 High-Throughput Agarose Gel Electrophoresis

Once a set of 96 DNA extractions has been processed, the researcher should proceed on to evaluating the quality and concentrations of all DNA samples on an agarose gel. Using a single channel pipet to load this many samples onto a single gel is not practical. Thus, using a multichannel pipet can greatly improve the speed of loading samples into a gel while reducing the incidence of errors. Anyone who has painstakingly pipetted a large number of samples—one sample at a time—into an agarose gel will appreciate the simplicity and effectiveness of loading eight samples at a time onto a gel.

Unfortunately, the agarose electrophoresis boxes and combs most often found in molecular ecology laboratories are inadequate for accommodating this high-throughput approach. Thus, a special agarose electrophoresis apparatus for high-throughput must be purchased. Be careful to obtain combs that match the spacing of the tips on the multichannel pipette.

After some practice, using a multichannel pipette can be easy. However, there are two important things to consider. First, you need to learn how much even pressure to put on the pipette while affixing pipette tips. Too little pressure and one or more tips may not form a good seal and either fall off or be incapable of holding the specified volume of liquid. That is, after drawing liquid into the tips you will see that the volumes are uneven among tips, which is not good. The second important thing to learn is how to pipette small volumes (0.5–20 μL range) without adding air bubbles to the solution. Multichannel pipettes can be difficult to use in the beginning because it can be difficult to simultaneously and correctly affix all 8 (or 12) clean tips to the pipette. If all tips are not properly affixed, then one or more tips will likely withdraw air instead of liquid reagent. Obviously adding a 0.5 μL of air instead of enzyme will not lead to a successful reaction in that tube. If you have trouble with this, then either you need to improve your technique, change the brand of tips you are using, or check to make sure the pipette is not defective or broken. Mastering a multichannel pipette is essential for high-throughput molecular biology.

4.3.1.3 High-Throughput UV Spectrophotometry

Measuring the purity and concentration of 96 extractions one tube at a time using a standard UV spectrophotometer would be arduous. Thus, for measuring a large number of extractions, it is much easier to use a 96-sample cuvette-plate UV spectrophotometer. If your DNA extractions are already arrayed in a 96-well (8 × 12) storage plate, then it will be easy to use a multichannel pipette to transfer a small volume of each sample to the wells in the cuvette plate. The plate-reading spectrophotometer can then quickly read the 96 samples and the results are then sent to a printer or saved in a spreadsheet file.

4.3.1.4 Preparation of Diluted DNA Templates for High-Throughput PCR

In preparation for high-throughput PCR, the spectrophotometric results for the original DNA extractions should be used to guide the making of appropriately diluted template in a clean 96 well (0.2 mL) microplate. Keeping in mind that the desired PCR template concentration should ideally be in the 10–100 ng/μL range, spectrophotometric data will allow you to make informed decisions on how to dilute the samples. Your original concentrated DNA extractions can be stored in the freezer for long-term storage. By having a separate plate of diluted templates, you will not only be ready to conduct PCR experiments in high-throughput fashion, but you will subject the concentrated DNA stock to fewer freeze–thaw cycles thereby better preserving the quality of those samples. The PCR template plate should be stored in the freezer when not in use. One last tip: always remember that whenever the original DNA extraction plate or the template plate are retrieved from the freezer, it needs to be spun in a centrifuge for a minute to force all the liquid to the bottom of each well, which will help minimize the risk of cross-contamination across wells and make it easier to pipet template for a new PCR. A nice alternative to using a 96 plate to hold your PCR templates, is to use 8-strip (0.2 mL) PCR tubes (i.e., 8 tubes connected together forming a strip). This approach gives you additional flexibility for managing your template DNAs for PCR. For example, if you don't have 96 extractions to load into a single microplate, then the 8-strip tubes allow you to still use the multichannel pipette for setting up PCR reactions in a quasi high-throughput manner. Of course, 12 strip-8 tubes is the same as a 96 well microplate.

Another aspect of the high-throughput approach to consider is the quality of the consumables. One of the keys to having consistent success in the extraction and sequencing lab is to use high quality consumables. Unfortunately, variation exists among products such as pipette tips, microcentrifuge tubes, and PCR tubes. For example, all the time and money invested to prepare a microplate full of PCR products for sequencing would be for naught if the plate is leaking! Be sure you are satisfied with all the products you are using. Also keep in mind that while kits often work well enough to save time and trouble, don't exclude the possibility of making some reagents yourself or using some "old fashioned" protocols. It is remarkable that gel electrophoresis, which was developed in the 1960s, continues to be an indispensable lab method in the age of genomics. Doing some things yourself can save you much money and in some cases can provide results that are comparable, and sometimes preferable, to kits.

REFERENCES

Carter, R. E. 2000. General molecular biology. *Molecular Methods in Ecology* 6:1.

Hillis, D. M., B. K. Mable, A. Larson, S. K. Davis, and E. A. Zimmer. 1996. Chapter 9. Nucleic acids IV: Sequencing and cloning. In *Molecular Systematics*, 2nd edition, eds. D. M. Hillis, C. Moritz, and B. K. Mable, 321–381. Sunderland: Sinauer.

Hykin, S. M., K. Bi, and J. A. McGuire. 2015. Fixing formalin: A method to recover genomic scale DNA sequence data from formalin-fixed museum specimens using high-throughput sequencing. *PloS One* 10:p.e0141579.

Life Technologies. 2014. Invitrogen Technical Note, Comparison of fluorescence-based quantitation with UV absorbance measurements. Qubit® fluorometric quantitation vs. spectrophotometer measurements.

Lowe, D. 2016. The myth of ethidium bromide. In the Pipeline, blog of *Science Translational Medicine*. http://blogs.sciencemag.org/pipeline/archives/2016/04/18/the-myth-of-ethidium-bromide (accessed April 18, 2016).

Miller, S. A., D. D. Dykes, and H. F. R. N. Polesky. 1988. A simple salting out procedure for extracting DNA from human nucleated cells. *Nucleic Acids Res* 16:1215.

Palumbi, S. R. 1996. Chapter 7. Nucleic acids II: The polymerase chain reaction. In *Molecular Systematics*, 2nd edition, eds. D. M. Hillis, C. Moritz, and B. K. Mable, 205–247. Sunderland: Sinauer.

Sambrook, J., E. F. Fritsch, and T. Maniatis. 1989. *Molecular Cloning: A Laboratory Manual*, 2nd edition. Cold Spring Harbor: Cold Spring Harbor Laboratory Press.

Shedlock, A. M., M. G. Haygood, T. W. Pietsch, and P. Bentzen. 1997. Enhanced DNA extraction and PCR amplification of mitochondrial genes from formalin-fixed museum specimens. *Biotechniques* 22:394–400.

Tang, E. P. ed. 2006. *Path to Effective Recovering of DNA from Formalin-Fixed Biological Samples in Natural History Collections: Workshop Summary*. Washington, DC: National Academies Press.

Yates, J. R., S. Malcolm, and A. P. Read. 1989. Guidelines for DNA banking: Report of the Clinical Genetics Society working party on DNA banking. *J Med Genet* 26:245.

CHAPTER FIVE

PCR Theory and Practice

Since the development of PCR during the mid- to late 1980s, this method has gone on to revolutionize biology, medicine, and forensics (Palumbi 1996). Nearly three decades later, PCR continues to be a workhorse in evolutionary genetics laboratories and it is an essential tool in phylogenomics. The major reason for this success is that PCR formed a synergy with Sanger sequencing because it allowed researchers to target and sequence particular genomic loci in a far more efficient manner than earlier cloning-based practices. PCR is invaluable because it enables researchers to easily and inexpensively (in time and money) generate sufficient amounts of target DNA templates for use by DNA sequencing methods.

As briefly mentioned in Chapter 1, the Sanger sequencing workflow (Chapter 6) first requires the use of PCR to copy or "amplify" a genomic target, which is usually <2 kb in length. When a PCR is finished, billions of copies of the target locus—called *PCR products* or *amplicons*—have been synthesized. In the second step, Sanger sequencing methods are used to determine the DNA sequences encoded by the PCR amplicons. Prior to the NGS era, the PCR-Sanger sequencing workflow represented the primary means for acquiring DNA sequence data and it continues to be important today. In addition to the standard PCR method used in most Sanger sequencing-based studies, a variety of other types of PCR methods such as *hot start PCR*, *long PCR*, and *reverse transcriptase* (RT)-PCR have also been developed. We will discuss each of these methods in this chapter. In Chapter 7, we will see that PCR continues to be a highly relevant methodology in NGS workflows as well. Like Sanger sequencing, some NGS approaches use PCR amplicons as input DNA templates for sequencing. However, several PCR variants such as *limited cycle PCR*, *suppression PCR*, and *bridge PCR* are also incorporated into NGS workflows. Thus, PCR remains a critically important technique in phylogenomic data acquisition.

5.1 HISTORICAL OVERVIEW

PCR was developed into a practical laboratory technique during the 1980s by researchers at the Cetus Corporation in Berkeley, California, though a concept for the method was sketched by Kleppe et al. (1971) a decade earlier (Kornberg and Baker 1992; Palumbi 1996). The initial papers, which described the modern PCR concept, provided the first empirical case studies and showed the promise of PCR (Saiki et al. 1985; Mullis et al. 1986; Mullis and Faloona 1987). In recognition for his contributions to these accomplishments, Kary Mullis was awarded the Nobel Prize in Chemistry. However, one of the major technical problems encountered during these early PCR experiments was DNA polymerase denaturation. Exposing the *Escherichia coli* Klenow Fragment of DNA polymerase I to high temperatures (95–100°C) caused loss of activity during each replication cycle (Saiki et al. 1988). The high temperatures were needed to melt hydrogen bonds of the double helix to generate single-stranded templates for replication. In order to drive a PCR of many replication cycles, fresh DNA polymerase had to be added to the reaction tubes each cycle (Mullis et al. 1986; Palumbi 1996). To solve this problem, David Gelfand tested DNA polymerases harvested from cultures of various thermophilic bacteria because these particular bacteria must have evolved high-temperature DNA polymerase enzymes (Brock 1997). What he discovered is that the DNA polymerase from *Thermus aquaticus*, a species of bacterium discovered in a hot springs pool in Yellowstone National Park (Brock and Freeze 1969; Figure 5.1), exhibited optimal DNA synthesis activity at 70°C and, importantly, could survive repeated exposure to the 94–95°C denaturation steps of PCR (Saiki et al. 1988; Brock 1997). This remarkable enzyme is now simply referred to as "*Taq*," which is a shortened version of the bacteria's scientific name. By using *Taq*, the specificity of PCR was also greatly increased because the higher reaction temperatures permissible with *Taq* polymerase enabled the desired reaction to proceed more efficiently. This gain in efficiency is partly explained by the decrease in production of nonspecific products, which reduces competition for DNA polymerases with target DNA (Saiki et al. 1988). *Taq* also enabled PCR to become a more rapid and automated process in that once started it could complete itself over the course of hours instead of days (Saiki et al. 1988). The discovery of *T. aquaticus* and the subsequent exploitation of its remarkable heat-tolerant DNA polymerase enzymes enabled PCR and, as we will see, DNA sequencing to achieve their potential.

Despite the PCR concept being seemingly simple to understand, beginners and even some experienced researchers nevertheless tend to regard PCR as a "black box" technique. This is because the reaction, which takes place in a small plastic tube containing only 10–50 μL of liquid, may often work as desired but fail at other times for no obvious reason. When PCR performs well it seems like an "easy" molecular method to practice, but when PCR fails it can lead a researcher to waste many hours in the laboratory with little or no results to show. Rather than merely hoping a PCR will work, the researcher should confidently *expect* it to work. If you learn the principles of PCR, then you can truly master PCR and enjoy consistent success. Part of this learning includes the importance of troubleshooting failed PCRs. Rather than viewing failed PCRs as a waste of time, it is better to learn why a given PCR failed. The rewards of such troubleshooting justify the additional effort expended.

Figure 5.1. Thermal hot springs in the vicinity of the Great Fountain Geyser in Yellowstone National Park. Brock and Freeze (1969) discovered the thermophilic bacterium *Thermus aquaticus* in Mushroom Pool, which is located off in the distance near the edge of the forest (the area in the foreground is closed to visitors because it is a sensitive ecological area and very dangerous). The original cultures of *T. aquaticus* used for testing *Taq* polymerase in PCR were obtained here. Photo taken by the author in August 2015.

5.2 DNA POLYMERIZATION IN LIVING CELLS VERSUS PCR

Regardless of whether DNA replication is accomplished in living cells or during *in vitro* experiments such as PCR, DNA is synthesized one strand at a time by DNA polymerase-catalyzed addition of 2′-deoxynucleoside triphosphates or what are generically referred to as "dNTPs." The four standard building blocks used to replicate genomes include: deoxyadenosine triphosphate (dATP), deoxythymidine triphosphate (dTTP), deoxyguanosine triphosphate (dGTP), and deoxycytidine triphosphate (dCTP). During DNA synthesis, bases exclusively pair with other complementary bases according to the "Watson–Crick base pair rules": guanine pairs with cytosine and thymine pairs with adenine. Although important differences exist between DNA synthesis in living systems and PCR, both forms of DNA replication generally require the same four major ingredients for proper function: (1) single-stranded RNA or DNA primers; (2) single-stranded DNA (ssDNA) template sequences; (3) dNTPs; and (4) DNA polymerase enzymes with cofactor Mg^{2+} cations.

In order to achieve a deeper understanding of PCR as well as DNA sequencing technologies, it is important to understand the details of DNA synthesis chemistry in living systems. Because the four different dNTPs represent the basic building blocks for DNA synthesis, we will begin with a look at the chemical structure of dNTPs. This remarkable molecule has several chemical components, each with a particular role during DNA synthesis. Figure 5.2 shows the basic structure of a representative dNTP—a dATP, which consists of three parts: the adenine base, a 2′-deoxyribose sugar, and a 5′-triphosphate group. The chemical groups found on one side of the base (not shown) play a critical role in the specificity of base pairing

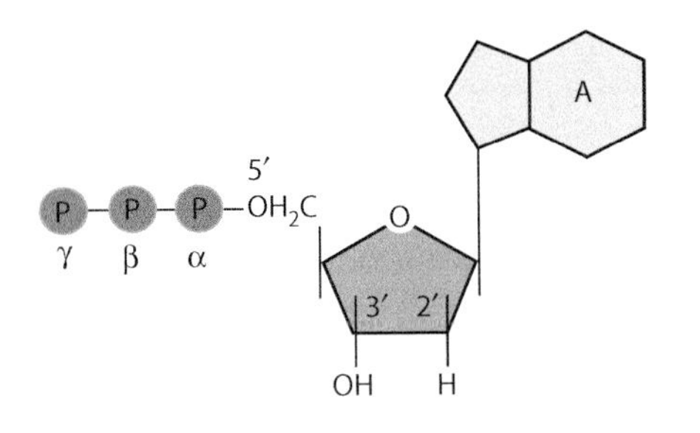

Figure 5.2. Chemical structure of a dNTP. The example shown here is dATP.

while the 3′–OH group on the sugar moiety and α-phosphate (Watson et al. 2014; Figure 5.2) are required for covalently linking together two bases on a DNA strand. The two remaining phosphates, which are labeled as γ- and β-phosphates (see Figure 5.2), interact with one of the metal ions in the active site of the DNA polymerase and later provide a source of energy to help drive the polymerization reaction. We will examine the chemistry of DNA polymerization in more detail in Section 5.2.1.

5.2.1 Brief Review of DNA Polymerization in Living Cells

Not only are many of the molecular components the same between DNA polymerization in a cell and in PCR, but the actual mechanism of polymerization is also the same. Certainly, many important differences exist between DNA synthesis in a living cell versus a man-made concoction in a plastic tube—and we will discuss these in this section. But first, let us not be distracted by these differences and instead focus on the basic polymerization reaction as it occurs in living systems because it is essentially the same as polymerization in PCR.

In a cell, after a stretch of chromosomal DNA has been separated into two single strands, DNA polymerization begins when an RNA primer binds with a ssDNA template (Figure 5.3a). Once this event occurs, a critically important *primer-template junction* is formed (Figure 5.3b). The primer-template junction enables DNA polymerase-catalyzed synthesis of DNA because it provides two critical entities: (1) a hydroxyl (-OH) chemical group (see Figure 5.2) available at the 3′ terminal nucleotide of the primer and (2) an adjacent template base. Although the correct base pairing between the template base (which is presently a thymine in Figure 5.3b) and its complement (adenine) is dictated by hydrogen bonding between the bases and their geometrical configurations relative to each other, DNA polymerase plays a critical role in helping to facilitate this pairing. After the successful base pairing between the thymine and adenine, the polymerase enzyme will then catalytically join the 3′ terminal nucleotide of the primer (a cytosine) to the adenine to complete the nucleotide addition. Note, that as each new nucleotide is added, the primer-template junction itself shifts one base along the template strand, or

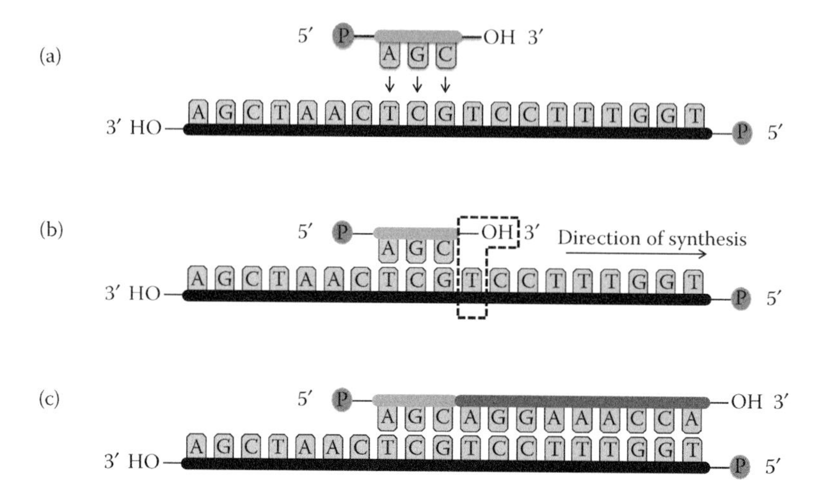

Figure 5.3. Overview of DNA synthesis. (a) A 3 bp RNA primer (light gray strand) hybridizes with the template (black) strand. (b) Primer-template junction formed (dotted box). (c) DNA polymerase synthesizes new strand (dark gray). Note, DNA polymerase and dNTPs not shown. (Modified after Figure 9-1, Page 258, of Watson, J. D. et al. 2014. *Molecular Biology of the Gene*, 7th edition. New York: Pearson Education, Inc.)

in a 5′ → 3′ direction with respect to the primer strand. Thus, each newly added nucleotide will, in turn, "prime" the addition of the next base because the 3′ end of the nontemplate strand will have the requisite 3′-OH thus allowing for further elongation of the nontemplate strand (Figure 5.3c).

Let's examine these chemical reactions in more detail. Figure 5.4 shows a 4 bp RNA primer that has just bonded to complementary nucleotides on a single-stranded template. Once the primer, a 5′–TGTC–3′ sequence in Figure 5.4, bonds to its complementary sites on the template, the DNA polymerase (not shown) will position the incoming dATP in the proper spatial orientation with respect to the template nucleotide (a thymine) and 3′-OH on the primer nucleotide (a cytosine). As soon as proper Watson–Crick base pairing occurs, the new base pair is in correct stereochemical configuration, which then allows the DNA polymerase to catalyze the reaction. In this critical step, a positively charged metal cation (Mg^{2+}) located in the active site of the polymerase enzyme removes a hydrogen atom from the 3′-OH group (Figure 5.4). This deprotonation of the hydroxyl group converts the oxygen into a negatively charged nucleophile that then attacks the α-phosphate on the nearby dATP (Figure 5.4). This action releases a diphosphate group (β- and γ-phosphates), which is eventually split apart by the enzyme pyrophosphatase. Breaking the bond linking the two phosphates releases energy that helps drive the formation of a new covalent bond linking together the cytosine and adenine nucleotides (Figure 5.4). These inorganic diphosphate molecules are called "pyrophosphates" (PPi) because when the covalent bond linking the two inorganic phosphates together is broken, a significant amount of energy is released. Inorganic pyrophosphates are natural byproducts of DNA polymerization in both living systems and in PCR. In cells these products are broken down by the enzyme pyrophosphatase. However, in PCR these byproducts accumulate with the target PCR products.

Another difference between DNA synthesis in living cells and PCR concerns the primers. In living systems, short RNA primers, which are only 5–10 bp long, bind to many complementary places along a template sequence simply because their targets are abundantly scattered throughout any given template sequence (Watson et al. 2014). Their function is to help initiate DNA synthesis. In contrast, PCR primers are different for two reasons. First, PCR primers are made of DNA and not RNA. Secondly, PCR primers are usually at least 18 bases long. The reason for this is that, in addition to priming DNA synthesis in the same manner as RNA primers, DNA primers have the added crucial property of being *highly specific* for certain genomic locations. This specificity property of PCR primers arises because of their length and sequence. Table 5.1 shows us that as primer

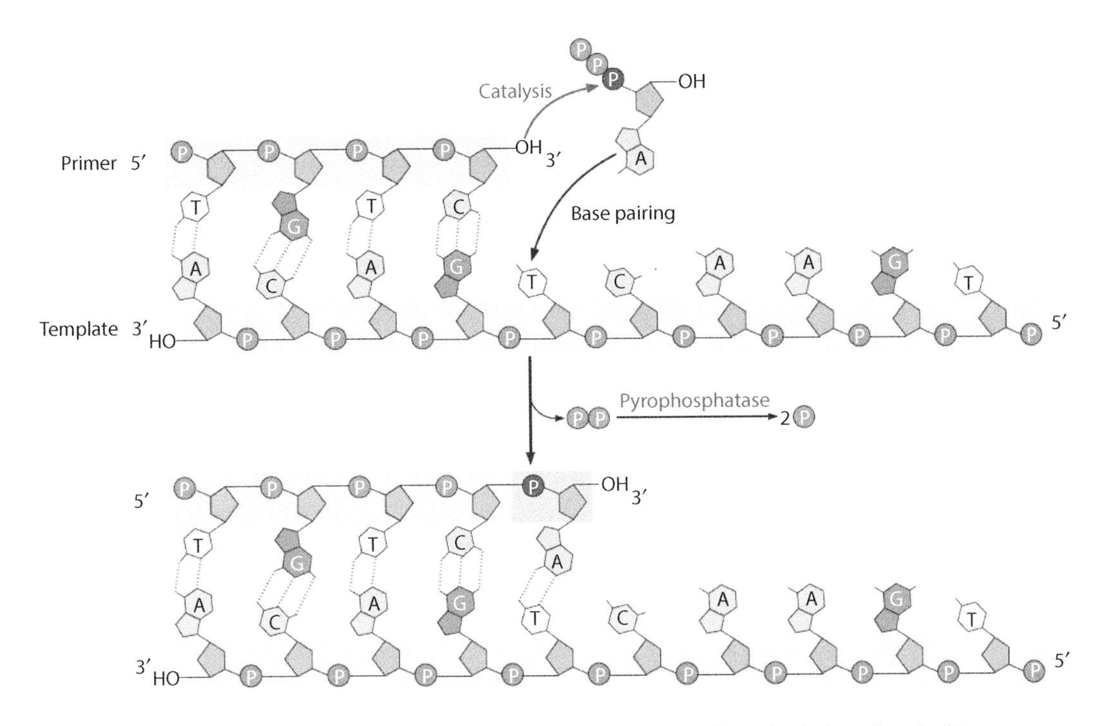

Figure 5.4. Mechanism of DNA synthesis in a living cell (Watson et al. 2014, fig. 9-2). The legend reads, "Diagram of the mechanism of DNA synthesis. DNA synthesis is initiated when the 3′-OH of the primer mediates the nucleophilic attack of the α-phosphate of the incoming dNTP. This results in the extension of the 3′ end of the primer by one nucleotide and releases one molecule of pyrophosphate. Pyrophosphatase rapidly hydrolyzes released pyrophosphate into two phosphate molecules." (Watson, J. D. et al. *Molecular Biology of the Gene*, 7th Ed., © 2014. Reprinted by permission of Pearson Education, Inc., New York, New York.)

length is increased the probability for a random match with a template declines rapidly. As primer length increases, the probabilities seem to become so small that they might lead someone to think that primers even as long as 10 bp may be sufficient for most PCR applications—after all, the probability that a primer this long would perfectly match a sequence of the same size is only 0.000000953674. However, as we saw in Chapter 2, genomes can be quite large. If we instead take the reciprocal value of this probability, assuming a random sequence of nucleotides, we can expect to find a perfect sequence match for our 10 bp primer, on average, once every 1,048,576 bp. Given that the human genome is over 3 billion bp, we could expect our 10 bp primer to anneal to roughly 2,800 different genomic locations! Perhaps, a primer this size may work fine for a genome that is on the order of millions of bases (a bacterial genome) but not for a larger genome such as ours. This is why PCR primers are usually at least 18 bases long; an 18-base primer is expected to find a random match only once every 69 *billion* bases—far larger than most eukaryotic genomes (Brown 2007)! This ability of longer DNA primers to target a single genomic location is one of the most desirable properties of PCR. As we will see in Section 5.2.2, a third major difference between primers in living systems versus PCR is that in PCR there are *two* primers of different sequences used together in a reaction to amplify a particular target sequence that usually ranges between 200 and 2,000 bases as opposed to the many primers needed to replicate an entire genome.

5.2.2 How the PCR Works

Now that our review of DNA synthesis and introduction of PCR primers is done, we can focus on the actual PCR mechanism and laboratory procedures for performing successful PCR. A PCR experiment begins when the researcher pipets the PCR reagents (template DNA, dNTPs, reaction buffer, water, and *Taq* polymerase) into 0.2 mL plastic reaction tubes. The tubes are then placed into

TABLE 5.1
Primer length and the probability of a match with a nonhomologous template sequence

Primer length (bp)	Probability of random template match	Genome size (bp) for one random template match
1	0.25	4
2	0.0625	16
3	0.015625	64
4	0.00390625	256
5	0.0009765625	1,024
6	0.000244140625	4,096
7	0.00006103515625	16,384
8	0.0000152587890625	65,536
9	0.000003814697265625	262,144
10	0.00000095367431640625	1,048,576
11	0.000000238418579101562	4,194,304
12	0.0000000596046447753906	16,777,216
13	0.0000000149011611938477	67,108,864
14	0.0000000037252902984619	268,435,456
15	0.0000000009313225746155	1,073,741,824
16	0.0000000002328306436539	4,294,967,296
17	0.0000000000582076609135	17,179,869,184
18	0.0000000000145519152284	68,719,476,736

NOTE: As the length of a primer increases the probability that the primer will match a random segment of DNA decreases according to the function $(0.25)^X$ (where X = primer length).

a *thermocycler*, a tabletop machine used to conduct the PCR. As its name implies, a thermocycler heats and cools the reaction mixtures. Once running, the thermocycler will typically undergo 30–40 *cycles* of heating and cooling that are controlled by a computer program. Each of these cycles, in turn, is comprised of three core steps: *denaturation*, *annealing*, and *extension* (Palumbi 1996). We will now discuss these temperature steps in a PCR cycle.

Step 1: Denaturation In this initial step the genomic DNA is heated to 94–96°C, which breaks all hydrogen bonds linking together the two complementary strands of DNA. This creates a mixture of ssDNA thus making available the primer targets on the genomic DNA templates.

Step 2: Annealing Next, the reaction mixture is gradually cooled down to a specific pre-selected temperature that depends on the optimal annealing temperatures of the primers being used, which is usually between 45°C and 65°C (we will discuss these properties of primers below and in Chapter 8). Upon arriving at the correct annealing temperature, the *forward primers* hybridize to their target locations on template strands. Meanwhile, the *reverse primers* also find and bind to their targets on the complementary template strand hundreds or thousands of bases distant from the forward primer. Note that the "forward" and "reverse" primer designations usually do not have any functional significance. The researcher usually designates which is which. What is important to keep in mind is that each pair of PCR primers usually consists of one forward and one reverse primer. The specificity of primer-template matches is influenced by the primer sequences, annealing temperature, and length of time spent at the annealing temperature.

Step 3: Extension After the forward and reverse primers bind to their targets the mixture is heated to a temperature of 70°C, which enables the *Taq* polymerases to incorporate free dNTPs at the optimal operating temperature for *Taq* and thus synthesizing new DNA strands.

After the first cycle, the targeted genomic template has, like DNA in living systems, been replicated in a semi-conservative manner meaning that for each original double-stranded DNA (dsDNA) template instead now exists two dsDNA molecules—each one comprised of an original or "parental" template strand plus a newly-synthesized complementary or "daughter" strand. Once the first cycle ends a second cycle immediately begins repeating steps 1–3 above. A diagram showing the steps during the first two PCR cycles is illustrated in Figure 5.5. During the first cycle the template consists of long strands of genomic DNA, each of which could span thousands if not millions of bases (even entire chromosomes). However, the synthesized DNA strands during the first cycle are far shorter in length because the time for extension is too short to allow the DNA polymerase to synthesize more than about 2,000 bp of DNA. Note, a technique known as "long PCR" can copy templates that span thousands or tens of thousands of bases but that is a more specialized form of PCR, which we will discuss later in this chapter. In Figure 5.5, notice that half of the newly synthesized strands at the end of the second cycle become even shorter in length. They span the exact distance between the 5′ ends of the forward and reverse primers with the target sequence located in between the primers (Mullis et al. 1986). This is how PCR can specifically target and copy a particular genomic locus. Within a few cycles these shorter double-stranded products will be far more involved in the synthesis of products than the original genomic DNA templates. Note also that the forward and reverse primers become permanently incorporated into each newly synthesized double-stranded product. Examining Figure 5.5 again, we not only see that the original template strands are available again for DNA synthesis for the start of the second cycle, but the new strands from the first cycle *also* contain target sequences for the forward and reverse primers. This means that the number of products at the conclusion of the second cycle will double to four. Figure 5.6 shows that the number of templates doubles after each cycle showing this to be a geometric growth process (i.e., $2 \rightarrow 4 \rightarrow 8 \rightarrow 16\ldots$ or 2^n, where n = number of cycles; Saiki et al. 1988), which is why PCR is also commonly referred to as DNA *amplification*. PCR is so sensitive that with just one double-stranded template DNA to begin a PCR, one can obtain 2^{35} or 34,359,738,368 copies after 35 cycles! If there are hundreds or thousands of genomic template copies present at the start of the reaction, then there will be trillions of copies when PCR is finished!

Let's now take a closer look at the synthesis of PCR products during the extension step of a PCR after a number of cycles have already been completed. Figure 5.7a shows two complementary strands of a product generated in some previous cycle and each strand already has a primer bound to the appropriate template sites. Thus, the denaturation and annealing steps have already been executed in this cycle. Once the primers become bound to their templates and the temperature is raised to 70°C the extension step commences—the DNA polymerases can start to synthesize new strands of DNA. Figure 5.7a shows how both primers become incorporated into the newly forming product. It is also important to notice that downstream from the bound primer the newly synthesized strand contains the complementary sequence to the other primer. This means that every new strand can itself be used as a template in a future cycle of synthesis (Figure 5.7b). After extension is complete, a 540 bp long product that is double-stranded from end to end is produced (Figure 5.7b). As you can see, each product consists of two original primers plus two newly made strands downstream of each primer.

It should now be clear that PCR can readily generate billions or more copies of a specific target locus. Interestingly, the accumulation of PCR products over the course of a typical PCR (e.g., 35 cycles) does not resemble an exponential growth process, but rather is better defined by an S-shaped or logistic growth curve (Saiki et al. 1988; Innis et al. 1990) as depicted in Figure 5.8. Thus, during the first ~20–25 cycles of PCR, the rate of amplicon production is exponential. However, after this point the rate of growth starts to slow down until the growth curve flattens out into a plateau of low growth (Saiki et al. 1988; Figure 5.8). A number of explanations have been advanced to explain this "plateau effect" in PCR including depletion

Figure 5.5. The first two cycles of PCR. Reaction starts with a genomic DNA template (black lines) and primers (short dark gray lines). D, denaturation step; A, annealing step; and E, extension step. Vertical lines depict hydrogen bonds. New strands of DNA are gray.

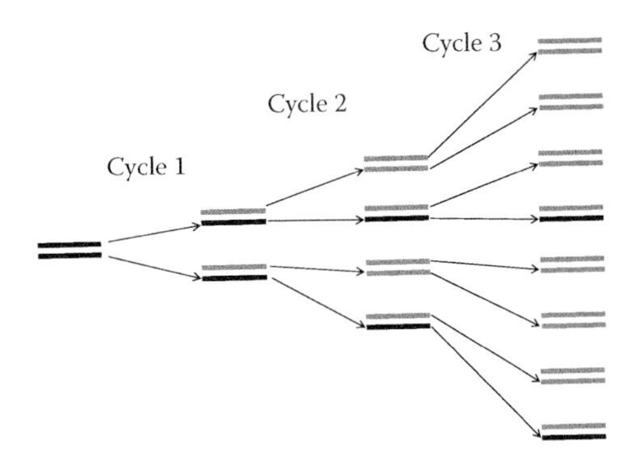

Figure 5.6. PCR replication from cycles 1 through 3 illustrating the geometric increase in dsDNA. Cycle 1 begins with one double-stranded template shown as two black (parental) strands and ends with eight after cycle 3. New (daughter) strands are gray.

of reagents such as dNTPs, primers, and Mg^{2+} cations; inactivation of *Taq* polymerases due to the high heating for many cycles; and inhibition of *Taq* caused by the accumulation of end products such as pyrophosphates and target products (Innis et al. 1990; McPherson and Moller 2006). Although all of the aforementioned possibilities could have some impact on amplicon production in PCR, the study by Kainz (2000) showed that this slowdown can simply arise due to negative feedback (i.e., density dependence) provided by the accumulation of the amplicons themselves. Regardless of its cause, however, the plateau effect in PCR seems to not be a concern. This does not mean that investigators should never consider fine-tuning their PCRs by fiddling with reagent concentrations or mixing in additives. Such fine-tuning, which is generally referred to as "PCR optimization," especially when new primers are being tested is actually a key part of the PCR experience. We will consider PCR optimization throughout the remainder of this chapter and in Chapter 8 when we consider primer design and loci development.

5.3 PCR PROCEDURES

Like agarose gel electrophoresis, the basic equipment and methodology of PCR has changed little

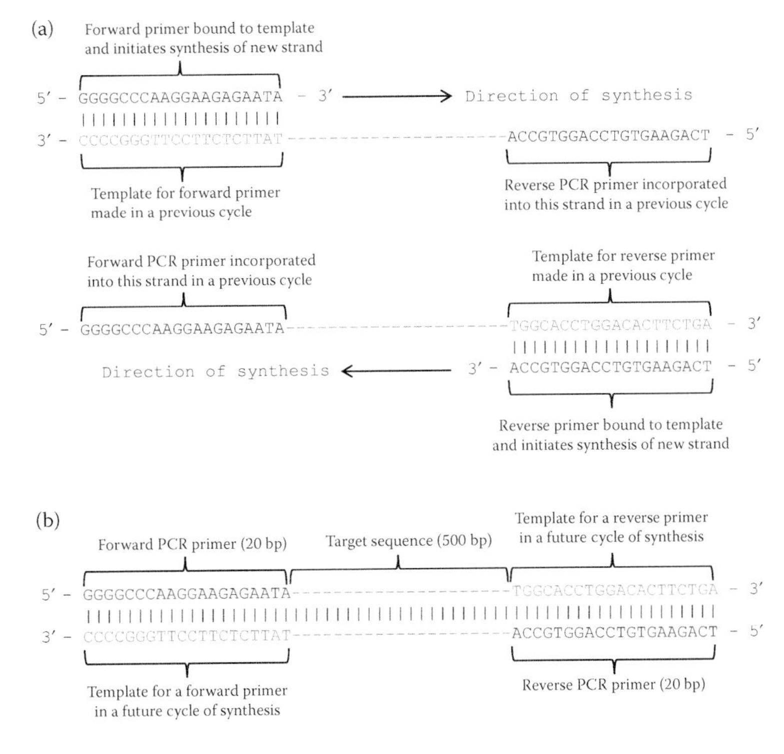

Figure 5.7. Synthesis and anatomy of a 540 bp PCR product. (a) Synthesis of two products from two complementary strands during the extension phase (after many cycles). (b) Parts of a PCR product. For clarity the target sequence is indicated as dashes. Note, most of target sequence not shown due to lack of space. Vertical bars indicate hydrogen bonding between complementary bases. Gray represents strands of DNA synthesized during PCR and primer bases are shown as black letters.

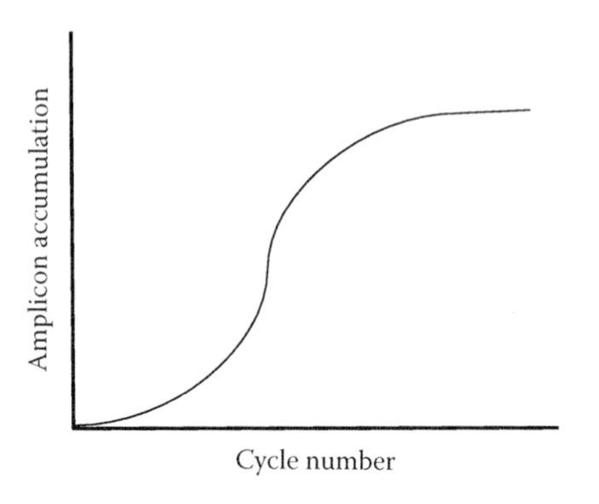

Figure 5.8. Growth curve depicting the accumulation of PCR amplicons during a PCR.

over the past 30 years. From start to finish, a PCR experiment requires 3–6 hours of lab time. There are three steps to conducting a PCR experiment: (1) *setting up the reactions*; (2) *thermocycling*; and (3) *agarose gel electrophoresis*. If a PCR is found to have failed to some degree, then PCR troubleshooting will be needed.

5.3.1 Preparation of PCR Reagents and Reaction Setup

The first step to conducting a PCR experiment is to prepare the PCR reagents prior to setting up the reactions. Before we discuss the processes of combining all reagents into the reaction tubes, thermocycling, and electrophoresis we will now review basic details about preparing each PCR reagent.

5.3.1.1 PCR Reagents

Primers—We are now familiar with the function of forward and reverse PCR primers in the PCR, but where do these primers come from and how are they designed? This is a complicated subject thus for now we will assume that the appropriate primers are already in hand (previously published or designed by someone else). We will explore the topic of primer design in Chapter 8.

Commercial manufacturers of primers such as Integrated DNA Technologies (IDT) desalt and dry (i.e., lyophilized) newly synthesized primers prior to shipment to customers. When the new primers arrive in the lab, they must be prepared for use in a PCR. The first step typically involves resuspending the dried primer with molecular biology grade water or TE buffer to make a 100 μM "concentrated" stock solution. The next step is to use the concentrated stock to make a more dilute 10 μM "working" stock solution for direct use in PCR. Although 10 μM working stocks will work for most routine PCRs, some researchers prefer to use less concentrated working stocks; so different working stocks can be made accordingly using the concentrated stock solution. The main reason why a concentrated stock is necessary is that repeated freeze–thawing of working stocks will lead to degradation of the primers, which in turn will diminish the efficiency of PCR. Thus, fresh working stocks of primers should always be used in PCR. To avoid contamination of the primer stocks it is highly advisable that this work be performed in a relatively sterile space such as a laminar flow hood and using aerosol-prevention (usually called "barrier" or "filter" tips) pipette tips.

Although making concentrated and working stocks of primers are simple procedures, it is critically important that this work be correctly done. The following is an example protocol for mixing up concentrated and working stocks of primers manufactured by IDT that my students and I have successfully used:

Step 1: Calculate the volume of molecular biology grade water or TE that you will add to each tube containing dried primers. For each primer you will need to look up the amount of primer synthesized. On the datasheet that comes with each primer, under the heading "amount of primer," you will see the quantity of primer in units of nmoles that was synthesized. This number is often between 20 and 40 nmoles. Multiply this number by 10. This new number will be the volume of water or TE in microliter that you will add to the primer tube. For example, if the amount of primer is 25 nmoles, then you will need to add 250 μL of sterile pure water or TE to the tube of dried primer. However, before opening the new primer tubes to add the water or buffer, briefly centrifuge the tubes for 1 minute in the microcentrifuge at maximum speed in order to guard against losing dislodged primer pellets.

Step 2: Add the correct volume of water or TE to each primer tube using a pipet with a filter tip.

Step 3: Allow the tubes to sit for 10 minutes at room temperature so the DNA pellet goes into the solution.

Step 4: Use the pipet to gently mix the primer solution (by pipetting up and down slowly about 20 times) to make homogeneous. If the correct amount of water or TE was added, you should now have a 100 μM solution of the primer. This concentrated primer stock is now ready for use to make diluted primer for PCR (i.e., the primer working stock).

Step 5: Now label a set of clean and sterile 1.5 mL microcentrifuge tubes (one for each primer) by writing the name of the primer and concentration (10 μM) on the lid. Write the date on the side of the tube. These new tubes will be the diluted primer ready for PCR.

Step 6: Add 90 μL of molecular biology grade water to each of the newly labeled microcentrifuge tubes. Next, add 10 μL of concentrated primer to each microcentrifuge tube. Use the pipet to gently mix the solution to make a homogenous concentration of the primer. You now have 100 μL of 10 μM primer ready for PCR. Store the primer working stock tubes at −20°C when not in use.

dNTPs—Deoxynucleoside triphosphates or "dNTPs" (i.e., dATP, dTTP, dGTP, and dCTP) are required for a PCR. As with primers, fresh dNTPs perform better than well-used solution of dNTPs due to the negative effects of repeated freezing and thawing of the nucleotides. A 10 mM solution containing equal molar amounts of each nucleoside triphosphate is often used in a PCR setup. Ready-to-use mixes containing all four dNTPs can be purchased from biotech supply companies or you can obtain the four types of dNTPs in separate tubes and then easily make your own mix.

Taq polymerase—In addition to adding the *Taq* enzymes to the reaction mixture, a 10× magnesium-containing buffer (i.e., MgCl) must also be added because, as we saw earlier, *Taq* polymerase requires free Mg^{2+} cations in order to catalyze the nucleotide polymerization process.

Molecular biology grade water—Critical to the success of PCR is the use of sterile, pure, and deionized water in all lab procedures otherwise problems can arise later. For example, DNA can be degraded or destroyed by DNA-destroying enzymes or chemicals may inhibit PCR. Such "PCR" water can be made in a lab by proper distillation and autoclaving procedures or can be purchased from life science or chemical supply companies as "Molecular Biology Grade Water," "HPLC Water," etc.

Template DNA—For most PCRs, DNA extracted from tissues is later used as template for the PCRs. Often this extracted DNA will be too concentrated for use in PCR so a sample should be diluted in the reaction. There are several advantages to diluting a sample of extracted DNA for use in PCR. These include diluting any PCR inhibitors (e.g., DNA polymerases are sensitive to salt) present in the original extraction; avoiding overamplification of the target locus; avoiding amplification of nontarget regions ("PCR artifacts"); and conserving DNA template by not using more than is required for a successful amplification. The old adage "less is more" applies well to the problem of how much template to add to a PCR. Although using a sample of template that is too dilute might lead to poor amplification of the target gene, the advantages to diluting the original template outweigh the disadvantages. So, how much template DNA is needed for each reaction? Since most PCR protocols require only 1.0 μL, the important thing to consider is the concentration of the extracted DNA. Although PCR can be robust to template concentration, 10–100 ng/μL represents an ideal range. It is very advantageous to start with a concentrated solution of extracted DNA and then dilute it to 10–15 ng/μL for use in PCR, as more than enough template will be available for a strong amplification while any inhibitors present in the original solution are less effective because of the dilution.

5.3.1.2 Importance of Making Reagent Aliquots

Researchers should always make aliquots of all working stocks of PCR reagents. This includes primers, molecular grade water, master mix (see Section 5.3.1.3), and template DNA. Usually such aliquots are dispensed into 1.5 mL microcentrifuge tubes then are stored in a −20°C freezer. One reason this is a good idea is that some of these reagents (especially the primers, dNTPs, and template DNA) become more degraded each time they are frozen and thawed making them perform more poorly in PCR. A second reason is that making aliquots greatly minimizes the risk of a "global" PCR contamination event in a laboratory. By using aliquots rather than concentrated stocks, contamination outbreaks can be confined to particular aliquots or PCR experiments. It only takes one misplaced pipette tip to contaminate an entire concentrated stock rendering that reagent useless for future PCRs, not to mention the wasted lab resource and time lost doing contaminated PCRs! Thus when new reagents arrive to a laboratory, a good strategy is to immediately go to a clean space such as a laminar flow hood and use filter tips to make aliquots. Afterward, the aliquots should be transferred to a freezer where they are stored. Be sure to also write the date the aliquot was made on the side of the tubes so that each lot can be monitored for quality control.

5.3.1.3 Setting Up PCRs

Typically PCRs are performed in 0.2 mL plastic PCR tubes or 96-well plastic PCR plates (0.2 mL wells). While performing this step, it is good to keep all reagents and tubes (or plates) on ice, which will help preserve the reagents. Once the reagents are thawed, they can be pipetted into the reaction tubes.

Negative and positive PCR controls—In addition to setting up a PCR for each of your different template DNA samples you should also set up a *negative control* and a *positive control*. The negative control is a PCR setup that contains all reagents except any template DNA—water is substituted for DNA. This control is important to every PCR experiment because it will indicate a DNA contamination problem when it exists. If all reactions containing template DNA plus the negative control amplify a product, then this indicates that foreign DNA (usually PCR products) contaminated one or more of the PCR reagents. The positive control contains all the reagents plus a template DNA sample that has successfully worked in prior PCR experiments with the same primers that are being used. Thus the positive control uses a proven DNA template sample. If the positive control amplifies correctly but the other samples do not, then you can safely conclude that the other DNA templates are somehow defective. If none of the samples containing DNA including the positive control amplify, then there are multiple possible reasons why the PCR failed. We will discuss this again when we consider troubleshooting failed PCRs later in this chapter.

Making a PCR master mix—When you are ready to set up the PCRs the first thing you will do is to retrieve aliquots of each reagent from the freezer and then thaw them. It is best to thaw the reagents and perform the pipetting steps inside a laminar flow hood because working in a relatively sterile environment will help to reduce the chances of contamination. Once the reagents are thawed, the next step is to make a "master mix," which is a mixture prepared in a 1.5 mL microcentrifuge tube containing exactly the right amounts of primers, dNTPs, enzyme buffer, water, and *Taq* (i.e., all components except the DNA template). Such master mixes are useful because they save you lots of time and pipette tips and reduce the likelihood of pipetting errors. Although micropipettes tend to perform well, you should expect that one or more of your pipettes, especially the larger volume pipettes (e.g., 200 μL and 1,000 μL models), may not deliver the exact desired volume to your master mix or PCR tubes. This potential problem is easily overcome by adding in extra reagents when making your master mix. For example, by making 10% more than you theoretically need to accommodate all your samples, you ensure that you will not run short on master mix. An example recipe for making a master mix for a PCR experiment that includes 30 different template DNA samples plus negative and positive controls is shown in Table 5.2.

Note, you can save even more time (fewer pipetting steps) and consumables (use fewer pipette tips) by using a so-called "pre-made master mix" not to confuse with the "master mix" discussed earlier. A *pre-made* master mix is usually a large stock solution that contains dNTPs, enzyme buffer, *Taq*, and water to set up hundreds or thousands of PCRs; thus it is a large stock for a great many PCR setups. Pre-made master mix usually does not include primers, though if your laboratory primarily uses a single primer pair (e.g., for DNA barcoding), then making a pre-made master mix that includes primers may be worthwhile.

There are two options for obtaining pre-made master mixes: purchase from a biotech supply company (e.g., Promega) or make your own. Commercially available pre-made master mixes are convenient because they save you time and trouble, are easy to use, and usually produce good PCR results. Also, some companies offer pre-made master mixes that already contain a loading dye,

TABLE 5.2

Example showing the setup for a PCR experiment that includes a total of 32 PCRs, which includes 30 DNA samples plus positive and negative controls

Master mix (MM) reagents	Volume/reaction (μL)	# Reactions[a]	MM volume (μL)
Water (molecular biology grade)	10.5	35.2	369.6
Taq buffer (10×)	1.5	35.2	52.8
dNTP mix (10 mM)	0.9	35.2	31.7
Forward primer (10 μM)	0.5	35.2	17.6
Reverse primer (10 μM)	0.5	35.2	17.6
Taq polymerase (5 U/μL)	0.1	35.2	3.5
Totals[b]	14.0	35.2	492.8

[a] Signifies that the # of reactions to use in the calculations = actual # PCRs to be set up (i.e., 32 reactions) + extra reagent (0.1 × total number of reactions) to account for minor pipette errors. Thus enough reagents to accommodate an additional 3.2 PCRs should also be included to ensure that enough master mix is made.

[b] Total reaction volume is 15 μL. Thus DNA or water (− control) must also be added to tubes.

which simplifies the loading of PCRs into agarose gels. A good alternative to the commercial pre-made master mixes, is to purchase the individual reagents separately and make your own pre-made master mix. The main advantage is you can lower your reagent costs by buying the components (dNTPs, *Taq*) separately. Another great advantage is that you have more flexibility on which type of DNA polymerase you can use. Biotech supply companies offer a variety of DNA polymerases suitable for PCR to choose from, ranging from the inexpensive (standard) to the expensive high-performance polymerases. My students and I have found that making our own pre-made master mix using a higher quality *Taq* polymerase produces better PCR results at a lower cost than buying the commercially available pre-made master mix. Testing a variety of different formulations of pre-made master mixes involving different *Taq* polymerases could provide a big payoff to the individual or laboratory by reducing reagent costs while improving PCR performance at the same time.

Returning to the act of setting up the reactions, once the reagents are thawed and ready to be used be sure to invert each reagent tube several times (or gently pipet) to mix the contents—this is especially critical for the pre-made master mix. You are now ready to begin pipetting the reagents into the PCR tubes. Use extreme care when pipetting to avoid cross contaminating samples, reagents, etc. When you are finished pipetting and have sealed the PCR tubes with their caps, it is time for the thermocycling step.

5.3.2 Thermocycling

Once the reactions are set up they are then ready to be placed in the thermocycler to start the PCR. Thermocyclers have computer touch pads so the operator can specify the actual PCR profile to be used. An example profile would be:

Step 1: 94°C for 30 seconds (initial denaturation step)

Step 2: 94°C for 30 seconds (denaturation step)

Step 3: 50°C for 30 seconds (primer annealing step)

Step 4: 70°C for 60 seconds (extension step)

Repeat Steps 2–4 total of 35 times

Although innumerable varieties of PCR profiles have been successfully used, the basic one shown above can work well for many applications assuming the correct annealing temperature is specified. This is because the temperature for denaturing the DNA (Steps 1–2) and extension of newly created DNA strands (Step 4) are standard for the vast majority of PCR experiments. On the other hand, the annealing temperature (Step 3) can vary between approximately 45–65°C depending on the optimal annealing temperatures for the primers being used. Here, we are assuming that 50°C is the optimal temperature for the primers. Other important variables in this profile include the time spent at Steps 2–4 as well as the total number of cycles to be run. For many PCRs a 30-second denaturation time is adequate to convert the double-stranded template into single-stranded DNA. However, some researchers prefer to include one long (2–5 minutes) denaturation only during the first step to ensure that all genomic DNA has been converted to single-stranded form thereby generating the maximum amount of single-stranded templates for the primers. Regarding the extension step, 30 seconds will be adequate time for the polymerase to complete a new strand of DNA provided the product size is less than about 500 bases, but one minute should be used for products that are 1,000 to 1,500 bp (Palumbi 1996). Typically, 35 cycles are adequate. However, if in doubt about which program to use, then you should consult the guidelines established by the manufacturer(s) of the PCR reagents you are using or see Palumbi (1996) for optimization strategies.

5.3.3 Checking PCR Results Using Agarose Gel Electrophoresis

In Chapter 4 on DNA extraction, we learned about using agarose gel electrophoresis to check the results of a DNA extraction. This same method also plays a key role in PCR because it is used to determine whether or not the PCRs functioned as expected. If the PCR worked well, then the investigator will only see the ladder and a single bright band in each well of the gel corresponding to the expected product sizes and with no evidence of contamination (Figure 5.9). If success is achieved, then the PCR products will be ready to sequence. However, if some degree of failure is experienced, then the investigator must perform some troubleshooting in order to hopefully fix the problem(s).

The same reagents and methods used to make agarose gels discussed in Chapter 4 remain the same for evaluating PCR products. The main difference between extraction and PCR gels, is that expected PCR product sizes are usually less than 1 kb, therefore a more concentrated gel (2%) is often used. This is because the more concentrated gel will provide greater resolving power for smaller linear fragments of DNA than a less concentrated one. Also, as before, the ladder functions as a positive control for the electrophoresis steps as well as enable the sizing of products. Typically, 5 μL of PCR product plus 1 μL of 6× loading dye are loaded into each gel well.

5.4 PCR TROUBLESHOOTING

When examining the results of a PCR experiment on an agarose gel the first thing you will be eager to look for will be the single bright bands that correspond to your expected product sizes, such as in the example shown in Figure 5.9. The presence of bands corresponding to their expected sizes usually means that specific amplification of the target locus has been achieved. If your sample lanes show good products and the negative control lane shows no evidence of DNA contamination (i.e., no band in this lane), then you are ready to prepare the products for sequencing. However, if the gel reveals any problematic issues with your PCRs such as missing target bands, weak target bands, multiple bands, smears, etc., then you will need to perform a troubleshooting analysis to determine the causes of the failures.

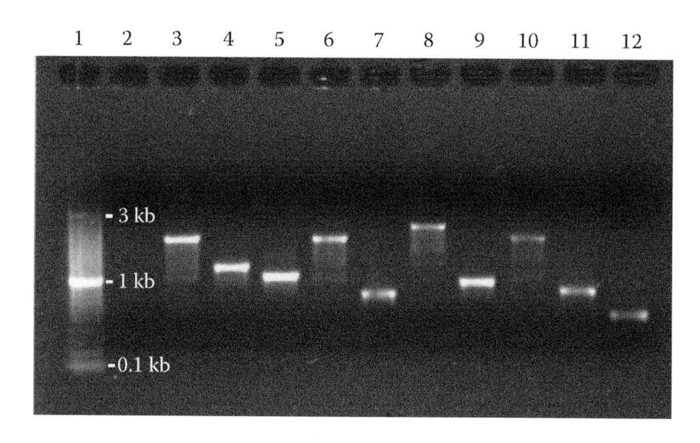

Figure 5.9. Photo of 2% agarose gel stained in ethidium bromide. Lane 1 is ladder; lane 2 is a negative control; and lanes 3–12 are PCR products for ten different target loci. Dark oval areas (below each number) are sample-loading wells in the gel.

Given that the entire process beginning with DNA extraction and ending with agarose gel electrophoresis of your target PCR products involves a number of procedures and intricate chemical reactions, numerous opportunities exist for one or more problems to arise causing the PCR experiment to fail in one way or another. Although the basic PCR procedure is straightforward, sooner or later everyone obtains gel results that indicate something went wrong somewhere during the PCR or electrophoresis procedures.

PCR gels can have many different appearances depending on the nature of the PCR or electrophoresis problems and they can therefore reveal important clues for troubleshooting problems. The first thing to rule out is a problem with the gel itself because if something malfunctioned with the gel electrophoresis procedure, then you will not be able to evaluate your PCRs. Nucleic acid stains such as ethidium bromide tend to lose their activity over time, especially if they are exposed to light. Thus, whenever a PCR gel does not show any DNA bands—not even the ladder, then the possible culprits include defective electrophoresis reagents or procedures. Other gels may show ladder and PCR bands but the bands are feint and difficult to see. This latter gel result suggests that the PCR worked but that the weak stain makes the bands difficult to visualize. A new gel with fresh nucleic acid stain should be prepared and run to retest the PCR products. If you can rule out gel problems and the PCR results are (still) disappointing, then you will need to troubleshoot the PCR itself. A summary of the most common PCR problems is shown in Table 5.3. Each of these scenarios is discussed in more detail including troubleshooting tips.

Consider another possible PCR gel result in which the only DNA bands on the gel are within the ladder lane—no bands in the positive control or sample lanes. In this scenario, there is sufficient evidence to conclude that the PCRs must have failed. Although the template DNA could be to blame (of course if any had worked successfully before, then this can be ruled out), because the positive control also did not work indicates that a more global problem exists (e.g., defective master mix or thermocycler). Although the thermocycler could be the problem, this is rarely the case as it is far more likely that the problem exists with one or more components in the PCR. If using a commercially made master mix, then you could set up a new PCR experiment but this time substitute in a new master mix. If the master mix was made in the lab then you need to troubleshoot the exact cause (bad *Taq*?, bad dNTPs?, etc.). This can be done by setting up a new PCR experiment in which all tubes get the *same* template DNA (use the proven positive control DNA), but each one differs by a single experimental variable:

Tube 1: New Taq, old dNTPs, old water, old forward primer, old reverse primer

Tube 2: Old *Taq*, *new dNTPs*, old water, old forward primer, old reverse primer

TABLE 5.3

Troubleshooting PCR experiments using information obtained from agarose gels

Ladder visible	Band(s) in (−) control	Band(s) in (+) control	Band(s) in sample lanes	Likely cause(s)
No	No	No	No	Defective ladder, stain, or gel[a]
Yes (D)	No	Single, correct size (D)	Single, correct size (D)	Weak stain
Yes	None	None	None	Defective master mix or thermocycler
Yes	Yes	Yes	Yes	Contaminated master mix
Yes	None	Single, correct size	None	Problem with template DNA
Yes	None	Single, incorrect size	Single, incorrect size	Nonspecific amplification
Yes	None	Multiple bands	Multiple bands	Nonspecific amplification
Yes	None	Smear	Smear	Over-amplification of target

NOTE: Each row summarizes observations in the form of presence or absence of amplified DNA ("bands") from a particular gel and identifies the likely cause(s) for those results. "D" means band was difficult to see on the gel.

[a] Indicates that problem could be electrophoresis conditions.

Tube 3: Old *Taq*, old dNTPs, *new water*, old forward primer, old reverse primer

Tube 4: Old *Taq*, old dNTPs, old water, *new forward primer*, old reverse primer

Tube 5: Old *Taq*, old dNTPs, old water, old forward primer, *new reverse primer*

In this case, a "new" reagent is either a new unopened product or one that is freshly remade using a concentrated stock solution, whereas the "old" reagent is the same reagent stock used in the prior failed PCR experiment. This is an easy way to troubleshoot many PCR problems because it is usually only one defective reagent that causes PCR failures. If none of these tests reveals the problem, then you can make a sixth tube that is made using all new reagents. Although it is quicker and less trouble to just throw away all old PCR reagents and start again with fresh ones, taking the time to troubleshoot these problems may help you avoid the same problems in the future thus saving you more time in the long run.

On other gels you might see that all of the samples including the positive control (if one is used) as well as the negative control amplified, a result that implicates DNA contamination. Pipetting errors or using pipettes contaminated with DNA (especially PCR products) is the usual culprit. If a contamination problem exists, then sterilizing the pipette(s) using UV light and following the approach outlined above for troubleshooting failed PCRs can help you understand the nature of the contamination problem. Taking the time to identify the source of contamination is a key step forward toward preventing similar problems in the future.

Although a positive PCR control is optional—provided the PCRs have been working as expected in previous experiments, this control can nonetheless be useful if the sample tubes show little or no sign of amplification. Remember, if the positive control amplifies, then this immediately allows you to rule out many possible causes for the failed PCRs—master mix, thermocycler, and gel electrophoresis. If the positive PCR control sample is successfully amplified, then this tells you that everything worked as expected except for the DNA templates in the samples. This would then point to the DNA obtained from the extractions. Perhaps those samples were too degraded to be used in PCR or something else was wrong with the template DNA (e.g., inhibitors or low template concentration). Occasionally, DNA templates contain too much salt or ethanol leftover from the extraction process, which can lead to inefficient or no amplification because these are inhibitors of the *Taq* enzyme. Recheck the template concentration and quality (extraction gel and UV spectrophotometry results). If a DNA extraction gel shows that the DNA is both concentrated enough for PCR and mostly undegraded, then diluting the template further and/or desalting the template using ethanol precipitation (with at least two 70% ethanol wash steps and ensuring that all ethanol is removed prior to the last step) can lead to dramatically better PCR results.

Other problems can include smearing of DNA or multiple bands within gel lanes. If the positive control amplified correctly but the samples did not, then this is likely a template problem; specifically the template could have been too concentrated leading to over-amplification of the target band. If the positive control and sample lanes all show smears or multiple bands, then this suggests that nonspecific amplification occurred during the PCR leading to multiple genome targets being amplified. One possible remedy for this is to increase the annealing temperature in 1–2°C increments until the optimal annealing temperature is identified. This can be accomplished in a single PCR experiment using a "gradient thermocycler," in which case a range of annealing temperatures can be simultaneously used or, if using a nongradient thermocycler, you can perform a number of separate experiments, each one using a different annealing temperature. Sometimes this problem cannot be resolved by trying different annealing temperatures. If multiple bands including a single discrete target band are evident in a gel, then you can use a gel purification protocol to obtain the desired PCR products. Alternatively, another method is to rerun the PCR products on a "low-melt" agarose gel. Then use a sterile pipette tip to obtain an agarose plug taken from the center of the target band on the gel. Next, place the agarose plug into a 1.5 mL microcentrifuge tube containing 1 mL of molecular biology grade water then heat the tube to a high enough temperature (i.e., 65.5°C) to melt the agarose plug. Use 1 μL of this solution as your DNA template (instead of the original DNA extraction) in a new PCR. This can be a surprisingly simple and fast method for acquiring PCR products though safe procedures (especially for eye protection) for working with a UV illuminator should always be carefully followed. Note that these gel-based methods for obtaining target PCR products are labor intensive and slow and thus they are not practical methods for obtaining many PCR products. If large numbers of products are needed, a better strategy is to find alternative primers or redesign them to function better.

5.5 REDUCING PCR CONTAMINATION RISK

A negative PCR control only tells you whether a contamination problem exists. Good preventative measures in the laboratory are needed in order to minimize DNA contamination risks. Contamination in the DNA laboratory increases the amount of time and costs of consumables not to mention the frustration on the part of the researcher. Establishing preventative measures in the lab can dramatically diminish the incidence of contamination.

DNA contamination often occurs when concentrated DNA such as freshly extracted genomic DNA and *especially* PCR products contaminate some component of the PCR (reagents, pipette, etc.). Although a pipette can become contaminated by direct contact with concentrated DNA while the pipette is being used to transfer DNA from one tube to another, concentrated DNA can also become *aerosolized* thus making cross-contamination of samples in adjacent tubes even more likely. DNA contamination can afflict a variety of things in the lab including DNA extractions, plastic reaction tubes, pipette tips, pipettes, reagents for PCR, etc. Although the source of a DNA contamination problem can be from genomic DNA, PCR products and plasmids represent the usual sources of contamination. This is because PCR products and plasmids tend to be highly concentrated, whereas genomic DNA is far less concentrated. PCR products and plasmids are often referred to as *high copy* DNA while genomic DNA is called *low copy* DNA. Given that PCR can amplify DNA even from only a few copies of template DNA, it should be easy to see that this process is quite vulnerable to DNA contamination and therefore implementing measures to prevent contamination of a PCR *before* conducting PCR is of critical importance.

By recognizing the threat posed by high copy DNA, a number of things can be done to prevent PCR contamination. Earlier we already discussed the importance of making aliquots of PCR reagents, which can greatly reduce the incidences of contamination. However, there are additional measures that can be taken. Many DNA laboratories maintain separate "Pre-PCR" and "Post-PCR" spaces so that DNA extractions, PCR setup, and other pre-PCR procedures are performed away from areas that are likely already contaminated with high-copy DNA such as thermocyclers and gel electrophoresis stations (see Figure 1.3).

Pre-PCR area—This area of the lab should have its own dedicated set of pipettes and consumables. Filter tips can further reduce the contamination risk because the filter contained within the plastic tip acts as an effective barrier to DNA thus helping to avoid cross contamination. A laminar flow hood can also help reduce contamination problems by providing a well-lit clean space where DNA extractions and PCR setup can be performed. As we discussed in Chapter 4 a 10% bleach solution will destroy DNA so occasionally wiping down all counter-tops can help maintain a hygienic work environment. Many plastic consumables such as microcentrifuge tubes and PCR plates have been pre-sterilized in the factory so you may not need to autoclave them. However, if you have any doubts about the sterility of your plastic consumables, then you should autoclave them or use a UV cross-linker to be sure (except filter tips). Although DNA contamination can arise despite using all precautions (e.g., pipetting error), the incidence of contamination can be greatly contained. This will help prevent a more global contamination problem that would lead to the loss of valuable DNA extractions and other reagents. In addition to protection from contamination, making aliquots can reduce the freeze–thaw damage to the original or concentrated stocks caused by repeated (and needless) retrieval of solutions from freezers. Concentrated stocks of extracted DNA, primers, etc. should be stored and used with utmost care.

Post-PCR area—DNA contamination becomes less of a worry once PCR has already been performed because the remaining steps toward obtaining sequence data are not likely to be compromised by template DNA contamination. Although Sanger sequencing is a PCR-like procedure (Chapter 6), the cycle sequencing reaction is robust to minor PCR-product contamination because the presence of the correct PCR template will outnumber the foreign template and therefore good quality sequences can still be obtained assuming the primers work equally well on all templates. The main requirements of a post-PCR area include a dedicated lab space where the thermocycler(s) are located as well as a dedicated set of pipettes, and both filter and nonfilter tips. The filter tips should be used when performing agarose gel electrophoresis to check the results of genomic DNA extractions, whereas nonfilter tips (which are less expensive than filter tips) can be used for loading samples of PCR products on gels.

5.6 HIGH-THROUGHPUT PCR

High-throughput PCR often involves the use of plastic "microplates" with 96 separate 0.2 mL wells arranged in 8 rows by 12 columns; the positions of samples in the plate can be noted using the row letter (A to H) × column number (e.g., C7), which is important when dealing with 96 samples on one plate. Moreover, it is important to distinguish between two varieties of microplates: "un-skirted" and "skirted." Laboratories may need both types of microplates because some thermocyclers only accept the former, whereas it is safer practice to use the latter type of microplate to send PCR products by mail (though you should read the recommendations given by the sequencing laboratory that will be receiving your plate). As an alternative to using plates, some flexibility in the numbers of samples processed can be added simply by using individual 0.2 mL PCR tubes or especially 0.2 mL "strip-8" tubes. If single tubes are arranged in "8s" on a plastic 96 tube rack, or if strip-8 tubes are similarly arranged on a rack, then an 8-channel pipette can be used to great advantage. Such an approach can be considered "quasi high-throughput."

5.6.1 Setting Up PCRs in a 96-Sample Microplate Format

The master mix should be poured into a clean (sterile) multichannel plastic tray or distributed among eight 0.2 mL PCR tubes situated in a rack. Doing this will allow you to use the 8-channel pipette to quickly transfer the master mix to each well in the new PCR microplate. Next, the same pipette is used to transfer the diluted DNA template from the template plate (or tubes) to the PCR plate. When the master mix and template have been added to the PCR plate, great care should be used to seal the plate so that liquid cannot escape. Some labs prefer to use silicone rubber "mats," but be careful that the lid of your thermocycler can apply sufficient pressure to seal each well in the plate otherwise some of the reaction mixtures may escape from the wells during the thermocycling process. An excellent alternative method to seal the plate is to use strip-8 tube caps (i.e., each 8-sample column is sealed with a different strip of plastic caps). Following thermocycling be sure to perform the following two procedures with your PCR plate before unsealing the wells. First, cool the plate in a refrigerator or freezer. Hot samples have a greater chance of volatilizing PCR products, which not only results in losing a portion of your samples, but also can cause cross-contamination problems. Secondly, spin the plate in the tabletop centrifuge for 1 minute in order to push all liquid to the bottom of each well in the plate, which will minimize the chance of liquid escaping from the plate upon opening.

By now you have realized that anytime you can switch from a single-channel micropipette to using a multichannel micropipette to handle a large number of samples you are going to greatly economize your lab work. As we will see in Chapter 6, there are many advantages to conducting your PCRs in the 96-sample format, as not only does it reduce the time and per sample costs for PCR, but it facilitates downstream applications such as DNA sequencing.

5.7 OTHER PCR METHODS

We conclude this chapter by considering several other types of PCR, which have important applications in phylogenomics. The first variant of PCR we will discuss is called *hot start PCR*. This procedure can improve the specificity of problematic PCRs by preventing the formation and coamplification of short nontarget PCR products as a result of primer–primer interactions, which can lead to poorer DNA sequencing results. Another variant of PCR, called *long PCR*, is a technique for amplifying loci up to 40 kb long. Although long PCR has thus far had limited applications, it has proved useful for amplifying and sequencing whole mitochondrial genomes (Zardoya and Suárez 2008). This technique may also be useful for avoiding mitochondrial pseudogenes or "numts" (Sorenson and Quinn 1998). The third variety of PCR we will discuss is called *RT-PCR*, which has been an important method for phylogenomic studies concerned with the evolution of protein-coding genes.

5.7.1 Hot Start PCR

The best-performing PCR primers only amplify the target product, a result that usually guarantees the acquisition of high-quality DNA sequence data. However, occasionally researchers use primer pairs that amplify the target product plus unwanted *primer dimers*.

Primer dimers are short (~40–60 bp) double-stranded PCR products that arise because of design defects in the primers that cause them to hybridize to each other during PCR. Because duplexed primers can serve as both template and primer, this can result in the accumulation of a very large number of these products. Primer dimers interfere with the sequencing of the target products and therefore diminish the quality of sequence data. We will take a closer look at why primer dimers form and how to prevent their occurrence in routine PCR when we discuss primer development in Chapter 8.

Although it is possible to obtain useable DNA sequences from PCR products containing primer dimers, it is preferable to use some method to remove and discard them from already-completed PCR products or to prevent their formation during the PCR. We will discuss PCR product purification methods in Chapter 6 but the best strategy for dealing with the primer dimer issue is to prevent their formation in the first place. This can be accomplished through careful primer design or by using a method known as "hot start PCR" (Chou et al. 1992).

Chou et al. (1992) realized that primer dimers initially form during the first cycle of a PCR—as the thermocycler begins increasing the heat toward the initial denaturation step, the problematic primers begin to anneal to each other, which can allow the polymerase to start synthesis using primers as templates. This creates a problem because during the first annealing step in cycle 1 not only will the primers anneal to the correct target sequences, but also to a far larger number of primer-based templates. The result of such a reaction will be to coamplify both target and primer dimers. In other words, the primer dimers form independently of the genomic DNA templates also present in the mixture. To address this problem, Chou et al. (1992) proposed a simple method for preventing the formation of primer dimers. The method basically involves combining all PCR reagents only after the thermocycler has reached a temperature in the 60–80°C range prior to the first denaturation step. Chou et al. used wax as a barrier to separate the enzyme (*Taq* polymerase) from the other reagents in the PCR tubes. Thus, when the high temperature is reached, the wax melts allowing for mixing of all reactants at a temperature above which short DNA molecules can anneal to other DNA molecules. The wax-based method allows for the hot start method to be more automated. More recently, biotech manufacturers have developed another such automated hot start format that uses a specially designed "hot start DNA polymerase." The hot start polymerase has an antibody attached that prevents it from synthesizing DNA at lower (i.e., below-annealing) temperatures; when the thermocycler reaches the initial denaturation temperature, the antibody is inactivated and the polymerase becomes activated. It is important to carefully read the manufacturer's instructions for hot start polymerase because longer-than-normal exposure times at 94–95°C are required to activate the polymerases. If the researcher wishes to only use the hot start method for one to several samples, then a "manual hot start" method can be implemented. This approach involves combining all reagents in the PCR tubes except for the *Taq* polymerase. The tubes are then placed into the thermocycler and heated up to the initial denaturation temperature. When the machine reaches a temperature between 60 and 80°C the researcher pauses the thermocycler, opens the tubes, and pipettes the *Taq* polymerase into each tube one at a time. The thermocycler program is then resumed and allowed to complete its normal run. Note that there are safety risks involved when attempting to access the PCR tubes while they are inside the running thermocycler because the metal plate containing the PCR tubes (or microplate) will be hot enough to cause severe burns if touched by the researcher.

The hot start method is effective at preventing the formation of primer dimers. However, if after using hot start short double-stranded "primer dimers" are still observed in the agarose gel, then the "primer dimers" are likely not primer dimers. Instead, the nontarget band(s) are likely short nonspecific targets amplified from the genomic DNA template. If the small nontarget band is also observed in the negative control, then this is also evidence for a primer dimer problem (i.e., because primer dimers form independently of genomic DNA templates).

5.7.2 Long PCR

The vast majority of PCRs involve amplifications of loci that are <2 kb in length. One reason for this is simply due to the limitations of Sanger sequencing. As we will see in Chapter 6, it becomes more complicated and difficult to sequence PCR products that are longer than 1–1.5 kb. Another reason why most PCRs involve shorter loci is because longer loci tend to be more difficult to amplify than shorter loci. This difficulty can be encountered when using degraded genomic templates. However, even if you are using undegraded templates, the maximum length locus that can be reliably amplified using *Taq* polymerase is around 5 kb (Barnes 1994).

Unlike a number of other DNA polymerases, *Taq* polymerase lacks detectable $3' \rightarrow 5'$ exonuclease or "proofreading" activity (Tindall and Kunkel 1988; Korolev et al. 1995). Consequently, when *Taq* commits a misincorporation (base substitution) error, which on average occurs once every 4,000–5,000 bases synthesized (Innis et al. 1988; Saiki et al. 1988; Keohavong and Thilly 1989), these enzymes are unable to repair the mistakes themselves. Accordingly, one problem with using *Taq* is that PCR products generated with this enzyme can have base substitutions. However, regarding attempts to amplify longer fragments, the primary significance of these errors is that after such misincorporation events occur, *Taq* polymerases are unable to continue extending those particular DNA strands and so synthesis abruptly terminates (Innis et al. 1988; Barnes 1994). The production of these truncated PCR products, in turn, does not allow for a true "chain reaction" to take place because the number of templates for both primers does not exponentially increase. Thus PCRs using *Taq* as the sole DNA polymerase are incapable of generating a sufficient yield of accurately made PCR amplicons that are >5 kb in size.

A solution to the fidelity problem of *Taq* was proposed by Lundberg et al. (1991) who showed that, by replacing *Taq* in PCR with thermostable $3' \rightarrow 5'$ exonuclease (i.e., proofreading) DNA polymerases called *Pfu* (obtained from the thermophile ***Pyrococcus furiosus***), they could improve the fidelity of final products by an order of magnitude. Because *Pfu* was evidently able to detect and excise its own nucleotide misincorporations, it might have seemed reasonable to believe that *Pfu*-driven PCRs might be a way to also solve the long PCR problem. Unfortunately, however, Barnes (1994) showed that such proofreading DNA polymerases are by themselves unable to sufficiently amplify loci longer than about 2 kb. Barnes hypothesized that the reason for these failures may be (oddly) due to the proofreading enzymes themselves. He reasoned that the addition of a sufficient amount of proofreading polymerase to amplify longer fragments may have the unfortunate side effect of rapid enzymatic degradation of the primers (i.e., exonucleases destroy single-stranded DNA). Therefore the presence of degraded primers in a running PCR may not only hinder the production of correct PCR products, but they can also lead to the production of PCR artifacts, which could adversely affect a DNA sequencing reaction. Interestingly, in the end, the use of such proofreading DNA polymerases proved to be a critical part of the solution to the long PCR problem.

The key was to use a *mixture* of two different types of thermostable DNA polymerases in order to take advantage of *Taq*'s high processivity along with the proofreading ability of $3' \rightarrow 5'$ exonuclease polymerases such as *Pfu*, Vent, or Deep Vent (Barnes 1994; Cheng et al. 1994a,b). The relative amount of each enzyme in the reaction was also important. The experiments by Barnes (1994) showed that by using a high level of *Taq* polymerase as a "parent" enzyme together with a low level (160–640-fold lower in terms of polymerase incorporation units) of a proofreading DNA polymerase, the *Taq*-caused nucleotide misincorporations could still be efficiently excised by the proofreading enzymes—thus allowing *Taq* to complete synthesis of those strands—but without suffering from excessive primer degradation caused by an abundance of proofreading polymerases. Using these enzyme mixes together with longer primers (27–33 bp) and modified thermocycling profiles, Barnes' approach provided an elegant resolution to the long PCR problem. This method, which is now commonly referred to as "long PCR" or "LA PCR" (for long and accurate), is capable of providing high product yields for loci as long as 35 or 40 kb (Barnes 1994; Cheng et al. 1994a,b).

In addition to the two-enzyme mix, the methodology of long PCR differs from standard PCR in several other ways. For example, a long PCR thermocycling profile can use the format:

Step 1: 94°C for 30 seconds (initial denaturation step)

Step 2: 94°C for 30 seconds

Step 3: 55–68°C for 30 seconds

Step 4: 68°C for 1 min/kb of target sequence

Repeat Steps 2–4 total of 30–35 times

You should notice several aspects of this profile that will differentiate it from a typical PCR profile. First, long PCR requires a significantly longer extension time. It is important to realize that *Taq* enzymes in commercially available long PCR "kits" have variable elongation rates (e.g., 1 kb per 15 seconds to 1 kb per minute) depending on the kit. Thus, extension times need to be calculated based on the kit specifications as well as the expected product length. Another difference is the primer annealing temperature. Long PCR tends to work best with PCR primers that are 30–33 bp (Barnes 1994; personal observation) and thus the optimal annealing temperatures for these longer primers will likely be higher than for more typical PCR primers. Some long PCR thermocycler profiles use the same temperature for both primer annealing and extension steps (e.g., Cheng et al. 1994a,b; Zardoya and Suárez 2008). Another difference between short and long PCR is the extension temperature. Notice that the extension temperature of 68°C is a little lower than the usual 70–72°C used in routine PCR. Although the optimal temperature for *Taq* polymerase is 70–74°C, Barnes (1994) found that extension temperatures above 68°C resulted in diminished long PCR product yields. Another factor limiting product yields is the duration of the denaturation step. Barnes (1994) conducted an experiment, which varied the time of the denaturation step and he observed that denaturing the DNA at 94°C for 20, 60, and 180 seconds resulted in progressively lower yields of an 8.4 kb product. He argued that the longer heat treatments stressed the DNA template enough to cause a higher incidence of depurination of sites, which can create problems because *Taq* polymerase stops synthesis at depurinated sites. Depurination of a site occurs when the glycosidic bond connecting the base with the sugar moiety is cleaved, which results in loss of the adenine or guanine base. Barnes also pointed out that a lower pH environment can worsen the depurination rate. Accordingly, based on these results as well as the study by Cheng et al. (1994a) it appears important to keep the denaturation step as brief as possible with the minimum temperature while also increasing the reaction pH to 8.8–9.2 (optimum of 9.1). Fortunately, a variety of long or LA PCR enzyme mixes with optimized buffers are commercially available so the researcher does not need to be concerned about fiddling with the pH of a long PCR. However, even with a proven commercially available long PCR reagent mix, there are additional "fine-tuning" procedures that the researcher can perform in order to improve the quality and yield of products. One variable is concentration of Mg^{2+} in the reaction mixture. As some product manuals will suggest, you should optimize the amount of free Mg^{2+} that is available for the polymerases. Secondly, improved specificity and target product yields can be achieved by hot starting the reactions (Cheng et al. 1994b; personal observations). This can be done by either purchasing a hot start long PCR kit or by performing a manual hot start.

5.7.3 Reverse Transcriptase-PCR

In phylogenomic studies concerned with the evolution of gene families it is essential to obtain DNA sequences for all homologous copies of a coding gene of interest that are present in a genome. However, because eukaryotic genes are usually split into different genomic segments consisting of exons and introns, this complicates the acquisition of full-length coding DNA sequences from genomic DNA templates. To address this challenge a variant of PCR called "RT-PCR" can be employed.

A summary of the RT-PCR method is as follows. The enzyme "RT" is used to synthesize DNA strands that are complementary to mature mRNA strands. As you will recall from Chapter 2, RT is a special type of DNA polymerase obtained from retroviruses that uses mRNA molecules as templates. Because most mature mRNAs have polyA tails, these stretches of sites can serve as annealing sites for polyT or "oligo(dT)" DNA primers, which, in turn, supplies a primer-template junction for the RT to begin synthesis of a DNA strand. These synthesized DNA strands are called *complementary DNAs* or "cDNAs." Normal PCR is then used to amplify these cDNAs into double-stranded cDNAs. However, because DNA polymerases cannot synthesize DNA using the mRNA template strands, the mRNA strands must first be eliminated using the enzyme *RNase* H followed by a round of ssDNA synthesis using a DNA polymerase. But how does the DNA polymerase synthesize the strand complementary to the first cDNA strand without adding a primer? After removal of the mRNA strand, the 3′ end of the remaining single-stranded cDNA strand spontaneously forms a hairpin by looping back on itself to generate a short stretch of dsDNA. This hairpin structure creates the needed primer-template junction from which the DNA polymerase can synthesize a complementary strand resulting in fully double-stranded cDNAs. A special endonuclease called S1 nuclease, which excises single-stranded loops in DNA, is then used to cut the loop thereby creating double-stranded but separate cDNA molecules. This step completes the construction of a *cDNA library*, which can later be used with standard PCR to target and amplify specific genes using gene-specific primer pairs. The next major step is to use standard PCR to amplify these double-stranded cDNAs. Note that because the cDNA products of RT-PCR are derived from mature mRNA molecules they will not contain introns and thus they provide the uninterrupted gene sequence needed for phylogenomic analyses.

In the laboratory, the first step is to acquire high-quality (i.e., undegraded) RNA that has been purified. RNA is much more vulnerable to degradation than DNA via enzymatic destruction by *RNases* or chemical disintegration due to alkaline hydrolysis. Note that some DNA extraction kits include *RNase* A and therefore careful consideration should be given whether to destroy the RNA at the time of extraction because the RNA may be needed later for RT-PCR. Most standard PCRs will function fine in the presence of RNA and thus it may not be necessary to do the *RNase* treatment during a DNA extraction. *RNase* can be used later on the purified DNA in case the DNA is needed for an RNA-free application such as making some types of NGS libraries (Chapter 7). RT-PCR is done either in a "one-step" procedure in which the RT and PCRs are carried out in the same PCR tube or in a "two-step" process whereby the RT reaction is performed first in one tube then the reaction products are transferred to a second (or more) tubes for PCR amplification. Because the RT reagents are expensive compared to normal PCR reagents, the two-step method is good for the preparation of a cDNA library, which can then be aliquoted as templates for multiple different PCRs. It is critically important to take measures to avoid *RNase* contamination such as wearing gloves and using *RNase*-free filter tips. Regardless of whether a one- or two-step RT-PCR procedure is used, the reactions are driven by a thermocycler. The first cycle of an RT-PCR thermocycler program, which is specifically for the RT reaction, consists of three parts: (1) the temperature is set to about 50°C so that oligo(dT) primers can anneal to the polyA tails of the mRNAs; (2) in the next step, the temperature is lowered to the preferred temperature of the RT enzyme (37–42°C). This allows the RT to extend the cDNA strand along the RNA template starting at the primer-template junction; (3) in the last step of this first cycle, the temperature is raised to 92°C to inactivate the RT enzymes. The second cycle starts the PCR process using the newly formed double-stranded cDNAs as the DNA templates for amplification. At this time, forward and reverse gene-specific primers are needed to target particular cDNAs for amplification. The PCR uses a typical master mix containing *Taq* polymerase and runs for 30–40 cycles. After the reactions are completed, the success of the amplifications can be evaluated on a 1%–2% agarose gel. In the past, RT-PCR products would have been sequenced via laborious vector-cloning methods, but now NGS-based methods can sequence far larger numbers of such products in a more efficient manner.

REFERENCES

Barnes, W. M. 1994. PCR amplification of up to 35-kb DNA with high fidelity and high yield from λ bacteriophage templates. *Proc Natl Acad Sci USA* 91:2216–2220.

Brock, T. D. 1997. The value of basic research: Discovery of *Thermus aquaticus* and other extreme thermophiles. *Genetics* 146:1207–1210.

Brock, T. D. and H. Freeze. 1969. *Thermus aquaticus* gen. n. and sp. n., a non-sporulating extreme thermophile. *J Bacteriol* 98:289–297.

Brown, T. A. 2007. *Genomes* 3. New York: Garland Science/Taylor & Francis.

Cheng, S., S.-Y. Chang, P. Gravitt, and R. Respress. 1994b. Long PCR. *Nature* 369:684–685.

Cheng, S., C. Fockler, W. M. Barnes, and R. Higuchi. 1994a. Effective amplification of long targets from cloned inserts and human genomic DNA. *Proc Natl Acad Sci USA* 91:5695–5699.

Chou, Q., M. Russell, D. E. Birch, J. Raymond, and W. Bloch. 1992. Prevention of pre-PCR mis-priming and primer dimerization improves low-copy-number amplifications. *Nucleic Acids Res* 20:1717–1723.

Innis, M. A., D. H. Gelfand, J. J. Sninsky, and T. J. White. 1990. *PCR Protocols: A Guide to Methods and Applications*. New York: Academic Press.

Innis, M. A., K. B. Myambo, D. H. Gelfand, and M. A. Brow. 1988. DNA sequencing with *Thermus aquaticus* DNA polymerase and direct sequencing of polymerase chain reaction-amplified DNA. *Proc Natl Acad Sci USA* 85:9436–9440.

Kainz, P. 2000. The PCR plateau phase—Towards an understanding of its limitations. *Biochim Biophys Acta* 1494:23–27.

Keohavong, P. and W. G. Thilly. 1989. Fidelity of DNA polymerases in DNA amplification. *Proc Natl Acad Sci USA* 86:9253–9257.

Kleppe, K., E. Ohtsuka, R. Kleppe, I. Molineux, and H. G. Khorana. 1971. Studies on polynucleotides: XCVL repair replication of short synthetic DNA's as catalyzed by DNA polymerases. *J Mol Biol* 56:341–361.

Kornberg, A. and T. Baker. 1992. *DNA Replication*, 2nd edition. New York: W. H. Freeman and Company.

Korolev, S., M. Nayal, W. M. Barnes, E. Di Cera, and G. Waksman. 1995. Crystal structure of the large fragment of *Thermus aquaticus* DNA polymerase I at 2.5-A resolution: Structural basis for thermostability. *Proc Natl Acad Sci USA* 92:9264–9268.

Lundberg, K. S., D. D. Shoemaker, M. W. W. Adams, J. M. Short, J. A. Sorge, and E. J. Mathur. 1991. High-fidelity amplification using a thermostable DNA polymerase isolated from *Pyrococcus furiosus*. *Gene* 108:1–6.

McPherson, M. J. and S. G. Moller. 2006. *PCR*, 2nd edition. Boca Raton: Taylor & Francis.

Mullis, K. B. and F. Faloona. 1987. Specific synthesis of DNA in vitro via a polymerase catalyzed chain reaction. *Methods Enzymol* 155:335–350.

Mullis, K. B., F. Faloona, S. Scharf, R. Saiki, G. Horn, and H. Erlich. 1986. Specific enzymatic amplification of DNA in vitro: The polymerase chain reaction. *Cold Spring Harb Symp Quant Biol* 11:263–273.

Palumbi, S. R. 1996. Chapter 7. Nucleic acids II: The polymerase chain reaction. In *Molecular Systematics*, 2nd edition, ed. D. M. Hillis, C. Moritz, and B. K. Mable, 205–247. Sunderland: Sinauer.

Saiki, R. K., D. H. Gelfand, S. Stoffel et al. 1988. Primer-directed enzymatic amplification of DNA with a thermostable DNA polymerase. *Science* 239:487–491.

Saiki, R. K., S. Scharf, F. Faloona et al. 1985. Enzymatic amplification of β-globin genomic sequences and restriction site analysis for diagnosis of sickle cell anemia. *Science* 230:1350–1354.

Sorenson, M. D. and T. W. Quinn. 1998. Numts: A challenge for avian systematics and population biology. *Auk* 115:214–221.

Tindall, K. R. and T. A. Kunkel. 1988. Fidelity of DNA synthesis by the *Thermus aquaticus* DNA polymerase. *Biochemistry* 27:6008–6013.

Watson, J. D., T. A. Baker, S. P. Bell, A. Gann, M. Levine, and R. Losick. 2014. *Molecular Biology of the Gene*, 7th edition. New York: Pearson Education, Inc.

Zardoya, R. and M. Suárez. 2008. Chapter 12. Sequencing and phylogenomic analysis of whole mitochondrial genomes in animals. In *Phylogenomics*, ed. W. J. Murphy, 185–200. Totowa: Humana Press.

CHAPTER SIX

Sanger Sequencing

In Chapter 5, we saw that a single PCR is capable of generating a very large number of copies for a locus of interest. Why then are so many replicate templates needed for the Sanger sequencing method? The reason is because enough copies of a particular locus are needed in order to generate sufficient signal for the laser detection system of a modern DNA sequencing machine to record the sequence of a PCR product. Of course in order for the sequencer to accomplish this imaging task, the templates must be labeled with a fluorescent dye, which can be detected by the machine's laser.

In recent years, it has become cheaper and easier to use Sanger sequencing thanks to further methodological improvements, enhancements to sequencing machines, and especially the availability of outsourcing of DNA sequencing. These factors have simplified the sequencing process to the extent that a researcher can now generate a wealth of DNA sequence data while having only a fuzzy understanding about the molecular mechanisms underlying Sanger sequencing. This is remarkable considering that the modern Sanger methodology is a medley of molecular biology techniques. However, as with other molecular methods such as PCR, a researcher who has a solid knowledge of the details underlying each of these components to the Sanger methodology can achieve a higher level of success in obtaining the desired sequence data.

6.1 PRINCIPLES OF SANGER SEQUENCING

6.1.1 The Sanger Sequencing Concept

The ingenious DNA sequencing method developed by Frederick Sanger and colleagues capitalized on an earlier discovery made by other molecular biologists. Atkinson et al. (1969), who were looking at how nucleotide analogs might be used to investigate the functions of DNA polymerases, demonstrated that an analog to the naturally occurring 2′-dTTP known as a 2′-,3′-dideoxythymidine triphosphate (ddTTP) could terminate the synthesis of a DNA strand after it had been incorporated into the same strand. Accordingly, they described ddNTPs as being *chain growth terminators*. An example of a ddNTP is shown in Figure 6.1. If you compare the usual dNTP (Figure 5.2) with the ddNTP shown in Figure 6.1 you will see that these nucleoside triphosphates differ only by the latter lacking a 3′-hydroxyl group; that is, there is only a hydrogen atom located at the 3′ carbon position on the sugar moiety of a ddNTP. However, as Atkinson et al. (1969) pointed out, this minor structural difference can have a dramatic functional consequence on DNA synthesis: a ddNTP *can* be added to a growing strand of DNA in normal fashion (like dNTPs) but it *cannot* act as a primer for another ddNTP or dNTP therefore synthesis of that strand abruptly terminates.

Sanger et al. (1977) exploited the chain terminating property of ddNTPs to develop their method of DNA sequencing. Their method consisted of two main steps. First, they performed *in vitro* DNA synthesis using a single primer, template DNA, DNA polymerase (Klenow fragment), and importantly, a *mixture* of dNTPs and ddNTPs. By using a mix of dNTPs and ddNTPs in the same reaction they could generate a distribution of variably sized but nested primer *extension products* (Watson et al. 2014). In the second step, they fractionated the extension products on a denaturing acrylamide gel. Because they had labeled some ddNTPs with a radioactive isotope (^{32}P) prior to

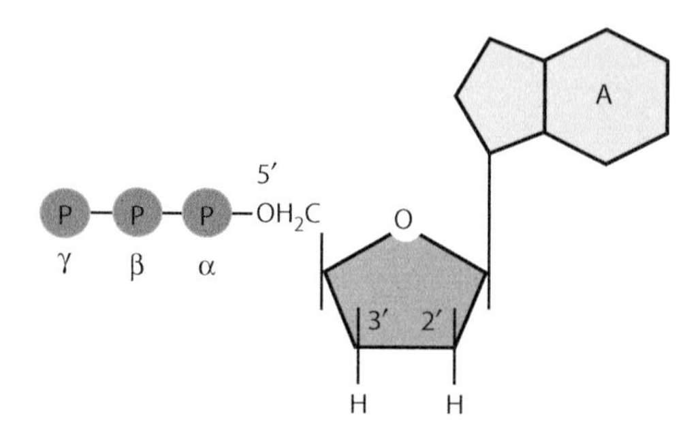

Figure 6.1. Chemical structure of a 2′-, 3′-dideoxynucleoside triphosphate (ddNTP). The example shown here is 2′-, 3′-dideoxyadenosine triphosphate (ddATP). The lack of an oxygen atom at the 3′ position on the ribose sugar represents the only difference between this ddATP and the dATP shown in Figure 5.2.

the sequencing reactions, they could visualize banding patterns in the gel that corresponded to the relative positions of a particular base type. Four separate sequencing reactions—one for each base type (A, G, C, and T)—had to be performed in order to elucidate the desired sequence.

Although this new chain termination method of DNA sequencing was superior to other proposed methods for DNA sequencing at that time (e.g., the method of Maxam and Gilbert 1977), the Sanger method would not come of age until more than a full decade later. Similar to the modernization of PCR, the 1980s were a period of tremendous innovation of new DNA sequencing refinements many of which dramatically improved the Sanger methodology. What were these advances that improved the Sanger method? A timeline for the modernization of the Sanger method is shown in Table 6.1. One of the difficulties with the Sanger method in its earliest implementation was that radioactive isotopes had to be used to "label" the ddNTPs to allow for visualization of the DNA sequence via autoradiographs. However, in 1985 researchers from Caltech published a paper (Smith et al. 1985) showing how sequence visualization could be accomplished using fluorescent-labeled primers (or "dye-primers") and a laser detector. This advance not only removed the health hazard posed by working with radioactive isotopes, but it enabled the production and first trial of a semi-automated DNA sequencing machine in 1986 (Smith et al. 1986). A year later, another research group unveiled the first use of fluorescent-labeled ddNTPs (or "dye-terminators") in DNA sequencing (Prober et al. 1987). Although the use of radioactive-labeled nucleotides and primers and dye-primers would continue to be used in Sanger sequencing into the mid-1990s, they would all be superseded by dye-terminators in DNA sequencing within a few years thereafter (Table 6.1).

The critical contributions from PCR to Sanger sequencing cannot be overstated. First, PCR provided a simple method for generating template DNA for Sanger sequencing that was far superior to cloning-based methods (Innis et al. 1988). Secondly, researchers dramatically improved the methodology for generating the Sanger sequencing extension products by co-opting PCR components including *Taq* polymerase and thermocyclers (Innis et al. 1988; Carothers et al. 1989; Murray 1989). This PCR-like sequencing reaction, which is called *cycle sequencing* or "sequencing PCR," uses a linear polymerase reaction to generate the extension products (Hillis et al. 1996). We will soon see why this is a linear rather than an exponential production process. The development of fluorescent dye-terminators

TABLE 6.1
Timeline for the modernization of Sanger sequencing

Year	Advance	References
1977	Chain-termination method for DNA sequencing published	Sanger et al. (1977)
1985	Fluorescent-labeled primers introduced	Smith et al. (1985)
1986	First semi-automated DNA sequencer introduced	Smith et al. (1986)
1987	Fluorescent-labeled ddNTPs introduced	Prober et al. (1987)
1989	Cycle sequencing using *Taq* polymerase introduced	Murray (1989); Carothers et al. (1989)
1996	ABI 377 automated "slab gel" DNA sequencer introduced	Applied Biosystems Wikipedia
1999	ABI 3700 automated "capillary" DNA sequencer introduced	Applied Biosystems Wikipedia

and cycle sequencing together with advances in computers collectively made possible the existence of the first fully automated DNA sequencing machines by the middle 1990s and a short time later the more advanced capillary sequencers (Table 6.1). The capillary DNA sequencer has remained largely unchanged since then and they are still being used despite the increasing popularity of newer NGS platforms. Readers interested in learning more about the history of DNA sequencing as well as the principles and methods not discussed here should read the comprehensive account in Hillis et al. (1996).

6.1.2 Modern Sanger Sequencing

6.1.2.1 Cycle Sequencing Reaction

Recall that the first essential principle of the Sanger method is to use a single primer, a DNA template, a mixture of dNTPs and ddNTPs, and DNA polymerase to produce a nested set of extension products. Accordingly, before the cycle sequencing reaction can be set up, the researcher must first use PCR to obtain good quality templates for sequencing. However, before PCR products can be sequenced they must be "cleaned," which means that unincorporated dNTPs and primers must be removed or inactivated otherwise they will interfere with the sequencing reaction. We will see why this cleaning step is necessary as well as learn about the techniques for cleaning PCR products later in this chapter. The cleaned PCR products are then used as templates in a cycle sequencing reaction.

Although regular PCR and cycle sequencing share many similarities, important differences between the two methods also exist. For example, the cycle sequencing reaction requires a mixture of dNTPs and fluorescent-labeled ddNTPs (dye terminators) in approximately a 100:1 ratio and only a *single* primer, which is called the *sequencing primer*, is used in the reaction. The sequencing primer, which directs the sequencing of one strand of the PCR product, is typically one of the original PCR primers used to generate the PCR product that is being sequenced. However, other types of sequencing primers also exist, which we will see later in this chapter. The computer program that the thermocycler uses to execute a cycle sequencing reaction is similar to a typical PCR program, as it subjects the reactants to the same denaturation, annealing, and extension steps repeated for 35 or more cycles. However, cycle sequencing only requires a brief annealing time of 5 seconds because a large concentration of the template, which perfectly matches the sequencing primer, is present. A longer extension time of 4 minutes per cycle is also used to help ensure that the extension products are properly formed (i.e., terminated by a ddNTP).

Let's now take a detailed look at the cycle sequencing reaction once it has begun in a thermocycler. After each denaturation step the double-stranded PCR products become single stranded. As the single-stranded amplicons cool down to the set annealing temperature (most often at 50°C), the sequencing primers bind to their complementary sites on one of the PCR strands as is shown in Figure 6.2a. The temperature of the reaction mixture is then raised to 60°C so that *Taq* polymerase can begin incorporating nucleotides starting at the 3′ end of the primer where the primer–template junction is initially located. Because there are 100 times more dNTPs than ddNTPs, the former type of nucleotide is more likely to be incorporated into the growing strand. However, synthesis stops once a ddNTP becomes incorporated into the strand (Figure 6.2b). In this example, synthesis of the strand was terminated after a ddGTP was added. The strand that includes the sequencing primer plus the newly synthesized DNA including the ddNTP at the 3′ end is defined as the extension product. This is a good moment to reflect on why unincorporated dNTPs leftover from the PCR must be removed prior to cycle sequencing. If unused dNTPs are not removed, then they will dilute the ddNTPs to the point that far fewer extension products will be made leading to a low quality sequence.

In our earlier discussion, we learned that cycle sequencing generates a linear increase in products, which is in contrast to the exponential process of PCR. Let's now see why this must be the case. During the first cycle of a cycle sequencing reaction, a single PCR product first denatures into two single strands. Only one of these strands will be used as template throughout all 35 cycles because only a single sequencing primer is used. The other PCR strand is not used in this reaction. If the researcher wishes to sequence the other strand, then a second cycle sequencing reaction must be performed in a separate PCR tube that includes the appropriate sequencing primer.

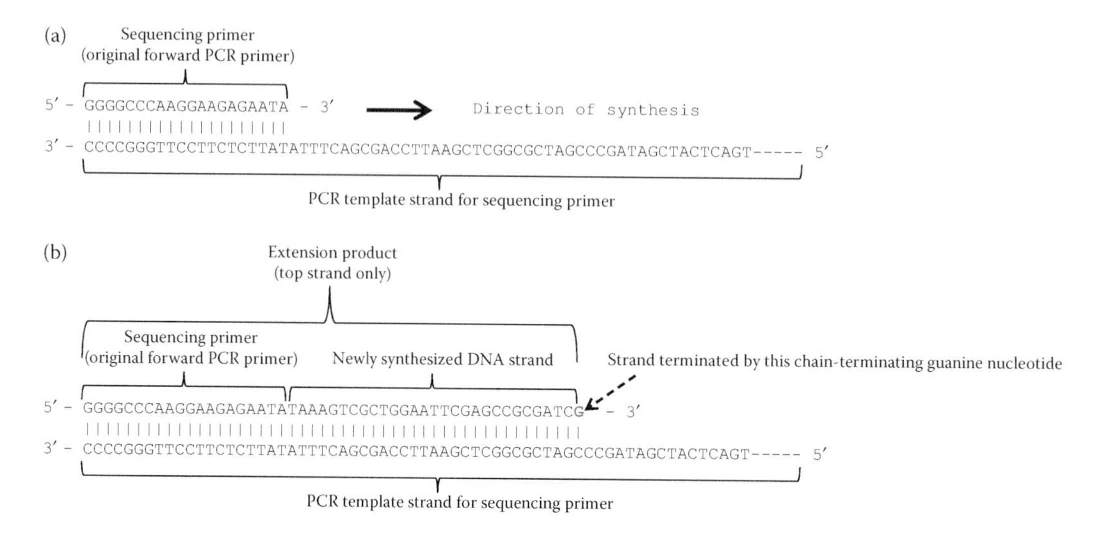

Figure 6.2. Annealing and extension steps in a cycle sequencing reaction. (a) Annealing step: forward sequencing primer (original forward PCR primer) anneals to the stretch of complementary sites on the PCR template strand. (b) Extension step: *Taq* polymerase (not shown) extends the newly synthesized strand until a ddNTP is incorporated into the strand. Once incorporated into the nontemplate strand, the newly added nucleotide (now ddGMP), which is shown in gray, terminates synthesis. Vertical dashes between bases indicate hydrogen bonding between Watson–Crick base pairs. The dashes at the 3′ end of the PCR template (lower) strand indicate that the PCR template strand is truncated on the right due to lack of space in the figure.

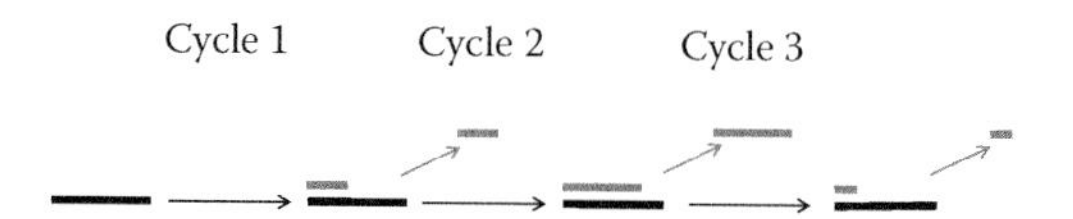

Figure 6.3. Synthesis of extension products during the first three cycles of a cycle sequencing reaction. After the extension step in each cycle, a new extension product (gray) can be formed from each PCR amplicon though the number of templates (black) does not change. The lengths of the extension products are variable owing to the random nature of ddNTP incorporation.

Following the first cycle, one extension product per PCR amplicon will be produced as shown in Figure 6.3. In the second cycle, the denaturation step denatures the PCR template-extension product duplex, which makes the PCR template available for a new sequencing primer during the annealing step. After the second cycle, a second extension product is made from a PCR template, and so on. Thus, each PCR template can potentially be reused in each cycle and the number of templates remains fixed throughout the process (Figure 6.3).

6.1.2.2 Gel Electrophoresis of Extension Products

The second essential principle of the Sanger method is to separate the extension products by their size using high-resolution gel electrophoresis. Polyacrylamide is used as the separation matrix in sequencing, rather than agarose, because it has the ability to separate DNA fragments differing in length by a single nucleotide. In order for this to perform as desired, the extension products must travel through the gel matrix in a single-stranded and completely linear state. The mobility of extension products through a gel will be adversely affected if the extension products are either duplexed with a PCR template or if they have any secondary structures (e.g., hairpin loops, etc.). Thus, immediately prior to electrophoresing the extension products, they are denatured into single strands using high heat (>80°C) and mixed with chemicals (e.g., formamide) to prevent the formation of secondary structures.

Once the extension products made in a sequencing reaction are loaded into a capillary sequencer, they are then fractionated by length in the gel-filled capillaries; that is, they are size-sorted using electrophoresis. As the extension products pass by the machine's laser in order of their length with shorter products appearing first, the fluorescent-dyed ddNTP of each extension product is recorded. Figure 6.4 illustrates a simplified example of capillary sequencing. In Figure 6.4,

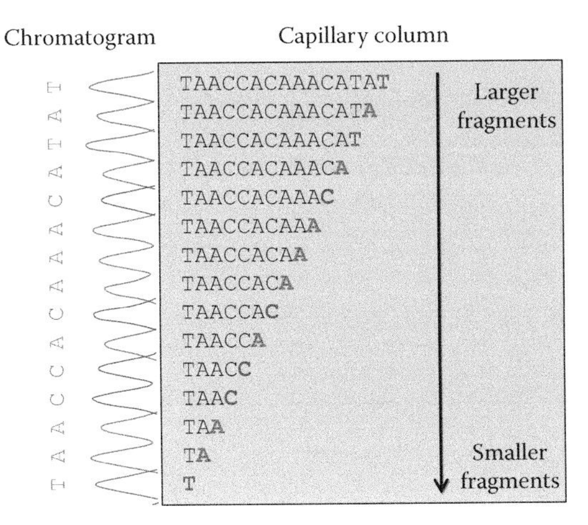

Figure 6.4. Electrophoresis of extension products in a capillary sequencer and recording of the sequence. Extension products are separated according to their length in the capillary column with smaller fragments moving through the gel matrix at a faster rate than longer fragments. The sequencer's laser detects the fluorescence emitted by the ddNTP located at the 3′ end of each extension product and displays the DNA sequence in a chromatogram. Extension products are shown in a 5′ → 3′ orientation.

each sequence in the capillary column represents a group of many thousands of identical extension products that pass by the laser over a short time interval thereby reporting the same type of ddNTP (i.e., A, C, G, or T base). When a group of such fragments passes by the laser a signal of sufficient strength is detected by the machine and recorded as a colored peak corresponding to one of the bases (Figure 6.4). As a consequence of the nested nature of the extension products, during electrophoresis the size-specific groups travel in an orderly progression through the gel matrix with the shortest products at the front owing to their rapid mobility through the matrix. After the last (and longest) of the extension products passes by the laser, the sequencer's computer software assembles the data into a DNA sequence. Looking at Figure 6.4, it should be apparent why only a single sequencing primer can be used in a cycle sequencing reaction: if two primers were used (one to sequence each opposing strand), then two sets of nonoverlapping extension products would be produced thereby producing ruined sequence data.

Capillary sequencers are capable of outputting high-quality sequence data between 700 and 1,000 bp per *read*. A raw sequence "read" is the length of the sequence in bp that a sequencer can accurately generate. As you can see in Figure 6.4, the sequencer determines the sequence using the information gleaned from laser-detection of the fluorescent-extension products. The name of the output file generated by an automated sequencer is called a "*chromatogram*," "*electropherogram*," or "*trace*," an example of which is shown in Figure 6.5. The chromatogram shows the sequence or *base calls* (A, T, G, and C) from one DNA strand. Each base call is based on a distinct colored peak in the chromatogram with the four base types being distinguished by color: adenine (green), thymine (red), guanine (black), and cytosine (blue) as is shown in Figure 6.5. For example, as a group of identical-length extension products passes by the laser, the instrument records each product's fluorescent ddNTP label and then tabulates the hundreds or thousands of independent values into distribution; a peak that resembles a single symmetrical distribution signifies that all the products of that length shared the same type of ddNTP and therefore the base call is correct. When the sequencing computer is unable to call a particular base due to weak or conflicting signals, an "N" symbol will be used to indicate an unknown base.

6.1.2.3 Sequence Data Quality

Chromatogram files represent an important milestone not only because they contain the raw sequence data needed for analyses, but because they also provide the researcher with an effective means for determining the quality of each base call in a sequence. It is crucial for a researcher to

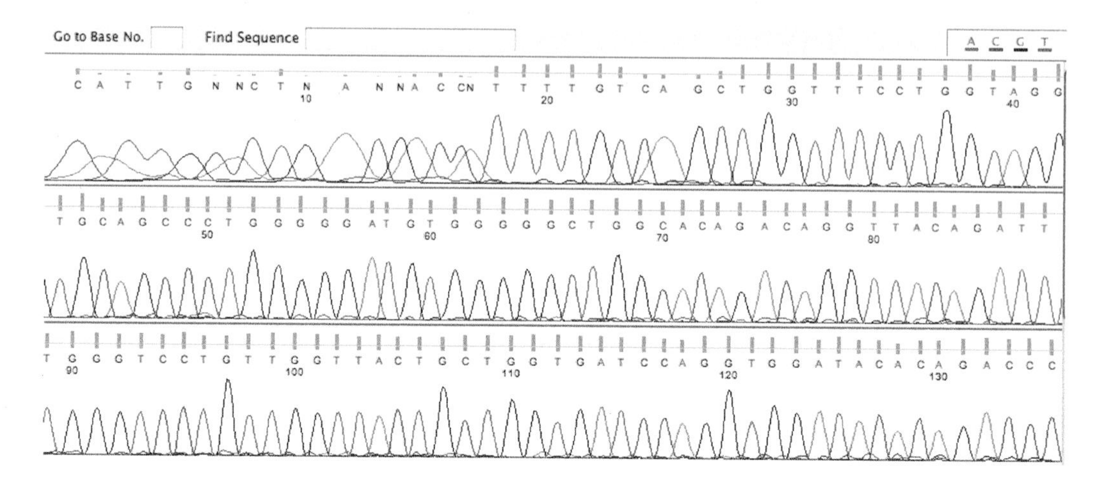

Figure 6.5. Chromatogram of a sequenced PCR product. Each colored peak represents a different base with thymine (red), guanine (black), cytosine (blue), and adenine (green). The sequencer records an "N" for ambiguous base calls when the quality of the sequence is poor. The vertical gray bars above each peak are the quality scores. Vertical gray bars that rise above the horizontal line are considered good quality base calls, whereas bars remaining below the line are judged poor quality.

distinguish poor versus good quality sequence data before proceeding with any computer analyses of those data. Chromatograms are rich in information, which can be used by the researcher to troubleshoot problems. Numerous software packages are available that allow the user to edit base calls, reverse complement a chromatogram, compute quality scores, and export a text version of the sequence for use in downstream analyses.

How is the quality of each base call evaluated on a chromatogram? Both qualitative and quantitative methods have been developed for assessing the quality of base calls. First, let us consider how to evaluate base calls using qualitative criteria. In Figure 6.6 we see a sequence chromatogram, which shows the first 140 base calls. Notice that the first half of the sequence is comprised of messy-looking or indistinct peaks, whereas the latter half is represented by a sequence of symmetrical peaks. Good quality sequence is generally comprised of *single distinct peaks* that are *evenly spaced from each other*, which is what we see from base calls #66–140. In contrast, base calls #1–65 are of poor quality. Another measure used to judge base calls is the *quality score*, which is shown on the chromatogram as a vertical gray bar above each peak. Quality scores that rise above the thin horizontal line are deemed to be good (acceptable) quality base calls, whereas bars that lie below the line indicate poor (unacceptable) quality base calls. In Figure 6.6 notice that the quality scores before base call #66 are consistently poor while the latter bases are all good—as we would expect had we only examined the colored peaks. Another characteristic you see in chromatograms concerns the unevenness of peak height, as some peaks are higher than others. This facet of a chromatogram should be ignored because this pattern reflects the idiosyncrasies of the cycle sequencing chemistry rather than any underlying reality about the organism's DNA.

Because visually scanning chromatograms to evaluate the quality of base calls is a tedious and slow process, researchers developed a bioinformatics approach that can evaluate the quality of each base call in an automated fashion. This was accomplished with the development of the program *Phred* (Ewing and Green 1998; Ewing et al. 1998), which computes a *Phred* quality score or "Q-score" for each base call on individual chromatograms. The Q-score reflects the probability of an incorrectly called base. For example, a Q-score of 20 or "Q20" is a commonly used threshold for discriminating poor versus acceptable base calls: scores below Q20 are considered poor quality, while at or above this value are deemed as good quality base calls. A score of Q20 means there is a 1 in 100 chance that the called base is incorrect. A Q-score of 10 means there is a 10% chance that a base call was incorrect, a Q30 is 1 in a 1,000 chance of such an error, and

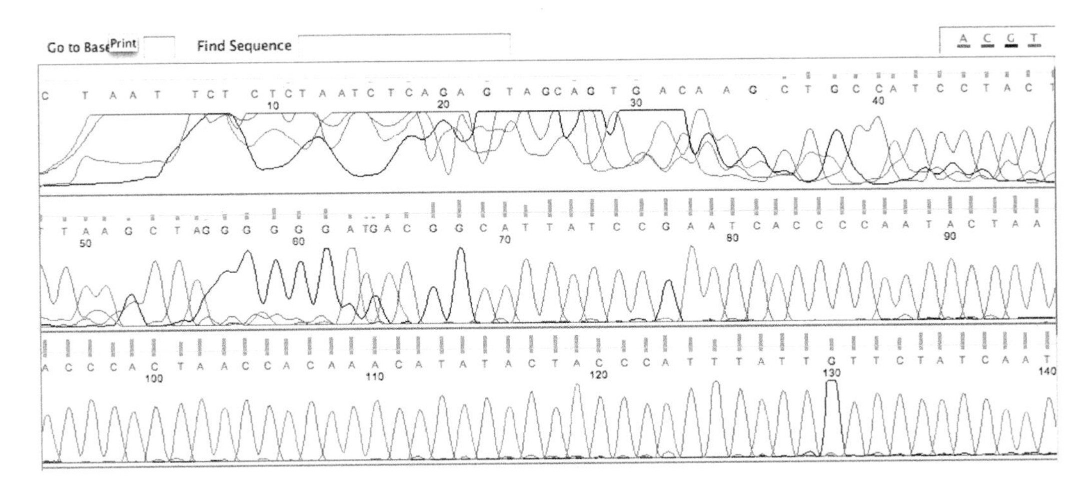

Figure 6.6. Chromatogram of a sequenced PCR product showing low-quality base calls in the first ~65 base positions of the sequence. Although the sequencer called a number of bases in this beginning part of the sequence, these base calls cannot be trusted because of the lack of evenly spaced single peaks, which is also reflected by the low quality scores.

Q60 implies there is a 1 in a million chance that a base was not correctly called.

It is not unusual for a chromatogram to show poor base calls for the first ten to thirty bases as well as the bases beyond approximately the 700th base call (e.g., Figure 6.5). Even in the highest quality sequences the first ~10 base calls are usually of poor quality simply due to the vagaries of the Sanger sequencing technology. If the first part of a sequence is needed for a study, then the same PCR product should be sequenced from the other direction (using the other primer) so that a reverse complement of the second sequence can be compared to the first sequence. Conducting a second cycle sequencing reaction of the same PCR product for purposes of sequencing the other strand is also desirable if the PCR products are longer than 700 bp or if there are concerns about quality of the first sequence. It is common practice for researchers to sequence each PCR product from both directions. Sequencing from both directions may be required to obtain whole sequences particularly if a long microsatellite region occurs in the sequence, as such repeats tend to cause problems when sequencing.

Chromatograms can also exhibit poor base calls beyond the norm just described, which can result in sequence data being either unusable or partially usable at best. There are various causes for such poor results. For example, problems can arise if unpurified PCR products (i.e., containing unincorporated primer and unincorporated dNTPs) or products containing nontarget amplicons are used in a cycle sequencing reaction. Primer dimers are one common type of PCR artifacts that can interfere with sequencing and unfortunately they are difficult to remove from PCR products. Inclusion of primer dimers in a cycle sequencing reaction results in poorer quality sequence data, as the chromatogram will show too much signal during the first 30–60 bases (i.e., poor base calls) and less signal downstream resulting in shorter read lengths. Other types of PCR artifacts include nonspecific PCR products caused by having the primers annealing to nontarget areas of the genome. PCR artifacts are generally caused by poorly performing primers (Chapter 8).

The use of uncleaned extension products can also lead to poor sequence data because unincorporated dye-terminators will interfere with the sequencing process. If dye-terminators are not removed following the cycle sequencing reaction, then so-called "dye blobs" can appear on the chromatogram. Dye blobs, which are colored lines that "float" above other (often distinct) peaks and span one or several (or more) base positions, can prevent the accurate calling of the first 30–100 bases of a sequence as well as in other places on the chromatogram. As we will see later in this chapter, there are other causes of problematic chromatograms that are not due to the factors just described but instead are due to limitations of the Sanger method.

6.2 SANGER SEQUENCING PROCEDURES

Following the successful amplification of a target locus the remaining steps toward getting those products sequenced will involve (1) PCR product purification; (2) cycle sequencing; (3) cleanup and resuspension of extension products; and (4) loading cleaned extension products into an automated sequencer. If you plan to outsource your DNA sequencing, then you can skip all or some of these steps. For example, one option is to simply send your unpurified PCR products by mail to the sequencing provider where their technicians will perform all four steps. Alternatively, many researchers prefer to only perform step 1 before sending their products to be sequenced or they complete steps 1–3 before delivering their products to be sequenced. If you do choose to outsource your sequencing, then be sure to check with the sequencing provider on their requirements for sample submission. Let's now take a closer look at each of these steps.

6.2.1 Purification of PCR Products

There are various methods available for cleaning up PCR products. It is important to note that each method has advantages and disadvantages and so the best method for a given situation will depend on one or more factors. These factors include the quality of the PCR results, cost of reagents or kits, and the lab time required to perform the procedures.

6.2.1.1 Exo-SAP Treatment of PCR Products

Werle et al. (1994) developed a simple PCR purification method that only consists of two steps: pipetting a two-enzyme mixture into original PCR tubes or plate followed by a brief thermal incubation period to allow the enzymes to perform their functions. Although the Exo-SAP method is considered a form of PCR product "purification," it does not actually purify anything. Instead, this method enzymatically destroys unused primers and inactivates unused dNTPs. One of the enzymes is **Exo**nuclease I, which destroys single-stranded DNA. This enzyme is thus used to degrade unincorporated primers. The second enzyme is **S**hrimp **A**lkaline **P**hosphatase or "SAP," which dephosphorylates the excess dNTPs thereby preventing these molecules from participating in DNA synthesis reactions.

Exo-SAP has significant advantages over other methods including its low cost/sample, fast preparation time without the need of centrifugation (a 96 plate of PCR products can be treated in much less than an hour), and some commercial Sanger sequencing facilities will perform the Exo-SAP procedure on a 96 plate for very low cost/sample—even as low as $10 USD per 96 sample plate ($0.10/sequence). The primary disadvantage is that Exo-SAP cannot destroy or inactivate primer dimers or other double-stranded PCR artifacts when they are also present with the target products. If these double-stranded nontarget products are much shorter in length than the target product (e.g., the artifacts are <100 bp and the target is >500 bp), then you can still acquire quality sequence data. The numbers of ambiguous base calls in the beginning of a chromatogram will be determined by the length of the nonspecific PCR products. For example, most primer dimers are ~40–60 bp long, thus the first 40–60 bp of the sequence will consist of unreadable base calls. If the reverse sequence can be obtained, and if it reaches the opposite end of the product, then you can usually recover the lost bases and obtain the entire target sequence. Also, if the nontarget products have both PCR primers incorporated at the ends, then both the forward and reverse sequences will be affected. However, occasionally such products are formed only from one of the primers and so in those cases only one sequence will suffer while the other will be unaffected. The Exo-SAP enzyme mixture can be purchased as a ready-to-use "kit" or you can save money and buy the enzymes and chemicals as separate components and prepare the mix yourself.

6.2.1.2 Spin Column and Vacuum Manifold Kits for PCR Product Purification

One of the earliest available "user friendly" methods for preparing PCR products for Sanger sequencing consisted of the "spin column" and "vacuum manifold" kits. The former method consists of adding the PCR products to membrane-containing plastic spin columns, which are subsequently treated using various proprietary buffers and multiple rounds of centrifugation at high speed (~14,000 rpm). The final round of centrifugation leads to the recovery of the purified PCR products. The vacuum method also uses plastic membrane-containing columns but the buffers and DNA are drawn through the column using vacuum suction instead of centrifugation. Both types of kits yield PCR products that are free of unused primers, dNTPs, and, depending on the kit, also primer dimers and short nontarget products. These kits can be used in either single tube or 96-sample formats. The main drawback is the higher cost/sample.

6.2.1.3 20% PEG 8000 Precipitation of PCR Products

Polyethylene glycol (PEG) has long been a useful chemical in molecular biology with many different applications. PEG acts as a "molecular crowding agent," which has been used to dramatically improve the efficiencies of enzymatic reactions (Zimmerman and Pheiffer 1983) and size-select DNA molecules in solutions (Lis and Schleif 1975; Lis 1980; Rosenthal et al. 1993). In phylogenomics, PEG is a simple and inexpensive method used to purify PCR products and, as we will see in Chapter 7, size-select DNA fragments in NGS libraries.

The PEG method purifies PCR products through precipitation of target PCR molecules. A 20% solution of PEG 8000 (PEG molecular weight = 8000) and sodium chloride is used to fractionate the DNA in a PCR according to molecular mass (length in bp) with DNA molecules longer than 100–200 bp precipitating out of solution while *all* types of smaller products—single- and double-stranded—remain in the supernatant (Lis and Schleif 1975; Lis 1980; Rosenthal et al. 1993). Thus, after the supernatant containing most of the unused primers and nonspecific products is discarded, the remaining DNA in the tube should only be the target product. The remaining steps consist of two separate ethanol washes of the PCR products, drying of the products, and resuspension of the products using pure water. Other than its negligibly low cost per sample, the main advantage of this method is that it not only removes unused primers and dNTPs, but it also works superbly to remove *all* nontarget nucleic acid molecules such as primer dimers and nontarget PCR products that are <200 bp long (when using 20% PEG). Another advantage of this method is that for weakly amplified PCR products the target PCR product can be further concentrated if a smaller volume of water or buffer is used to resuspend the DNA pellet in the last step. Thus one strategy for obtaining a sufficient amount of target PCR product for samples that are weakly amplified is to redo a PCR with a much larger volume. The PCR products are then cleaned using the PEG procedure and the desired DNA is resuspended with a smaller volume of water thereby yielding a higher concentration sufficient for Sanger sequencing. Although 20% PEG works well when purifying PCR products of typical size (>200 bp), this method can be modified to purify products that are in the 100–200 bp range. To do this, one must optimize the concentration of the PEG solution for a given PCR product size (i.e., a less concentrated solution will retain smaller sized nucleic acids).

Of the methods described so far, the PEG precipitation method is the most inexpensive in terms of cost/sample, but it is substantially more labor intensive unless a high-throughput protocol for 96 sample PCR/sequencing plates is followed. Although the PEG 8000 method may seem like an old-fashioned molecular biology protocol, this method enjoys some major advantages over more modern kit-based methods of PCR purification and it plays an important role in the following PCR clean up method.

6.2.1.4 Solid-Phase Reversible Immobilization Beads

Hawkins et al. (1994) reported that DNA can reversibly bind to the surface of paramagnetic beads in a solution with a high concentration of PEG 8000 and salt. When these carboxyl-coated beads (1 μm diameter) are bound to DNA and are later placed close to a magnet, they can effectively isolate the target DNA, which is then washed, dried, and eluted. This method was termed *solid-phase reversible immobilization* or "SPRI" because the beads, which are the "solid phase," and the DNA can be first bound together and then unbound simply by exchanging buffer solutions (Hawkins et al. 1994). Using this principle DeAngelis et al. (1995) developed an SPRI-based method for cleaning PCR products.

The SPRI method is easily applied to 96 PCR products in a microplate. Typically, SPRI beads and PEG buffer are added at 1.8× the amount of PCR products (e.g., 36 μL SPRI beads/buffer + 20 μL PCR product). A synopsis of the procedure follows. First, the paramagnetic beads are brought to room temperature before they are mixed with PCR products and a 20% PEG 8000/2.5 M NaCl buffer solution. During a brief incubation period the beads and DNA bind to each other in solution. The PCR plate is then placed onto a magnet, which "pulls" the bead/DNA complexes to the sides of the plate's wells nearest to the magnet. The supernatant, which still contains unused primers and dNTPs as well as primer dimers (if present), is then pipetted away and discarded. Next, the beads are washed twice with 80% ethanol (70% is often used as well). The beads and PCR products are allowed to air-dry for several minutes before the microplate is removed from the magnet. Water or elution buffer is added to the plate's wells, which liberates the PCR products from the beads. The microplate is then placed back onto the magnet, which immobilizes the beads thereby allowing the supernatant containing the purified PCR products to be saved for sequencing. Interested readers should consult the original step-by-step protocol in DeAngelis et al. (1995) for more details.

In order for the SPRI bead cleanup procedure to yield the best results it is essential to do the following. First, ensure that the beads are brought to room temperature and are fully resuspended (e.g., using a vortexer) before combining them with the PCR products. Second, use the pipette to thoroughly mix the beads and PCR products in order to obtain maximal binding efficiency. Third, use only freshly made ethanol in the wash steps (usually this will be at a 70% concentration). Fourth, while the beads are "on magnet," do not disturb them while adding the wash buffers; that is, carefully and gently pipette the liquid into each well and not directly onto the beads. Lastly, be sure that all ethanol has evaporated from the wells before adding the water or elution buffer otherwise the ethanol will inhibit subsequent enzymatic steps (e.g., PCR).

The SPRI method has many advantages. Hawkins et al. (1994) and DeAngelis et al. (1995) noted that SPRI-based methods are readily amenable to high-throughput sample workflows (i.e., highly automatable). Furthermore, there are no centrifugation or filtration steps and it is relatively inexpensive on a cost/sample basis. Another major advantage is that SPRI clean up of PCR products results in elimination of all small DNA molecules including primer dimers and other short nonspecific products (Lis and Schleif 1975; Lis 1980; Rosenthal et al. 1993; Quail et al. 2009).

Although the SPRI method was deemed early on to be an inexpensive method (DeAngelis et al. 1995), the high cost of commercially available SPRI beads has limited its use thus far especially given other alternatives such as Exo-SAP. However, Rohland and Reich (2012) showed that it was possible to prepare batches of "homemade" SPRI beads and PEG/NaCl buffer at low cost, which perform as well as AMPure XP beads (Agencourt, Beckman Coulter), the market standard (Faircloth 2012; Ford 2012; Rohland and Reich 2012). The high performance of homemade SPRI beads/buffer is not surprising given the early observations by Hawkins et al. (1994) that the DNA-binding efficiency is 100% when beads are in excess and remains high at around 80% following the washing steps. The Rohland-Reich (2012) SPRI preparation protocol, which was elaborated into a detailed step-by-step protocol by Faircloth and Glenn (2011), can be found on the following website: https://ethanomics.wordpress.com/2012/08/05/homemade-ampure-xp-beads/. Given the simplicity, effectiveness, and newfound availability of inexpensive SPRI beads/buffer, this method is certainly going to be used far more frequently in various phylogenomic methods (Faircloth 2012). Indeed, as we will see in Chapter 7, SPRI technology is being extensively used in next generation sequencing workflows.

6.2.1.5 Gel Purification of PCR Products

Agarose gel electrophoresis can also be used to isolate target PCR products, particularly those that are "messy" in that they contain many different nontarget products. Similar to the PEG 8000 method, though via a different mechanism, agarose electrophoresis can separate different DNA molecules by their length, which means unused primer, dNTPs, and nontarget PCR products can be physically separated from the target product. In this procedure, the entire PCR is first run through a 2% agarose gel. Following electrophoresis, the target bands are excised from the gel and the DNA fragments recovered using a gel purification kit (i.e., spin column format such as the QiaQuick kit, Qiagen). Although this method is quite effective at isolating the target DNA, it is not practical for general use owing to high costs in terms of purchasing gel purification kits, consumption of other lab supplies (e.g., plastics, agarose, etc.), and the amount of time the procedure requires.

6.2.1.6 *Which PCR Product Purification Method Is Best?*

If we look at Table 6.2 for a comparison of the aforementioned methods, we can immediately see why Exo-SAP has become the PCR purification method of choice: it is inexpensive and fast. The only drawback to Exo-SAP is that it is unable to destroy or remove small dsDNA artifacts such as primer dimers and PCR products that are less than 200 bp. However, if your target product is less than ~800 bp long, then you can overcome this problem simply by sequencing your product in both directions; that is, each sequence can fill in the correct base calls rendered ambiguous by the cosequencing of the artifacts. Given that the SPRI beads method is also quick and easy along with the relatively newfound protocol for making homemade SPRI beads and buffer, this method of PCR clean up now represents a highly attractive option especially when primer dimers are a concern and thus its use may surge in the future. Although some spin column and vacuum manifold kits can effectively remove these nonspecific artifacts, the cost per sample is far higher than using Exo-SAP or homemade SPRI beads and buffer. PEG 8000 and gel purification can effectively remove short PCR artifacts in a cost-effective manner but they are labor intensive compared to the other methods. For most routine PCRs, Exo-SAP followed by SPRI beads represent the best available options.

6.2.2 Setting Up Cycle Sequencing Reactions

Setting up cycle sequencing reactions is similar to setting up a PCR. You only need to combine purified PCR product template, sequencing primer, and a sequencing master mix containing dNTPs, dye terminators, buffer, and *Taq* polymerase. Note, that the concentration of the sequencing primer is usually less than the concentration of a PCR primer. In Chapter 5, we saw that a typical working concentration for a PCR primer is 10 μM, but for cycle sequencing it is usually in the 3–5 μM range. Some PCR or cycle sequencing protocols specify the primer concentration in units of picomoles/μL (pmol/μL). Or protocols may state that the total amount of sequencing primer should be X pmol per cycle sequencing reaction. Thus, it is useful to keep in mind the following conversion: 1 μM = 1 μmole/L = 1 pmol/μL. Cycle sequencing reactions can be performed in individual 0.2 mL plastic tubes or 96-well microplates. The sequencing master mix is sold by the manufacturer of the automated sequencer. Although some differences exist between thermocycling programs for PCR versus cycle sequencing, both involve the basic denaturation, annealing, and extension steps and a similar number of cycles (30–40 cycles). However, be sure to use the correct thermocycling program for a cycle sequencing reaction.

6.2.3 Purification of Extension Products

When the cycle sequencing reactions are completed, the extension products must be purified to remove unincorporated dye terminators and primers. Several methods are commonly used including ethanol (EtOH) precipitation and sephadex (G-50 grade) spin columns. The EtOH method is advantageous because of its lower cost per sample, however, the quality of sequence data tends to be more variable than using spin columns. Sephadex columns are more expensive

TABLE 6.2
Comparison of methods for purifying PCR products

Method of PCR purification	Reagent costs (per sample)	Labor cost (per sample)	Reduces or eliminates small dsDNA artifacts
Exo-SAP	Low	Low	No
Spin column	High	Low	Yes
Vacuum manifold	High	Low	Yes
PEG 8000	Low	High	Yes
SPRI beads[a]	Low	Low	Yes
Gel purification	High	High	Yes

[a] Indicates that the SPRI beads and associated PEG/NaCl buffer are homemade (see section 6.2.1.4).

than EtOH but this method consistently yields the best quality sequence data. Lab costs can be reduced if the sephadex powder is purchased separately and the plastic columns are reloaded for use. The sephadex method can be used for single tubes or in a 96 plate format. The final step in sample preparation prior to loading onto a capillary sequencer is resuspending the dried purified extension products with a resuspension or "sequencing" buffer.

6.2.4 Sequencing in a Capillary Sequencer: Do-It-Yourself or Outsource?

Once the extension products have been purified and resuspended in the appropriate sequencing buffer they are ready to be loaded into a capillary sequencer. After the sequencer door is shut, there is little more to do than specify a number of things on the desktop computer, which controls the sequencer. At the conclusion of a sequencing run, the computer creates a folder containing all sequence files (i.e., chromatograms). These machines are capable of sequencing 96 samples in less than a day.

Although capillary sequencers are easy to operate, they cost hundreds of thousands of US dollars (USD) to purchase plus demand yearly maintenance costs on the order of tens of thousands more USD. This means that many labs are simply unable to afford them. However, with the introduction of commercially available Sanger sequencing services starting in the early 2000s, this outsourcing option suddenly made the power of DNA sequencing available to anyone. This represented a key innovation because it liberated labs from having to purchase and maintain expensive sequencers and hire technicians to run and maintain the machines. In effect, a person could set up his or her own DNA lab in a small room for mere tens of thousands of dollars simply by creating a clean work environment equipped for DNA extractions and PCR; the PCR products can be sent by mail for sequencing with sequence data downloaded from the Internet. This is why today most DNA labs outsource their DNA sequencing needs—even sending their PCR products overseas to be sequenced. In addition to increased availability of DNA sequencing services, the cost per sample has dropped as well. For example, in 1997 the cost for one sequence, which included PCR cleanup, cycle sequencing, and running on a capillary sequencer, was approximately $16 USD, whereas by 2007 the price dropped down to at least $3 USD per sequence. If you choose to send your PCR products via international mail, then be sure to follow all rules and regulations for export and import of your samples as laws vary by country.

6.3 HIGH-THROUGHPUT SANGER SEQUENCING

In the DNA extraction and PCR chapters we discussed high-throughput strategies for increasing the number of samples that are processed at the same time. We will now consider two additional high-throughput strategies that can dramatically facilitate the sequencing of large numbers of PCR products and save money.

6.3.1 Sequencing 96 Samples on Microplates

Regardless whether you performed your PCRs in a single 96 sample plate or in individual tubes it is still worthwhile to submit your PCR products for sequencing on a 96-well microplate. This can be a particularly good strategy if you intend to outsource your samples for sequencing. Companies that offer a Sanger sequencing service often use robotic equipment to process samples for sequencing. Thus having your samples already arranged on a plate will greatly facilitate the handling of your samples. Note that usually "skirted" microplates are used in this case but check with the provider to be certain. The outsourcing sequencing option avails you with at least two choices. You can simply send a microplate full of unpurified PCR products, which their technicians can treat with Exo-SAP, perform the cycle sequencing reactions, and then run on a capillary sequencer. Alternatively, you can perform the PCR cleanup yourself then send a plate of cleaned PCR products for sequencing. In any case, sending your PCR products on a 96 well plate, instead of single tubes, should reward you with significant price discounts. A last important tip: before you send a plate by mail, be *extra careful* about sealing your plates—if in doubt, contact the sequencing lab for advice on which plate brands to use and proper shipping methods. Plates that are not properly sealed can leak during shipment resulting in partial or entire loss of your samples.

6.3.2 Adding Sequencing Primer "Tails" to PCR Primers

The plate-based approach to sequencing PCR products is easy and works well provided two conditions are satisfied: (1) the sequencing primer is known to work properly in the cycle sequencing reaction and (2) you use the same sequencing primer in all 96 cycle sequencing reactions that will be on the same plate. The first condition is satisfied only after a primer has been tested in a cycle sequencing reactions and proven to yield good quality sequences. Although perhaps the majority of typical PCR primers that function well in PCR may also work well in sequencing, this is not always the case. The second condition is easily satisfied if all the PCR products used the same pair of PCR primers. For example, if you are going to sequence 96 products with the forward PCR primer, then you can simply use a multichannel pipette to dispense the forward primer to all 96 cycle sequencing reactions. Alternatively, you can add the sequencing primer to a separate 96 well plate and send both the PCR and primer-containing plates to the sequencing provider (some sequencing providers even allow you to simply send the sequencing primer in a 1.5 mL microcentrifuge tube). However, if more than one pair of primers was used in the PCRs, then special care must be used to ensure that the correct sequencing primers are used in the cycle sequencing reactions. If you will be sending your PCR plate to a sequencing provider so that they can perform the cycle sequencing reactions, then you will need to prepare a special primer plate that has the sequencing primers placed correctly in the primer plate so that they can later be combined with their appropriate PCR templates in the sequencing reactions. If a given plate contains PCR products generated from many different primer pairs, then obviously this greatly complicates the process due to the primer plate preparation. Indeed, making a primer plate with multiple sequencing primers is a tedious and time-consuming task that can easily lead to errors that result in failed sequences.

To address these dilemmas, a nice trick was developed to make plate-based sequencing easy and fast. This method involves adding an actual *proven* sequencing primer to the 5′ end of a normal PCR primer at the time the primer is synthesized. This results in a compound or "tailed" primer that has a sequencing primer located at the 5′ half and a PCR primer located on the other (3′) half. Although in principle the sequencing primer could be any primer that is known to function without any problems during DNA sequencing, in practice most researchers have used so-called "M13 universal primers." M13 universal primers were originally developed for sequencing insert DNA in cloning vectors for shotgun DNA sequencing (Messing et al. 1981; Vieira and Messing 1982; Table 6.3). However, more recently these special sequencing primers have been co-opted for use as PCR product sequencing primers (e.g., Dinauer et al. 2000; Ivanova et al. 2007). Although any of the primers listed in Table 6.3 could be useful sequencing tails, it is recommended to carefully read Section 6.3.2.3, which explains potential pit

TABLE 6.3
List of M13 universal primers for DNA sequencing

Primer name	Sequence (5′ → 3′)
M13 forward (−20)	GTAAAACGACGGCCAGT
M13 forward (−21)	TGTAAAACGACGGCCAGT
M13 forward (−40)	GTTTTCCCAGTCACGAC
M13 forward (−41)	GGTTTTCCCAGTCACGAC
M13 reverse (−27)	CAGGAAACAGCTATGAC
M13 reverse (−29)	CAGGAAACAGCTATGACC
M13 reverse (−20)	GCGGATAACAATTTCACACAGG
M13 reverse (−49)	CAGCGGATAACAATTTCACACAGG

NOTE: Similar primers are grouped together and aligned to facilitate pairwise sequence comparisons. The original M13 universal sequencing primers are from the work of Messing et al. (1981) and Vieira and Messing (1982).

falls associated with using M13 tails and how to avoid them.

6.3.2.1 How an M13-Tailed Primer Functions in PCR

In order to better understand how an M13-tailed PCR primer can simplify Sanger sequencing, we must first consider how an M13-tailed primer functions in PCR. Recall the PCR primers we examined in Figure 5.7, which generated a 540 bp product. Let's say we wish to add the M13 forward (−21) and M13 reverse (−29) sequencing primers listed in Table 6.3 to these PCR primers. By convention the forward PCR primer will have an M13 forward (−21) tail in our example (Figure 6.7a), whereas the reverse PCR primer will have an M13 reverse (−29) tail (Figure 6.7b). Although adding such primer tails to proven PCR primers may lead to unwanted primer–primer interactions with the end result being poor PCR results, thankfully this seems to seldom occur.

Let's now look at what happens when these tailed primers are involved in synthesis during the extension phase of a PCR. Once the M13-tailed primers have annealed to their respective target templates (which were synthesized in some previous cycle), DNA polymerase begins synthesizing new strands of DNA starting at the 3′ end of the primers (Figure 6.8a). At the end of this extension phase, new double-stranded products have been made each of which is 576 bp long (Figure 6.8b). Thus when M13-tailed primers are used, the PCR should yield target products that are ~38–50 bp longer than products generated using the original (tail-less) PCR primers; the difference explained by the combined lengths of the forward and reverse tail primers. An important thing to notice is that not only have the entire M13-tailed primers been incorporated into each newly synthesized strand, but priming sites for the M13 sequencing primers are also present in the products, which will later be used during the cycle sequencing reaction. Accordingly, during the first several cycles of PCR when the reaction mixture is dominated by target templates located on genomic DNA, the true PCR primers (not the tails) drive the synthesis of new strands because the genomic templates naturally do not contain binding sites for the M13 tails. However, in later cycles when the mixture is dominated by synthetic target templates—which now contain the target templates for the original PCR primers + their attached M13-tails, the entire M13-tailed primer can now bind to the templates and drive synthesis.

6.3.2.2 Cycle Sequencing and M13 Primer Tails

From our earlier discussion of sequencing nontailed PCR products, remember that a nontailed PCR product will have, at each end of the double-stranded molecule, strings of sites that are complementary to the forward and reverse PCR primers (see Figures 5.7 and 6.2 for review). These primer-binding sites are then used in the cycle sequencing reaction when one of the PCR primers is used as a sequencing primer (see Figure 6.2). If M13-tailed primers are used to generate PCR products, then the double-stranded amplicons will have sites complementary to both the PCR primer and M13 primer tail, as is shown in Figure 6.9a. However, in the downstream

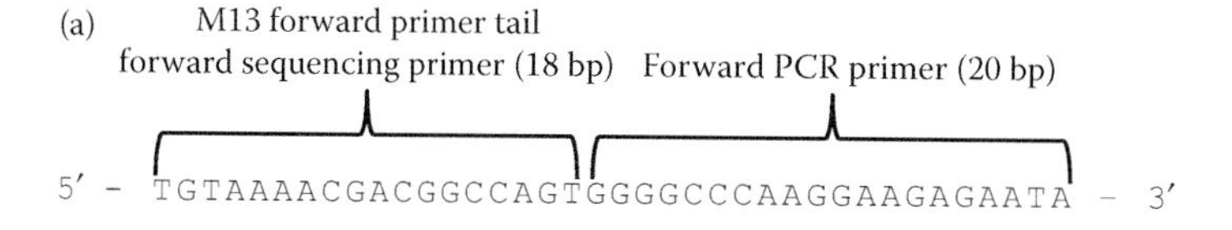

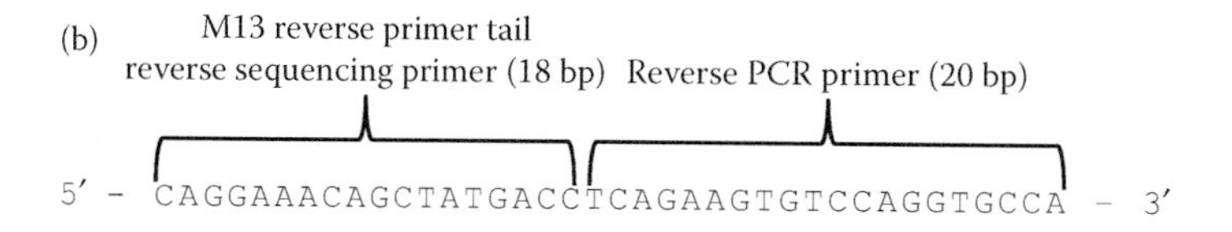

Figure 6.7. M13 universal primer tails added to normal PCR primers. (a) The forward sequencing ("tail") primer is added to the 5′ end of the forward PCR primer. (b) The reverse sequencing ("tail") primer is added to the 5′ end of the reverse PCR primer.

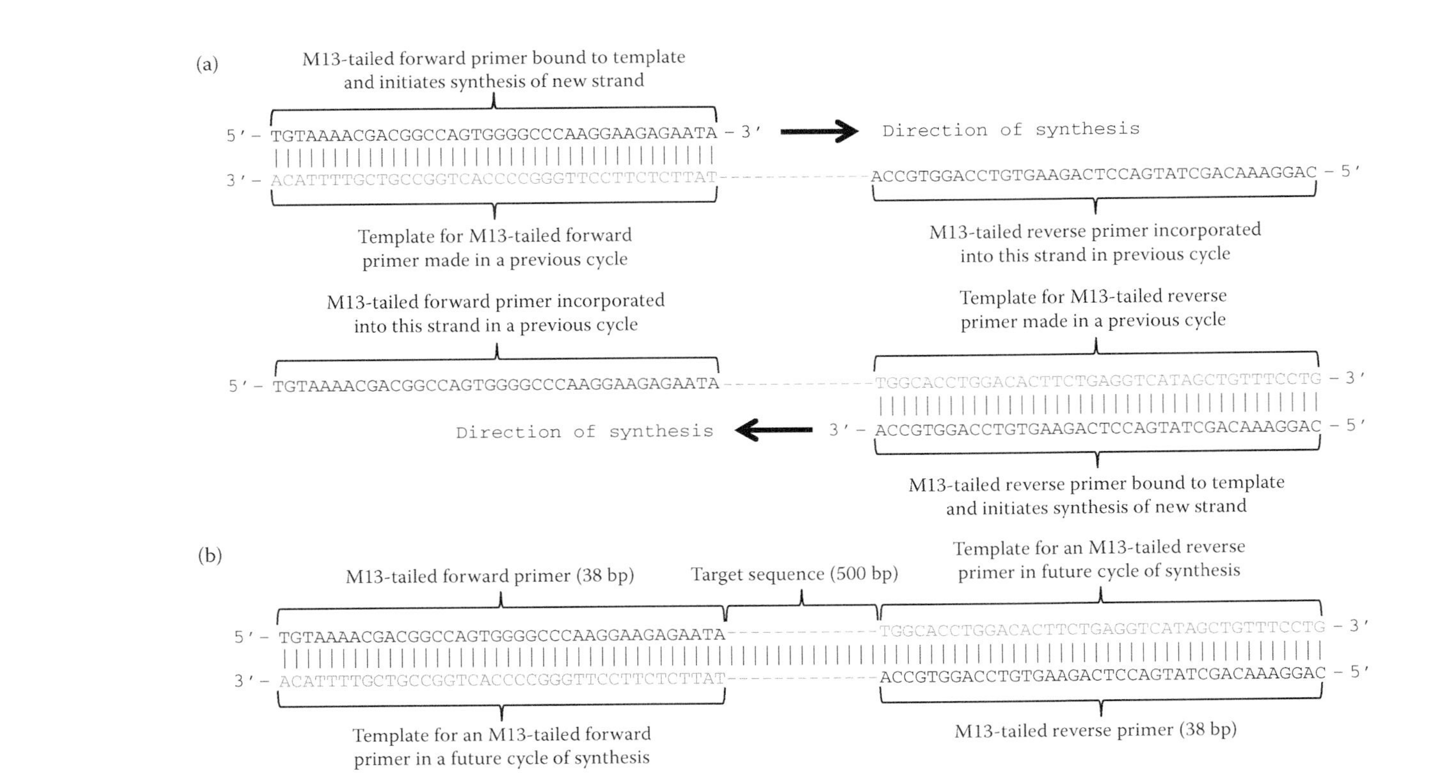

Figure 6.8. Synthesis and anatomy of a 576 bp PCR product using M13-tailed PCR primers. (a) Synthesis of two products from two complementary strands during the extension phase (after many cycles). (b) Anatomy of an M13-tailed PCR product. For clarity the target sequence is indicated as dashes. Note, most of target sequence is not shown due to lack of space in the figure. Vertical bars show hydrogen bonding between complementary bases. Gray represents strands of DNA synthesized during PCR and the M13-tailed forward and reverse primers are black. PCR primers are same as shown in Figure 6.7.

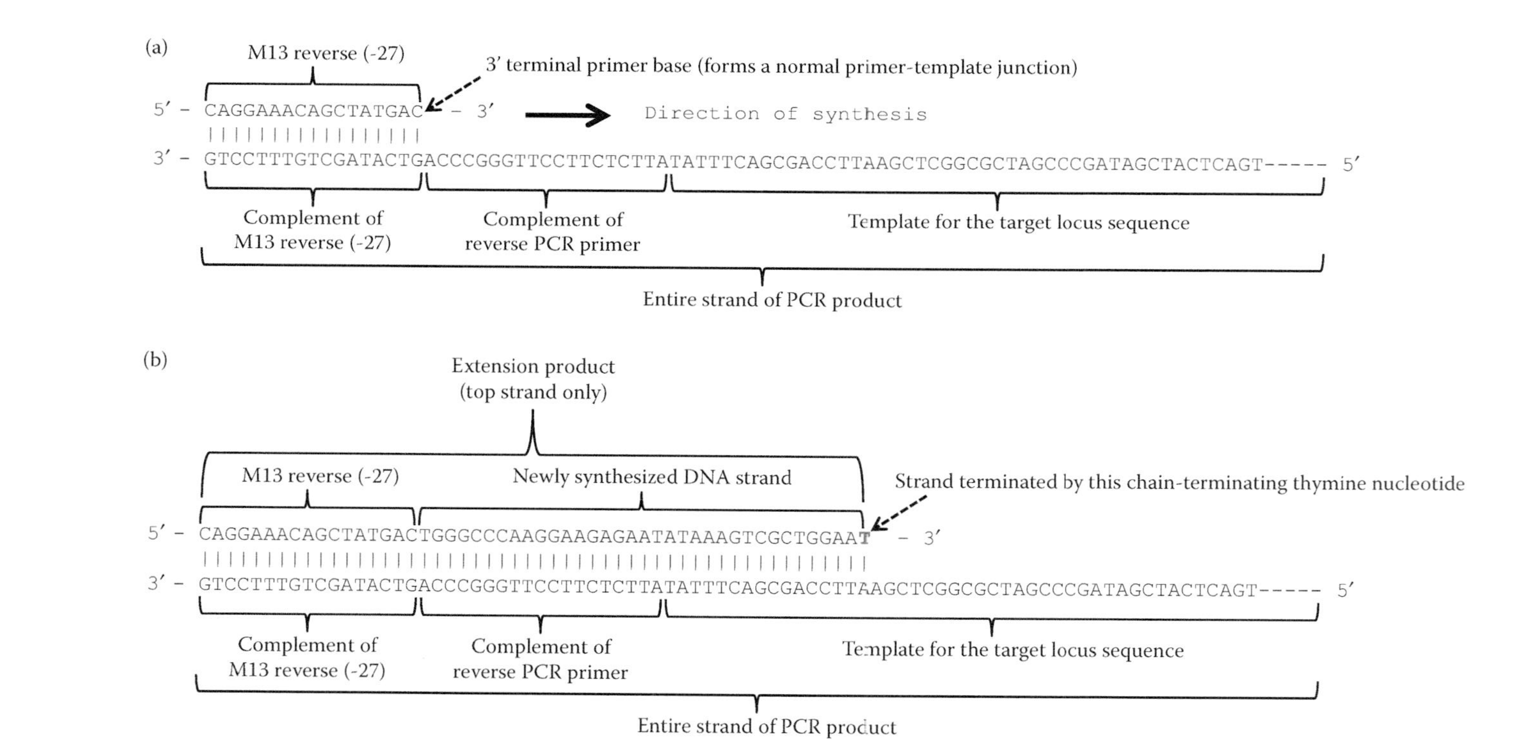

Figure 6.9. Synthesis of an extension product using an M13 universal primer. (a) An M13 reverse (−27) sequencing primer anneals to its complementary sequence on one strand of the PCR product. The match between primer and template is perfect because the same M13 primer was used as a "tail" in the PCR. Notice that the 3′ terminal base of the M13 primer forms a normal primer–template junction from which *Taq* polymerase can start synthesizing the extension product. (b) Formation of the extension product is completed once a ddNTP becomes incorporated into the new strand. Here, a chain-terminating thymine (shown in in gray) abruptly stopped synthesis.

cycle sequencing reaction, only the correct M13 primer should be used as the sequencing primer (e.g., Figure 6.9a). An interesting consequence of sequencing with an M13 primer is that the extension product will include the actual PCR primer sequence ahead of the target sequence (Figure 6.9b), and hence a chromatogram of this sequence may show part or all of the primer at the start of the sequence. Other than this difference between nontailed and tailed PCR products, the process of generating extension products is the same. Thus when the M13 sequencing primer anneals to its appropriate location on the PCR template strand (Figure 6.9a), *Taq* polymerase will start extending the new strand using dNTPs as substrates until one of the four possible ddNTPs (a ddTTP in this case) becomes incorporated at the 3′ end of the strand (Figure 6.9b). It is critical to use forward and reverse M13 tails for the forward and reverse PCR primers, respectively or only a single tailed primer per PCR. If instead the same M13 tail is used on both forward and reverse PCR primers, then two conflicting sets of extension products will be in the cycle sequencing reaction with the result being a ruined sequence.

6.3.2.3 On the Importance of Matching Sequencing Primers

As we saw in Table 6.3 there are a variety of M13 sequencing primers that could be used as PCR primer tails. Note that some of these sequencing primers only differ by one base such as the M13 forward (−20) and M13 forward (−21) primers. Although you may have some flexibility in which M13 primers you use as PCR primer tails, the important thing is for you to *exactly* match your M13 tails to the sequencing primers that you will use to sequence those same products. If you choose one type of M13 primer to be a PCR primer tail and later sequence your PCR products with a different M13 primer, then you run the risk of having a failed sequencing reaction due to mismatches between the PCR product and sequencing primer. Even sequencing primers that only differ by one base from your PCR tail primer *could* be the difference between success and failure. Needless to say, having all 96 of your sequences fail is a costly mistake to make. We will now take a closer look at the consequences of using a PCR primer tail that differs from the sequencing primer.

Let's say a researcher used an M13 reverse (−29) primer as a tail attached to a reverse PCR primer, but then used an M13 reverse (−27) primer in the cycle sequencing reaction to sequence this product. Notice in Table 6.3 that these two M13 primers only differ by a single base at the 3′ end. In this particular scenario, which is depicted in Figure 6.10a, despite the sequencing primer being one base shorter than the stretch of complementary sites created by the M13 reverse (−29) primer, a normal primer–template junction is formed. Assuming the sequencing primer is still able to firmly hybridize to the binding sites, the DNA polymerases in the reaction should have no problem generating extension products and thus the reaction is expected to proceed in normal fashion.

Now let's see what happens if we change the scenario by reversing the aforementioned tail and sequencing primers. As you can see in Figure 6.10b, the M13 reverse (−29) sequencing primer is one base longer at the 3′ end than the stretch of complementary binding sites generated by the M13 reverse (−27) PCR tail. Because the 3′ terminal primer base is a cytosine base, which is not complementary to the adenine base on the PCR template strand (Figure 6.10b), this creates a rather unfortunate mismatch. This is probably one of the worst kinds of mismatches at the primer–template junction because, as noted by Palumbi (1996), cytosine cannot form any hydrogen bonds with adenine. The consequence of this mismatch will be severe because the efficiency of primer extension by the DNA polymerases will be so low that the end result will be failed sequences (i.e., useless chromatograms). I have observed this same type of failure before. To confirm that the problem was indeed the wrong sequencing primer, the same plate of M13 reverse (−27) tailed PCR products was sequenced again but this time using the same M13 primer. Nearly all 96 of the resulting sequences were high quality in the second sequencing attempt thus verifying the cause of the earlier failure.

In the scenario we just looked at, the abnormal primer–template junction was due to a mismatch between the 3′ terminal primer base and the first template base. What if by chance the first template base was a guanine instead of an adenine? As you can see in Figure 6.10c, in this scenario the primer–template junction would of course be normal because of the Watson–Crick base pairing

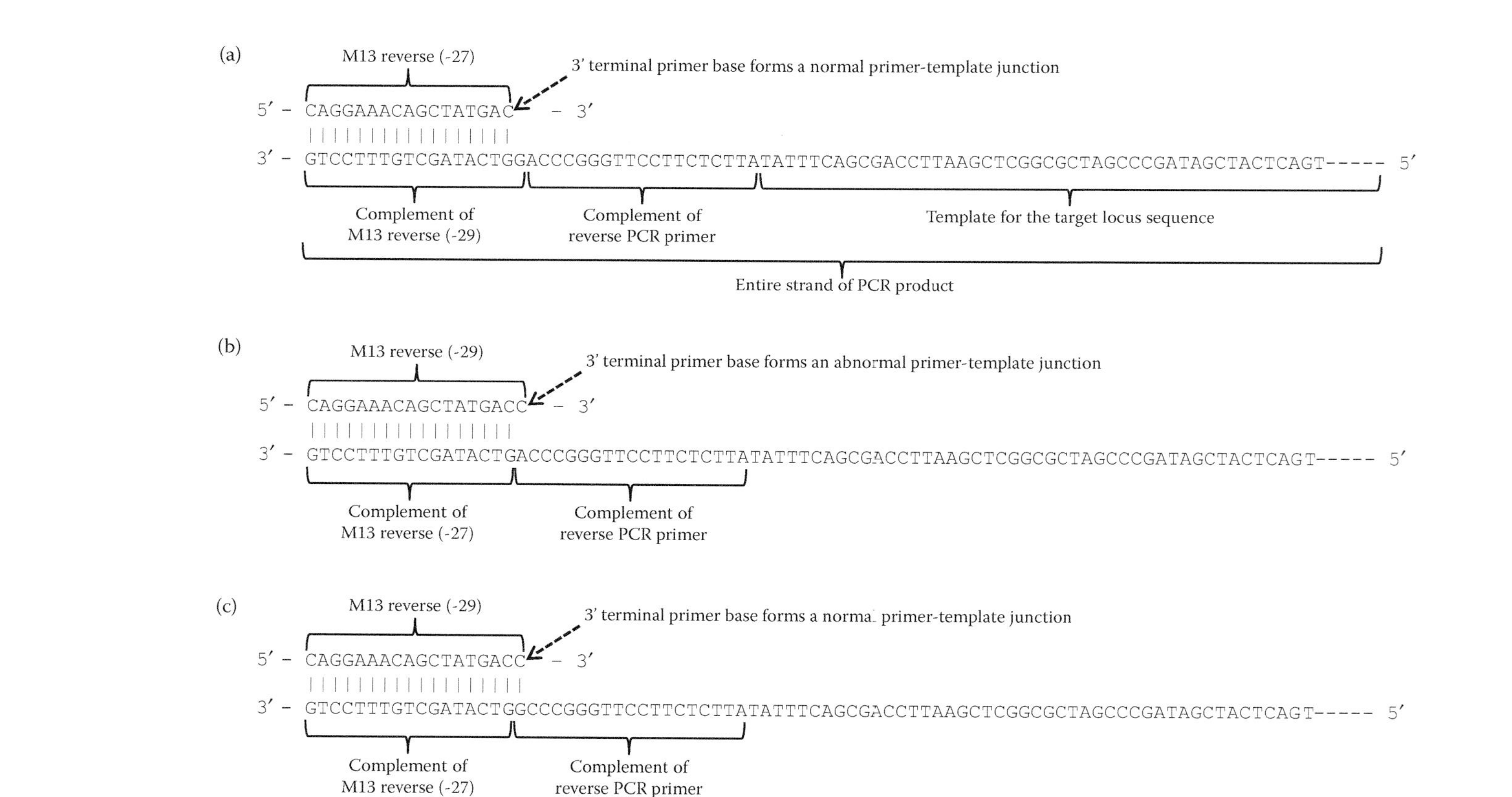

Figure 6.10. Consequences of not matching M13 primers in the PCR and cycle sequencing reactions. (a) Using an M13 reverse (−27) sequencing primer to sequence a PCR product containing an M13 reverse (−29) primer tail. Despite the sequencing primer having one less base at its 3′ end than the complement to the other M13 primer, a normal primer–template junction can form. (b) Using an M13 reverse (−29) sequencing primer to sequence a PCR product containing an M13 reverse (−27) primer tail. An abnormal primer–template junction arises between the cytosine base located at the 3′ terminal end of the primer and the adjacent adenine base in the template base position. (c) Same as previous but the template base is now shown as a guanine base. In this scenario, a normal primer–template junction can form. Note, the 5′ end of the PCR template strands are truncated (at the dashes) due to space limitation in the figure. Vertical dashes represent hydrogen bonding between complementary base pairs and M13 universal primers shown are from Table 6.3.

between cytosine and guanine. Thus, a researcher who uses a sequencing primer that differs from the tailed primer could still obtain excellent sequencing results but this would be by dumb luck. If the same researcher changes PCR primers and repeats the same tail-sequencing primer mismatch, then it is likely that eventually a major sequencing failure would occur. Note also that a primer tail and sequencing primer can differ from each other at the 5′ end and still produce the desired sequences provided the correct primer–template junction is formed at their 3′ ends.

The critical message here is to use the exact same M13 primers for both tailing your PCR primers and as the sequencing primers in order to avoid sequencing problems. If you are outsourcing your Sanger sequencing and you request that the sequencing provider uses their stock of M13 sequencing primers to sequence the PCR products, then it will be your responsibility to ensure that the M13 sequencing primers they use will exactly match your M13-tailed PCR products.

6.3.2.4 *Benefits of Using M13-Tailed Primers*

There are several advantages to using M13-tailed PCR products. First, M13 primers are proven to sequence well, whereas not all PCR primers sequence well even if they perform well in PCR. Secondly, PCR products that were generated using different pairs of PCR primers (i.e., multilocus study) can all be sequenced together on the same microplate as long as they have the same primer tails. Thirdly, if outsourcing an entire microplate of tailed PCR products, then many commercial sequencing facilities will supply the M13 sequencing *free of charge* thus saving you the expense of purchasing your own M13 primers. Lastly, recall that earlier in this chapter we saw that the first 10–30 bases of a Sanger sequencing chromatogram typically are of low quality and not reliable. However, if we instead sequence a tailed PCR product, then the extension products from the cycle sequencing reaction will represent the sequence of the PCR primer and not the locus of interest. Thus, the poor quality base calls in the first part of a chromatogram will represent primer sequence (which is unimportant) and thus the target sequence will likely be represented by higher quality base calls. This means that for PCR products that are less than about 700 bp and free of primer dimers and other coamplified products, it may be possible to obtain high quality sequences from just one sequence read per PCR product.

6.4 HAPLOTYPE DETERMINATION FROM SANGER SEQUENCE DATA

Obtaining DNA sequence data from haploid loci such as mitochondrial genes using the Sanger methodology is straightforward. However, Sanger sequencing of diploid (or worse, polyploid loci) presents a serious complication to the investigator. What is this complication? First, consider the simple case of sequencing any haploid locus such as a mitochondrial locus. In this scenario the output chromatograms (from forward or reverse sequencing primers) should show high-quality single peaks for each nucleotide site in the locus as in Figure 6.5. Now think about sequencing any diploid autosomal locus. If the individual being sequenced happens to be homozygous at all sites in this locus, then the resulting chromatogram should resemble the basic form observed in Figure 6.5. However, what if the individual is instead heterozygous for any of the sites? What are the implications for Sanger sequencing in such a scenario? Let's now look at this important issue in detail.

6.4.1 PCR Amplification and Sanger Sequencing of Diploid or Polyploid Loci

When a haploid locus from the mitochondrial genome or human Y-chromosome is amplified via PCR, millions of copies of a *single* allele or haplotype are made. This assumes, of course, that paralogous copies are not coamplified with the orthologous haplotype. However, it is important to realize that when a diploid or polyploid locus is similarly amplified, *all* orthologous haplotypes will likely be copied. This means that copies of *multiple haplotypes will be produced in a single PCR* (Clark 1990).

For example, let's consider two different human autosomal loci that were PCR-amplified in two separate reaction tubes. Again we will consider the simplest scenario in which no paralogous copies of these loci occur elsewhere in the genome. Note that without reference to comparable sequence data from the parents, we cannot hope to ascertain which alleles in our PCR tubes were inherited from the mother and which were from the father but these details are unimportant. However, given that each cell contains one genomic copy from each parent, we can assume that the two *parental haplotypes* each account for approximately 50% of the total PCR products in each reaction tube (Figure 6.11). In our example, the parental haplotypes for each locus are heterozygous but in different ways. In the first locus, the two haplotypes differ only by a single SNP site but are otherwise the same length (Figure 6.11a), whereas in the second locus the two haplotypes differ by their length due to the presence of an indel (Figure 6.11b). While it is generally not a problem to amplify each parental haplotype for a given locus in a PCR, such PCR products do present additional complications when they are sequenced via the Sanger method.

Recall from Chapter 5 we considered the simplest case of PCR and Sanger sequencing, which is to amplify and sequence a haploid locus such as a mitochondrial or human Y-chromosome locus. When such loci are sequenced, you expect to see chromatograms consisting of only single and well-defined peaks. Likewise, a diploid locus that is homozygous in an individual should also generate a chromatogram exhibiting only single peaks even though there were, in actuality, *two different sets of extension products* produced—one set from each parental haplotype. Such an occurrence is not a concern for us because the two parental haplotypes are the same.

Now let's return to our earlier example from Figure 6.11 and consider what would happen if we sequence each of the heterozygous PCR products. When the PCR products from Figure 6.11a are cycle sequenced we would expect the reaction to generate two sets of extension products reflecting each of the two parental haplotypes in a 50:50 manner as shown in Figure 6.12. As these extension products become sorted by their size in the capillary sequencer, the products of equal length will each have their terminal (dye terminator) nucleotide "read" at the same time leading to a base call for that site on the chromatogram. All homozygous extension products will have matching signals at each site and therefore only single peaks will appear on the chromatogram. However, for the extension products with the polymorphic site (#380) located at the terminal labeled nucleotides, two overlapping peaks (green and black) will show on the chromatogram at that position reflecting the 50:50 mixture of terminally located adenine and guanine bases (Figure 6.12). In this situation, the computer may specify the base reflecting the taller of the two peaks, as was the case here, or it may instead call an "N" for unknown base. We thus see that sequencing a PCR product that is heterozygous for a SNP site can still

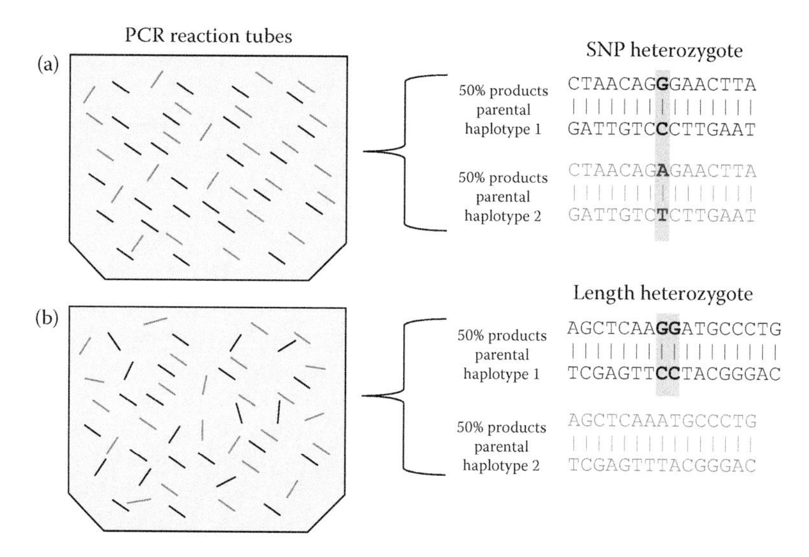

Figure 6.11. PCR of a diploid locus results in a mixture of parental products. (a) PCR involving a locus heterozygous for a single SNP site. One parental product (black) has a G:C base pair while the other parental product (light gray) has an A:T base pair. (b) PCR on a locus heterozygous for a 2-bp indel. The black parental product contains two consecutive G:C base pairs, which are missing from the light gray parental product. Polymorphic sites are shaded in dark gray and vertical dashes represent hydrogen bonding between complementary bases.

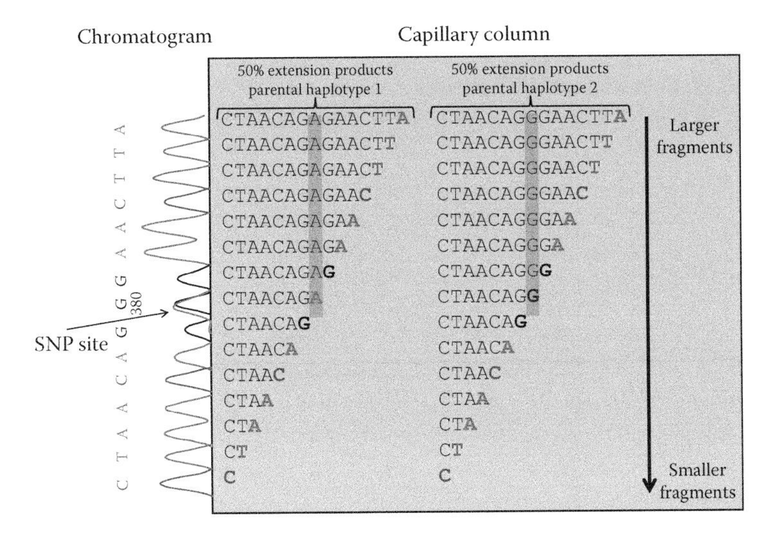

Figure 6.12. Sanger sequencing of a PCR product that is heterozygous at a single SNP site. The dark shaded rectangle shows the SNP site and represents the only polymorphic site on this sequence. SNP is located at site #380 on the adjacent chromatogram where a green and black double peak is located. A guanine base (black peak) was called but an adenine base (green peak) is also evident.

produce a high quality chromatogram. Indeed, even though quality scores for these heterozygous sites are usually low, the two overlapping peaks are usually distinct enough that the existence of each allelic base is not questionable.

When evaluating chromatograms generated for diploid loci one does not expect to see a large number of SNPs clumped together; instead they tend to be dispersed as single polymorphic sites that occur every tens or hundreds of sites along a sequence. Thus, if you observe two, three, or more double-peaks that occur on successive sites, then most likely they are not SNPs, but instead are artifacts of poor quality sequence. This is often observed on one chromatogram based on one sequencing primer, while the chromatogram containing the sequence generated from the other sequencing primer shows only sharp and distinct single peaks in the locations of the double-peaks seen on the other chromatogram.

A cycle sequencing reaction of the PCR products from Figure 6.11b will likewise produce two sets of extension products, which will be sorted by their length in the capillary sequencer (Figure 6.13). Notice that the extension products representing each parental haplotype are homozygous for the first seven sites (beginning with the smallest fragments) resulting in the formation of single peaks and unambiguous base calls on the chromatogram (Figure 6.13). However, most of the sites after the seventh site (#230) have double peaks and about half of the bases were called as "N." It might be tempting to explain the double peaks in this chromatogram as multiple heterozygous SNP sites but this would be unlikely because SNPs are usually infrequently encountered in sequences. This chromatogram pattern showing a number of high-quality single peaks suddenly changing to a long series of double peaks is the hallmark of an indel present in one parental haplotype but not in the other (i.e., a length heterozygote). If we compare the extension products for the two parental haplotypes, we see that the indel—a G–G present in the first parental haplotype, but not in the second parental haplotype—has shifted the two haplotypes out of alignment so that the extension products are out of phase with each other for the latter half of the sequence. If you imagine the G–G being removed from the first haplotype then the two sequences would be homozygous and a clean chromatogram with single peaks would result. How abundant are indels? For any given Sanger chromatogram of a diploid locus at most you are likely to see only one or two separate indels. A sequence read will be most affected if an indel is present early in the chromatogram. If this is the case, one strategy to recovering the sequence is to sequence the PCR product using the other primer. We have just learned that it is quite possible to obtain good quality PCR products (i.e., faithful copies of both parental haplotypes) but, if they are heterozygous for one or more indels, then these

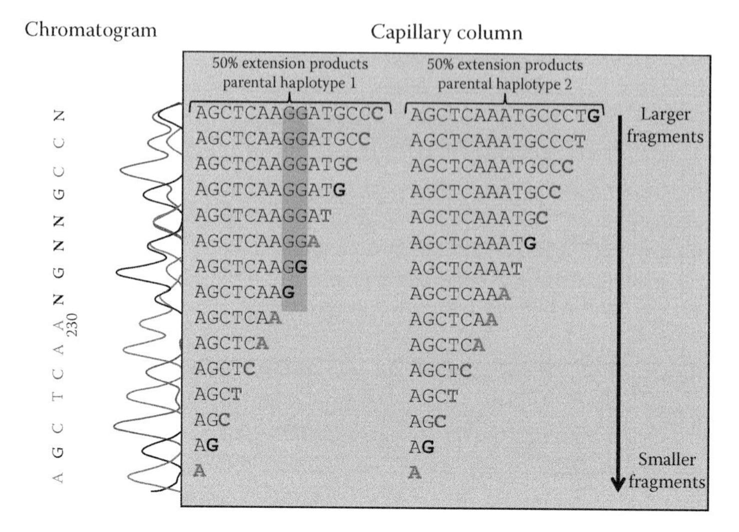

Figure 6.13. Sanger sequencing of a PCR product that is heterozygous for a 2-bp indel. Dark gray is the indel shown as two consecutive guanine bases present in the first haplotype but not the second.

products will not sequence well owing to current limitations of Sanger sequencing. Fortunately, newer NGS methods are able to circumvent this problem.

6.4.2 Multiple Heterozygous SNP Sites and Haplotype Sequences

When a single heterozygous SNP site or *singleton* is observed on a chromatogram, then the two parental haplotypes can be unambiguously determined. For example, in Figure 6.14 we see the sequence for a diploid PCR product that was recorded from a chromatogram (not shown). The chromatogram contained one site with a double peak corresponding to a guanine base and an adenine base. The presence of this singleton does not present a problem because the two parental haplotypes can be resolved with certainty—each is identical at every site except for the SNP site, which has an adenine base on one haplotype and a guanine base on the other. However, for some types of loci (e.g., anonymous loci) it is not uncommon for a chromatogram spanning hundreds of bases showing evidence of two or more heterozygous SNP sites. In these cases, the haplotypes from such PCR products cannot be determined without using outside information (Clark 1990). Figure 6.15 shows a case with two heterozygous SNP sites on the same sequence indicated by two sites that exhibited two overlapping and differently colored peaks. The base calls from the chromatogram (not shown) showed a heterozygous SNP site indicating that one haplotype has an adenine base (green peak) while the other has a guanine base (black peak). The second SNP site similarly showed two double peaks with one corresponding to a thymine base (red peak) and the

Base calls from chromatogram AGCCCAGATTACAGCTAGGTACTACTAAGGGATACATCAA
G

Haplotype 1 AGCCCAGATTACAGCTAGGTACTACTAAGGGATACATCAA

Haplotype 2 AGCCCGGATTACAGCTAGGTACTACTAAGGGATACATCAA

Figure 6.14. Haplotype determination from a Sanger sequence containing a "singleton." Sequence data were obtained from a chromatogram of a directly sequenced PCR product that was heterozygous for a single SNP. The top sequence, which was read directly from the base calls of a chromatogram (not shown), showed a single "double peak" indicating that both adenine and guanine bases were present at the same site. Because this is a singleton SNP, both haplotype sequences are easily resolved. The singleton and corresponding distinct haplotype bases are shaded by the dark gray rectangle.

Base calls from chromatogram	AGCCCAGATTACAGCTAGGTACTACTAAGGGATACATCAA G C
Haplotype ?	AGCCCAGATTACAGCTAGGTACTACTAAGGGATACATCAA
Haplotype ?	AGCCCGGATTACAGCTAGGTACTACTAAGGGATACATCAA
Haplotype ?	AGCCCAGATTACAGCTAGGTACTACTAAGGGACACATCAA
Haplotype ?	AGCCCGGATTACAGCTAGGTACTACTAAGGGACACATCAA

Figure 6.15. Haplotype determination from a Sanger sequence containing multiple heterozygous SNPs. The top sequence was obtained from a chromatogram of a directly sequenced PCR product that was heterozygous for two sites. One SNP is evident as an adenine and guanine double peak while the second SNP site was shown by a thymine and cytosine double peak. There are four possible haplotype sequences but only two are correct. In the absence of outside information it is impossible to unambiguously resolve the haplotype sequences.

other a cytosine base (blue peak). With no other information we cannot determine with certainty the two haplotype sequences because there are *four different possible haplotypes* that can be inferred from these data yet only two of them are correct (Figure 6.15). This problem becomes more complicated with an increasing number of heterozygous SNP sites: three SNPs = eight possible haplotypes; four SNPs = 16 possible haplotypes; and so on.

In order to circumvent this problem, it has been common practice by researchers to simply label each heterozygous SNP site with the appropriate IUPAC ambiguity code (Table 6.4) or eliminate those problematic sites from the dataset. However, doing this reduces the information content in the sequences, which can adversely impact downstream analyses especially in phylogeographic studies. As we will see in Section 6.4.3, methods are available that allow for recovery of the complete haplotype sequences from PCR products that are heterozygous for SNPs or indels.

6.4.3 Methods for Obtaining Nuclear Haplotype Sequences from Sanger Sequence Data

A number of methods have been developed to allow investigators to "resolve" or "phase" PCR haplotypes from Sanger sequence data. These methods fall into two general categories: (1) using laboratory procedures to physically isolate each of the unique amplicons *prior* to Sanger sequencing and (2) using bioinformatics algorithms to infer each PCR haplotype from a directly sequenced PCR product. Although the physical separation methods are capable of generating high-quality haplotype sequences, their laborious time-consuming nature presents a significant obstacle to obtaining large phylogenomic datasets. Thankfully, bioinformatic approaches

TABLE 6.4

IUPAC nucleic acid codes by the Nomenclature Committee of the International Union of Biochemistry

Bases	Symbols	IUPAC code
Adenine	A	A
Guanine	G	G
Cytosine	C	C
Thymine	T	T
Adenine or guanine (purine)	A or G	R
Cytosine or thymine (pyrimidine)	C or T	Y
Cytosine or adenine	C or A	M
Thymine or guanine	T or G	K
Thymine or adenine	T or A	W
Cytosine or guanine	C or G	S
Cytosine, thymine, or guanine	C, T, or G	B
Adenine, thymine, or guanine	A, T, or G	D
Adenine, thymine, or cytosine	A, T, or C	H
Adenine, cytosine, or guanine	A, C, or G	V
Adenine, cytosine, guanine, or thymine	A, C, G, or T	N

SOURCE: From Moss, G. P. 1984. Nomenclature for incompletely specified bases in nucleic acid sequences. http://www.chem.qmul.ac.uk/iubmb/misc/naseq.html (retrieved March 2, 2015).

provide a superior way to efficiently resolve PCR haplotype sequences.

6.4.3.1 Physical Isolation of PCR Haplotypes prior to Sequencing

Single-Strand Conformational Polymorphism—A variety of laboratory methods have been used to isolate the two parental haplotypes (when they are heterozygous) in a completed PCR mixture. One early method that has been successfully used is called *single-strand conformational polymorphism* or "SSCP" (Friesen 2000; Sunnucks et al. 2000). In brief, PCR product amplicons are first converted into single-stranded molecules via heat denaturation then are electrophoretically run through a "native" polyacrylamide gel. A native gel is simply a gel that does not contain any denaturant chemicals (e.g., those that are found in sequencing gels). Under these conditions single-stranded DNA has a tendency to form secondary structures such as loops and hairpins, which, in turn, influences their mobility through a gel. Heterozygous haplotypes usually form different secondary structures and hence each haplotype travels through the gel at a different rate. Thus, rather than separating DNA molecules by length, as is often done, amplicons consisting of heterozygous haplotypes are separated primarily by their two-dimensional secondary structures. At the conclusion of electrophoresis, the individual gel bands containing each amplicon haplotype are excised and sequenced. Although this method may well produce the best results of all haplotype resolving methods, a huge drawback is the large labor cost and use of hazardous chemicals.

PCR Cloning—Another method that has been successfully used for isolating amplicon haplotypes is PCR cloning, which is also called "TA cloning" (Jennings and Edwards 2005). In brief, PCR products are first "A-tailed" meaning that a single adenine is added to the 3′ ends of each double-stranded amplicon. This is accomplished by combining the PCR products from a single reaction with *Taq* polymerase and dATPs then incubating the mixture in a thermocycler at 70°C for about 10 minutes. The A-tailed amplicons are then efficiently ligated into plasmid cloning "T-vectors," which are plasmids having single thymine overhangs on the 3′ ends of the two vector strands. The recircularized plasmids are then transformed into *E. coli* bacteria and grown overnight into individual clone-colonies (i.e., a single bacterium containing a plasmid eventually gives rise to millions of other bacteria each containing the identical plasmid). The colonies containing recombinant plasmids are identified via blue-white screening and then the plasmids are isolated away from the individuals in each colony. The purified plasmids are then used as sequencing templates and vector sequencing primers (e.g., M13 universal primers) are used to sequence the inserted haplotype amplicons. One advantage of this method is that since each haplotype amplicon is separately sequenced, one does not have to be concerned about resolving haplotypes or about having indels wrecking all or part of a sequence (see Figure 6.2 in Jennings and Edwards 2005). If enough clones are sequenced, then paralogous loci can also be detected. Although TA cloning kits are available for this purpose, the amount of labor involved makes this an impractical method for generating large phylogenomic datasets especially with superior NGS-based approaches that are now available (Chapter 7).

6.4.3.2 Statistical Inference of Haplotypes from Sanger Sequence Data

Clark (1990) recognized the problem of inferring PCR haplotype sequences for diploid loci using Sanger sequence data. Moreover, he also suggested ways that haplotype sequences could be inferred. By locating sequences containing "singletons," a number of different haplotypes can be quickly identified. Using these known haplotypes, a process of subtraction can be used in which the known haplotypes are used to help resolve remaining ambiguous haplotypes. This method, which is done by the "eye" can work reasonably well provided a sufficient number of individuals have been sequenced and the number of polymorphic sites is not too large. Once the number of polymorphic sites becomes too large, the process of elucidating haplotypes by the eye becomes unwieldy. However, algorithmic methods for statistically inferring haplotypes from Sanger data have made this bioinformatics approach very popular. Stephens et al. (2001) developed such a Bayesian program, which they called PHASE, for inferring haplotypes from Sanger data. Other bioinformatics software developed to compliment PHASE, such as SEQPHASE (Flot 2010), have further enhanced the user-friendliness of this bioinformatics approach to phasing haplotypes. Recent examples of studies using this methodology include Reilly et al. (2012) and Gottscho et al. (2014). However, one problem this bioinformatics method cannot effectively solve is the indel problem. If only a single heterozygous indel is observed in a chromatogram, then it might be possible to elucidate the downstream "frame-shifted" haplotypes using some software programs that allow for editing chromatograms. For example, Lee and Edwards (2008) used this approach. However, if multiple heterozygous indels are present in one PCR product, then there is nothing that can be done to fix the sequence—it will likely be unusable. Another potential problem with the bioinformatics method is that incorrectly inferred haplotypes will artificially recombine the two haplotypes, which may lead the investigator to incorrectly conclude that one or more historical recombination events occurred and not realize that the cause was bioinformatics error. In other words, false positive results might be obtained from recombination tests such as the four gamete test (Hudson and Kaplan 1985; Chapter 3). Nonetheless, the bioinformatics method for inferring haplotypes is still the best available method for inferring haplotypes from Sanger sequence data. For additional discussion about the use of PHASE in population studies see Garrick et al. (2010).

REFERENCES

Atkinson, M. R., M. P. Deutscher, A. Kornberg, A. F. Russell, and J. G. Moffatt. 1969. Enzymatic synthesis of deoxyribonucleic acid. XXXIV. Termination of chain growth by a 2′,3′ dideoxyribonucleotide. *Biochemistry* 8:4897–4904.

Carothers, A. M., G. Urlaub, J. Mucha, D. Grunberger, and L. A. Chasin. 1989. Point mutation analysis in a mammalian gene: Rapid preparation of total RNA, PCR amplification of cDNA, and Taq sequencing by a novel method. *BioTechniques* 7:494–496.

Clark, A. G. 1990. Inference of haplotypes from PCR-amplified samples of diploid populations. *Mol Biol Evol* 7:111–122.

DeAngelis, M. M., D. G. Wang, and T. L. Hawkins. 1995. Solid-phase reversible immobilization for the isolation of PCR products. *Nucleic Acids Res* 23:4742.

Dinauer, D. M., R. A. Luhm, A. J. Uzhiris, D. D. Eckels, and M. J. Hessner. 2000. Sequence-based typing of HLA class II DQB1. *Tissue Antigens* 55:364–368.

Ewing, B. and P. Green. 1998. Base-calling of automated sequencer traces using phred. II. Error probabilities. *Genome Res* 8:186–194.

Ewing, B., L. Hillier, M. C. Wendl, and P. Green. 1998. Base-calling of automated sequencer traces using Phred. I. Accuracy assessment. *Genome Res* 8:175–185.

Faircloth, B. 2012. A penny for your method: AMPure Substitute. The Molecular Ecologist blog. http://www.molecularecologist.com/2012/01/a-penny-for-your-method-ampure-substitute/ (accessed May 11, 2016).

Faircloth, B. and T. Glenn. 2011. Serapure. Unpublished protocol based on the original Rohland and Reich (2012) homemade SPRI beads protocol (see also the Ford 2012 reference).

Flot, J. F. 2010. SeqPHASE: A web tool for interconverting PHASE input/output files and FASTA sequence alignments. *Mol Ecol Res* 10:162–166.

Ford, E. 2012. Homemade AMPure XP beads. Ethanomics, Everything Ethan knows about biology—the Ethan-ome. https://ethanomics.wordpress.com/2012/08/05/homemade-ampure-xp-beads/ (accessed May 16, 2016).

Friesen, V. L. 2000. Introns. In *Molecular Methods in Ecology*, ed. A. J. Baker, 274–294. Oxford: Blackwell.

Garrick, R. C., P. Sunnucks, and R. J. Dyer. 2010. Nuclear gene phylogeography using PHASE: Dealing with unresolved genotypes, lost alleles, and systematic bias in parameter estimation. *BMC Evol Biol* 10:118.

Gottscho, A. D., S. B. Marks, and W. B. Jennings. 2014. Speciation, population structure, and demographic history of the Mojave Fringe-toed Lizard (*Uma scoparia*), a species of conservation concern. *Ecol Evol* 4:2546–2562.

Hawkins, T. L., T. O'Connor-Morin, A. Roy, and C. Santillan. 1994. DNA purification and isolation using a solid-phase. *Nucleic Acids Res* 22:4543.

Hillis, D. M., B. K. Mable, A. Larson, S. K. Davis, and E. A. Zimmer. 1996. Chapter 9. Nucleic acids IV: Sequencing and cloning. In *Molecular Systematics*, 2nd edition, eds. D. M. Hillis, C. Moritz, and B. K. Mable, 321–381. Sunderland: Sinauer.

Hudson, R. R. and N. L. Kaplan. 1985. Statistical properties of the number of recombination events in the history of a sample of DNA sequences. *Genetics* 111:147–164.

Innis, M. A., K. B. Myambo, D. H. Gelfand, and M. A. Brow. 1988. DNA sequencing with *Thermus aquaticus* DNA polymerase and direct sequencing of polymerase chain reaction-amplified DNA. *Proc Natl Acad Sci USA* 85:9436–9440.

Ivanova, N. V., T. S. Zemlak, R. H. Hanner, and P. D. N. Hebert. 2007. Universal primer cocktails for fish DNA barcoding. *Mol Ecol Notes* 7:544–548.

Jennings, W. B. and S. V. Edwards. 2005. Speciational history of Australian Grass Finches (*Poephila*) inferred from thirty gene trees. *Evolution* 59:2033–2047.

Lee, J. Y. and S. V. Edwards. 2008. Divergence across Australia's Carpentarian Barrier: Statistical phylogeography of the Red-backed Fairy Wren (*Malurus melanocephalus*). *Evolution* 62:3117–3134.

Lis, J. T. 1980. Fractionation of DNA fragments by polyethylene glycol induced precipitation. *Methods Enzymol* 65:347–353.

Lis, J. T. and R. Schleif. 1975. Size fractionation of double-stranded DNA by precipitation with polyethylene glycol. *Nucleic Acids Res* 2:383–389.

Maxam, A. M. and W. Gilbert. 1977. A new method for sequencing DNA. *Proc Natl Acad Sci USA* 74:560–564.

Messing, J., R. Crea, and P. H. Seeburg. 1981. A system for shotgun DNA sequencing. *Nucleic Acids Res* 9:309–321.

Moss, G. P. 1984. Nomenclature for incompletely specified bases in nucleic acid sequences. http://www.chem.qmul.ac.uk/iubmb/misc/naseq.html (retrieved March 2, 2015).

Murray, V. 1989. Improved double-stranded DNA sequencing using the linear polymerase reaction. *Nucleic Acids Res* 17:8889.

Palumbi, S. R. 1996. Chapter 7. Nucleic acids II: The polymerase chain reaction. In *Molecular Systematics*, 2nd edition, eds. D. M. Hillis, C. Moritz, and B. K. Mable, 205–247. Sunderland: Sinauer.

Prober, J. M., G. L. Trainor, R. J. Dam et al. 1987. A system for rapid DNA sequencing with fluorescent chain-terminating dideoxynucleotides. *Science* 238:336–341.

Quail, M. A., H. Swerdlow, and D. J. Turner. 2009. Improved protocols for the Illumina genome analyzer sequencing system. *Curr Protoc Hum Genet* 18–2.

Reilly, S. B., S. B. Marks, and W. B. Jennings. 2012. Defining evolutionary boundaries across parapatric ecomorphs of Black Salamanders (*Aneides flavipunctatus*) with conservation implications. *Mol Ecol* 21:5745–5761.

Rohland, N. and D. Reich. 2012. Cost-effective, high-throughput DNA sequencing libraries for multiplexed target capture. *Genome Res* 22:939–946.

Rosenthal, A., O. Coutelle, and M. Craxton. 1993. Large-scale production of DNA sequencing templates by microtitre format PCR. *Nucleic Acids Res* 21:173–174.

Sanger, F., S. Nicklen, and A. R. Coulson. 1977. DNA sequencing with chain-terminating inhibitors. *Proc Natl Acad Sci USA* 74:5463–5467.

Smith, L. M., S. Fung, M. W. Hunkapiller, T. J. Hunkapiller, and L. E. Hood. 1985. The synthesis of oligonucleotides containing an aliphatic amino group at the 5′ terminus: Synthesis of fluorescent DNA primers for use in DNA sequence analysis. *Nucleic Acids Res* 13:2399–2412.

Smith, L. M., J. Z. Sanders, R. J. Kaiser et al. 1986. Fluorescence detection in automated DNA sequence analysis. *Nature* 321:674–679.

Stephens, M., N. J. Smith, and P. Donnelly. 2001. A new statistical method for haplotype reconstruction from population data. *Am J Hum Genet* 68:978–989.

Sunnucks, P., A. C. C. Wilson, B. Beheregaray, K. Zenger, J. French, and A. C. Taylor. 2000. SSCP is not so difficult: The application and utility of single-stranded conformation polymorphism in evolution biology and molecular ecology. *Mol Ecol* 9:1699–1710.

Vieira, J. and J. Messing. 1982. The pUC plasmids: An M13mp7-derived system for insertion mutagenesis and sequencing with synthetic universal primers. *Gene* 19:259–268.

Watson, J. D., T. A. Baker, S. P. Bell, A. Gann, M. Levine, and R. Losick. 2014. *Molecular Biology of the Gene*, 7th edition. New York: Pearson Education, Inc.

Werle, E., C. Schneider, M. Renner, M. Völker, and W. Fiehn. 1994. Convenient single-step, one tube purification of PCR products for direct sequencing. *Nucleic Acids Res* 22:4354–4356.

Zimmerman, S. B. and B. H. Pheiffer. 1983. Macromolecular crowding allows blunt-end ligation by DNA ligases from rat liver or *Escherichia coli*. *Proc Natl Acad Sci USA* 80:5852–5856.

CHAPTER SEVEN

Illumina Sequencing

Illumina sequencing is revolutionizing phylogenomics in the much the same manner as Sanger sequencing had spurred the growth of molecular phylogenetics. Although this technology was originally developed for purposes of obtaining whole genome sequences, phylogenomics researchers are primarily using it to target large numbers of specific genomic loci. This more focused approach enables researchers to acquire dozens to thousands of loci from genomes of many different individuals all in a single sequencing experiment (Lemmon and Lemmon 2013). Indeed, Illumina sequencing is proving to be a powerful means by which enormous phylogenomic datasets can be obtained.

One class of methods known as *hybrid selection* has effectively been coupled with Illumina sequencing in order to generate targeted genomic datasets (Turner et al. 2009). Hybrid selection, which is also commonly referred to as "target capture," involves the hybridization of sequence-specific oligonucleotide probes to a large number of target locations in genomes followed by isolation of these templates for a single mass sequencing run (Turner et al. 2009). For example, Faircloth et al. (2012) and Lemmon et al. (2012) developed probes for sequencing large numbers of UCE-anchored and AE-anchored loci, respectively. Large numbers (e.g., 25–100) of different PCR products from dozens to hundreds of individuals can also be sequenced using the Illumina platform (e.g., Barrow et al. 2014). These giant-sized datasets now routinely obtained by researchers would have been almost inconceivable a decade ago. Among these methods, hybrid selection represents the most powerful approach owing to its capability of generating datasets with thousands of DNA-sequence loci, which are long enough (i.e., ~250–1,500 bp) to allow for the robust inferences of gene trees. Although sequencing of PCR amplicons using Illumina sequencers is effectively limited to datasets with fewer than 100 loci, these datasets can still provide robust phylogenomic results. Other types of loci being used in phylogenomic studies such as RAD-seq loci and single-nucleotide polymorphisms (SNPs) generally provide poorly reconstructed gene trees. Thus, we will focus our attention on hybrid selection and PCR loci-based methods in this chapter because both approaches usually yield sequences long enough to produce well-constructed gene trees. For additional information about how Illumina sequencing is being used to generate other types of phylogenomic datasets the reader should consult reviews by Lemmon and Lemmon (2013), McCormack et al. (2013), and Toews et al. (2015).

7.1 HOW ILLUMINA SEQUENCING WORKS

The Illumina sequencing workflow consists of three basic steps: (1) construction of indexed sequencing libraries; (2) generation of clusters on a flow cell; and (3) sequencing of clusters. Like Sanger sequencing this workflow begins with the preparation of sequencing templates from input DNA samples such as genomic DNA, short or long PCR products, cDNAs, etc. However, while the former sequencing method uses PCR products directly as sequencing templates, the latter uses sequencing libraries. Furthermore, as organismal phylogenomic studies include DNA samples from multiple individuals or species, one library must be made for each input sample.

Thus, each library must be tagged with at least one individual-specific *index* sequence so that they can be sequenced together in a single Illumina sequencing run, a practice called *multiplexing* (Cronn et al. 2008). Library constructs may have a single index sequence, which ranges from 7 to 10 bp long (depending on the kit or protocol) or, more usually, they have dual indices. Later, during the bioinformatics processing of sequencing data, software is used to *deconvolute* the sequences (i.e., sequences are sorted by individual according to their indices).

Once the desired libraries are acquired, the next step is to generate *in situ* clonal colonies of target DNA templates called *clusters*. Each cluster comprises of ~1,000 ssDNA molecules derived from each target library fragment (Shendure and Ji 2008). As with Sanger sequencing, Illumina sequencing also relies on detecting and imaging base-specific light signals emitted from fluorescently labeled nucleotides during the sequencing process. However, because the imaging devices found on Sanger and second-generation NGS sequencers such as Illumina are not sensitive enough to detect light emissions from single molecules, large numbers of identical molecules must first be generated via PCR before they can be sequenced. Thus, when these populations of molecules are subsequently tagged with fluorescent bases and are illuminated by the sequencer's laser, the emitted light signals from each population are strong enough to be detected and imaged (Shendure et al. 2004).

The earliest version of the Illumina platform generated sequences only 25–35 bp long (e.g., Bentley et al. 2008; Cronn et al. 2008). Since then, however, Illumina has increased their read lengths up to 100–300 bp depending on the sequencer model. While the "short-read" nature of Illumina sequencing is one disadvantage compared to the longer Sanger reads, the hundreds of millions to *billions* of reads output per run by some Illumina platforms is nothing short of mind boggling. It is the massively parallel nature of this form of DNA sequencing that gives it tremendous sequencing power, which is why this and other NGS technologies are commonly referred to as "massively parallel sequencing" (Shendure et al. 2004; Mardis 2008).

In addition to increasing their read lengths, Illumina introduced an innovation called *paired-end sequencing* to facilitate the construction of *de novo* genomes particularly for organisms having large genomes (see Brown 2013 and Masoudi-Nejad et al. 2013 for further explanations). Paired-end or "PE" sequencing is a form of DNA sequencing whereby forward and reverse sequences of each target library fragment are obtained. For example, if a target library fragment is 500 bp long and 150 bp PE sequencing is performed, then a 150 bp read will be made from one end of a target library molecule while the other 150 bp read will be taken from the opposite end of the same molecule (but the latter sequence will reflect the complementary strand). In this example, a ~200 bp stretch of bases in the middle of the target would not be sequenced. However, given the massive number of reads obtained from a single Illumina run, a sufficient number of overlapping reads will usually be obtained to allow for the accurate reconstruction of each entire allele sequence regardless of their lengths (Lemmon and Lemmon 2013). Although this is similar to using forward and reverse primers to sequence a single PCR product, it is not the same thing. This is because PE reads are effectively obtained from single molecules reflecting actual chromosomal sequences, whereas forward and reverse Sanger reads based on a single PCR product are obtained from a population of molecules that often consist of multiple haplotypes (Lemmon and Lemmon 2013). This general property of NGS offers another major advantage over Sanger sequencing: nuclear haplotypes can be easily and unambiguously determined for typical phylogenomic loci (Brito and Edwards 2009; Edwards and Bensch 2009; Lemmon and Lemmon 2013; O'Neill et al. 2013). Both single-end and PE Illumina sequencing have been successfully used to generate large multilocus datasets. However, because most studies are now using PE sequencing we will only examine this type of sequencing here. The following sections in this chapter are meant to provide a simplistic overview of Illumina sequencing chemistry. Moreover, as Illumina owns the rights to this technology, they may alter any of the procedures and reagents at any time. Given the fast-paced nature of genomics, it is essential that you use procedures and reagents that are compatible with the Illumina sequencing facility you plan to use. If in doubt, contact the intended sequencing provider before a project is begun in order to minimize the risk of wasting thousands of dollars and many hours of lab time.

7.1.1 Construction of Indexed Sequencing Libraries

Many similarities exist between genomic library-making methods used in the pre-NGS era and the Illumina methodology. The key difference, however, is that the latter type of library requires special adapter sequences, which must be attached to the ends of each library DNA fragment (Figure 7.1a). The adapters contain a variety of oligo sequences required for subsequent PCR-amplification and sequencing steps (Figure 7.1b). Because of the extensive space needed to fully explain how these constructs are acquired in phylogenomic studies, we will defer this discussion until later in this chapter. However, at this time we are able to examine how these adapters function in the cluster generation and sequencing processes.

7.1.2 Generation of Clusters on a Flow Cell

The process of creating clusters is quite different from the standard PCR methods used to amplify templates for Sanger sequencing. Instead of using a thermocycler to perform the amplification step, the Illumina workflow uses a fluidics device called a *cluster station*. Furthermore, rather than conducting the amplification reaction inside plastic microtubes or PCR microplates, clusters are generated on the inside surfaces of a *flow cell* (Figure 7.2a). The flow cell, which is small enough to fit in your hand, is a glass slide that is subdivided into eight enclosed glass channels referred to "lanes" (Fedurco et al. 2006). Flow cells with fewer lanes have also been developed for some sequencing platforms but the eight-lane flow cell has been the workhorse in Illumina sequencing. Multiplexed sequencing libraries for a phylogenomics project are usually loaded into a single lane, which is not uncommonly shared with the libraries from other projects.

The top and bottom surfaces of the flow cell's lanes are carpeted by two types of oligos for PE sequencing, all of which had their 5′ ends covalently anchored or "grafted" to the flow cell while their 3′ ends remain free (Adessi et al. 2000; Fedurco et al. 2006; Bentley et al. 2008; Figure 7.2b). These oligos, which are randomly distributed across the flow cell's surfaces in a 1:1 ratio, are commonly referred to as the "P5" and "P7" flow cell oligos though they were originally named "oligo C" and "oligo D," respectively (Bentley et al. 2008). The sequence of the P5 flow cell oligo is 5′-TTTTTTTT TTAATGATACGGCGACCACCGAGAUCTACAC-3′ where U = 2-deoxyuridine, while the sequence of the P7 flow cell oligo is 5′-TTTTTTTTTT CAAGCAGAAGACGGCATACGAGoxoAT-3′ where Goxo = 8-oxoguanine (Bentley et al. 2008). The uracil and Goxo residues in the P5 and P7 flow cell oligos, respectively, are cleavage sites used during the sequencing process. An enzyme mix called uracil-specific excision reagent ("USER") is used to cut each P5 flow cell oligo at the appropriate time (Bentley et al. 2008; see below). The enzymes used in this mix are uracil DNA glycosylase (UDG) and DNA glycosylase–lyase endonuclease VIII (New England Biolabs). The former enzyme excises the uracil base while the latter breaks the phosphodiester bond thereby cleaving the oligo into two fragments minus the uracil. The enzyme used to cleave the Goxo cleavage site in the P7 flow cell oligo is called "Fpg" (Bentley et al. 2008), which stands for formamidopyrimidine DNA glycosylase (New England Biolabs). For further description about the development of the Illumina flow cell technology see Adessi et al. (2000) and Fedurco et al. (2006).

Let's now examine the process of cluster formation. First, notice that two of the adapter sequences embedded in the library construct shown in Figure 7.1b are complementary to the P5 and P7 flow cell oligos. Thus, the adapter sequences at the ends of the library strands are able to hybridize to both flow cell oligos where they are subsequently amplified into individual clusters that are affixed to the flow cell surface. Indeed, one major function of the flow cell oligos is to serve as PCR primers in a special type of PCR known as *isothermal bridge amplification* or "solid-phase DNA amplification" (Adessi et al. 2000).

At the start of the cluster generation process the double-stranded library constructs must be chemically denatured into single-stranded molecules using 0.1 M NaOH. The single-stranded molecules are then added to a flow cell lane where they are allowed to hybridize to their complementary flow cell oligos under stringent thermal conditions controlled by the cluster station (Figure 7.3a). For simplicity, Figure 7.3a shows both strands of a library construct hybridizing to their respective flow cell oligos in close proximity to each other. Note, however, it is extremely unlikely that both strands from the same construct will hybridize

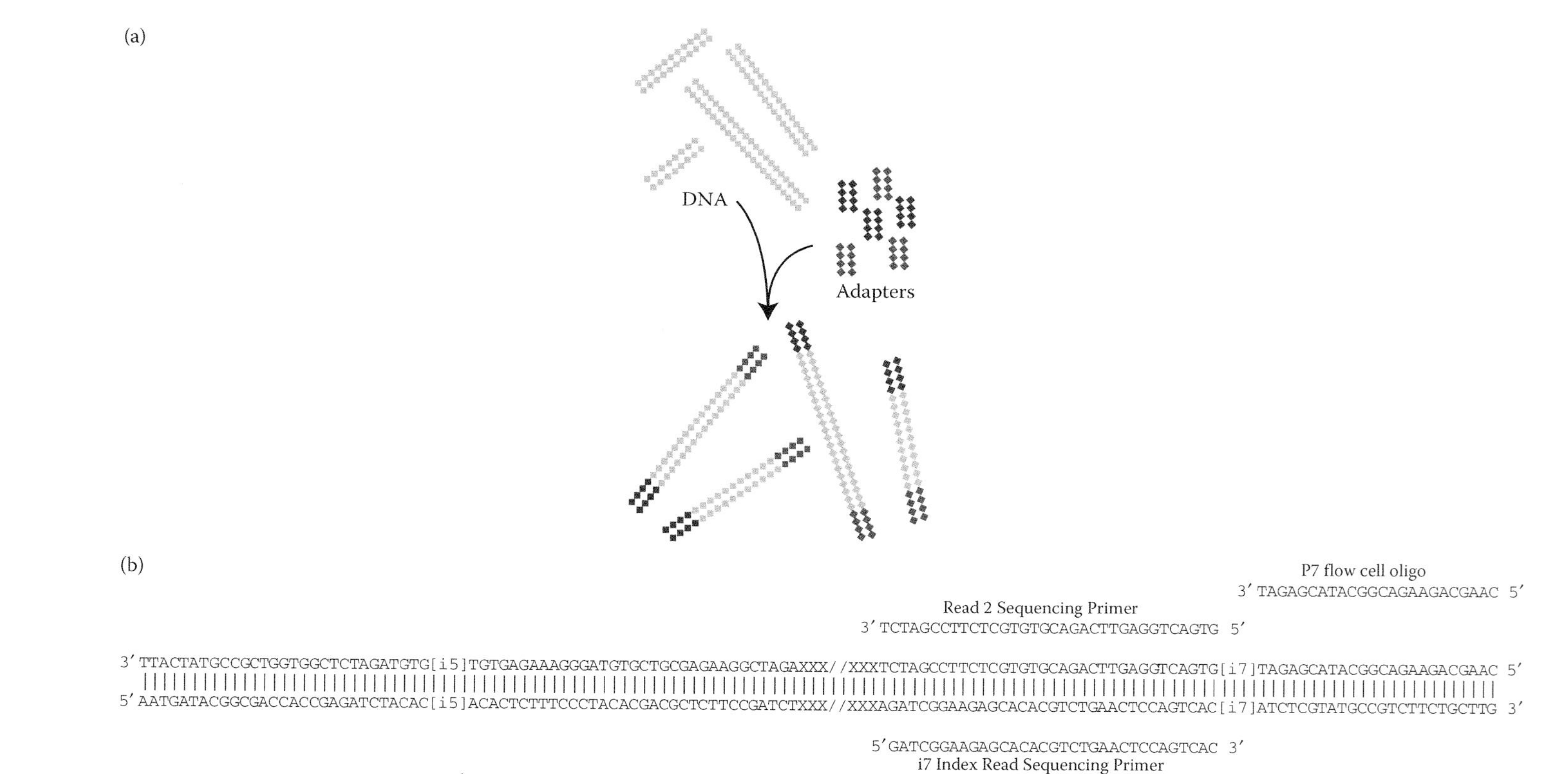

Figure 7.1. Construction of Illumina adapter–library fragment constructs. (a) Before target (library) DNA fragments can be sequenced, Illumina adapters must be attached to the ends of the fragments. Each library fragment must include both the P5 and P7 adapters (shown in different colors) in order to be sequenced. (Image courtesy of Illumina, Inc.) (b) Detailed look at a finished Illumina adapter–library fragment construct. The middle section of the construct contains the target DNA (a string of Xs). The // signifies that most of the target fragment length is not shown due to space limitations and the vertical dashes indicate hydrogen bonding between complementary bases. The [i5] and [i7] sequences represent two index sequences used to identify sequenced fragments after sequencing. Oligos shown above and below the construct are discussed in the main text and can be found in Table 7.1.

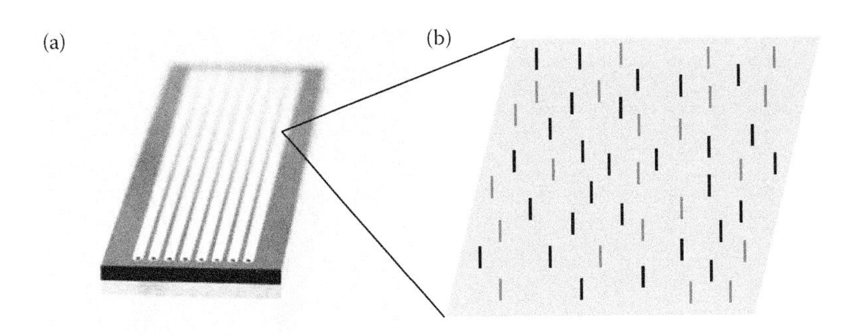

Figure 7.2. (a) An Illumina flow cell. This flow cell is a glass slide with eight lanes. (Image courtesy of Illumina, Inc.) (b) Magnified view of the flow cell surface in one of the lanes. Top and bottom (not shown) flow cell surfaces are populated with two different oligos, here shown as vertical black and gray lines, which have been covalently attached to the surface.

to adjacent flow cell oligos; in other words, each cluster generally originates from a single library strand. Once a library strand hybridizes to its complementary flow cell oligo, a primer template junction is created, which allows for synthesis of a new strand of DNA after unlabeled dNTPs and DNA polymerases are added to the flow cell (Figure 7.3b,c). Notice that the new strand, which is complementary to the original strand, is covalently attached to the flow cell surface (Figure 7.3c). Thus, when these double-stranded molecules are chemically denatured via treatment with formamide in the next step, the original library strands can be washed away leaving the new complementary strands firmly attached to the flow cell (Bentley et al. 2008; Figure 7.3d). Each flow cell-tethered strand will be the initial template molecule used to make each cluster. Next, grafted library strands fold over to hybridize or "bridge" nearby free flow cell oligos, which again creates primer–template junctions that are used in another round of DNA synthesis (Figure 7.3e). After synthesis, double-stranded bridges are formed in scattered locations across the flow cell surface (Figure 7.3f). The double-stranded bridges are then chemically denatured with formamide, which linearizes each library molecule (Figure 7.3g). After many additional synthesis cycles (35 cycles total; Bentley et al. 2008), the surface of the flow cell contains hundreds of millions of randomly distributed clusters, each of which will be sequenced (Figure 7.3g). As we will soon see, the fixed-location nature of these clusters represents one of the keys to the Illumina sequencing method. At this moment in the process, sequences matching both strands of the original library construct are now covalently attached to the flow cell; the strand anchored by the P7 flow cell oligo is hereafter referred to as the *forward strand*, whereas the stand connected to the P5 flow cell oligo is now the *reverse strand* (Figure 7.3g). In the last step before sequencing, the reverse strands are cut with USER enzymes (as explained before) and washed away leaving only the forward strands attached to the flow cell (Figure 7.3g). After removal of the reverse strands, notice that the P5 flow cell oligos remain anchored to the flow cell except that they are seven bp shorter at their 3′ ends. These P5 flow cell oligos will still be used during the PE sequencing process.

7.1.3 Sequencing of Clusters

The process of sequencing clusters is more complex than sequencing PCR products via Sanger sequencing. The Illumina PE adapters shown in Figure 7.1b contain sequences that function as annealing sites for three different sequencing primers: Read 1 and Read 2 Sequencing Primers generate sequences at each end of the library fragment (i.e., PE reads) while the i7 Index Read Sequencing Primer obtains the i7 index. A fourth sequencing primer is not needed for acquiring the i5 index sequence because the P5 flow cell oligo is used for this purpose (Figure 7.1b). Thus, there are four different sequencing passes made on the library construct in order to record the PE sequences and dual indices.

As we saw in Chapter 6 fluorescent-labeled chain terminator nucleotides are key components of the Sanger sequencing methodology. Similarly, dye-terminator nucleotides are also essential

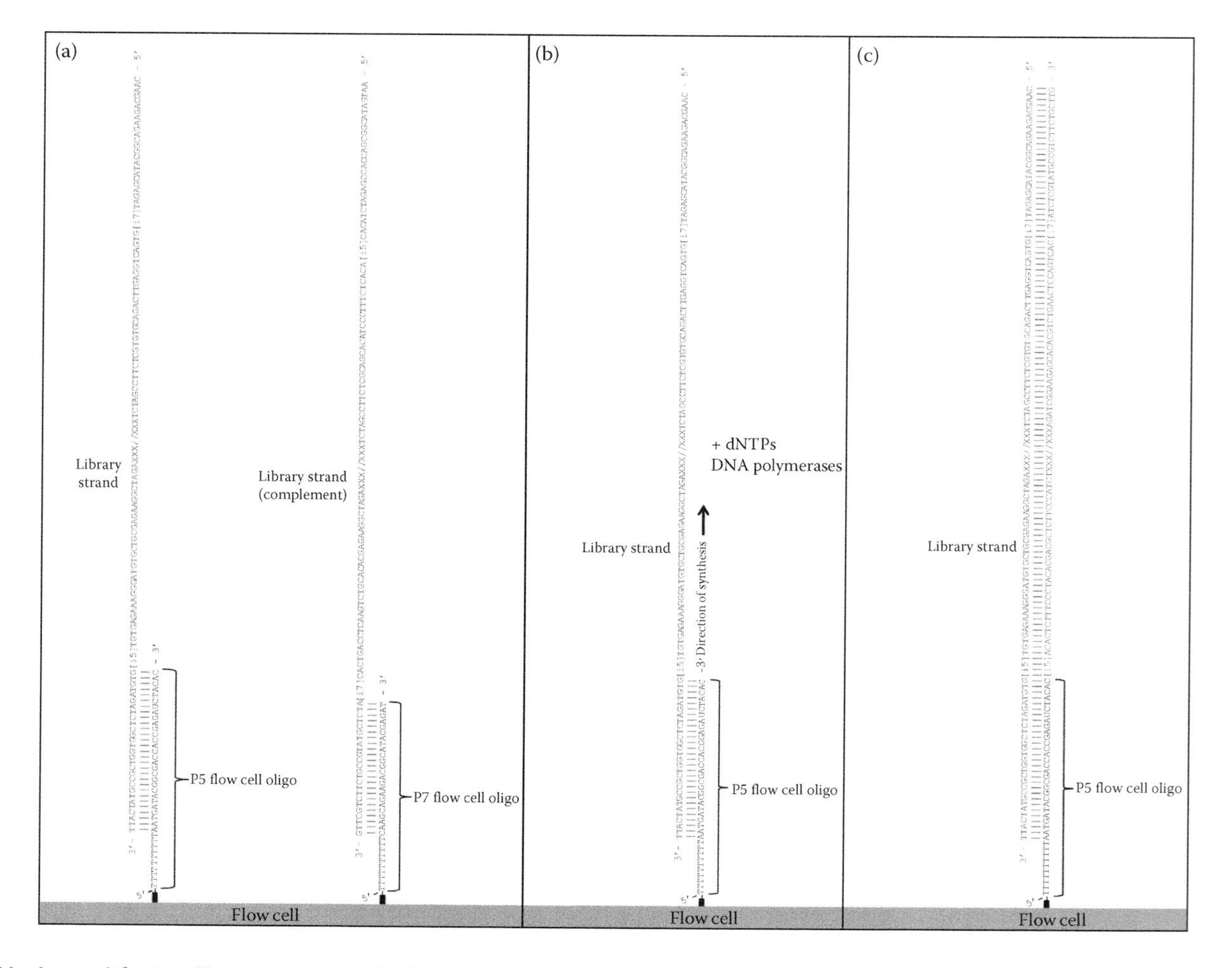

Figure 7.3. Isothermal bridge amplification. (a) Adapters at the ends of single-stranded library DNA hybridize to P5 and P7 flow cell oligos. Sequence of Xs in each strand represents target library fragments, // indicates that much of the library sequence is not shown due to space limitations, and vertical dashes signify hydrogen bonding between complementary bases. (b) Hybridization of a library strand to a flow cell oligo (only P5 flow cell oligo is shown) creates a primer–template junction, which is used for synthesizing a complementary strand. (c) A new strand is synthesized (gray bases).

(Continued)

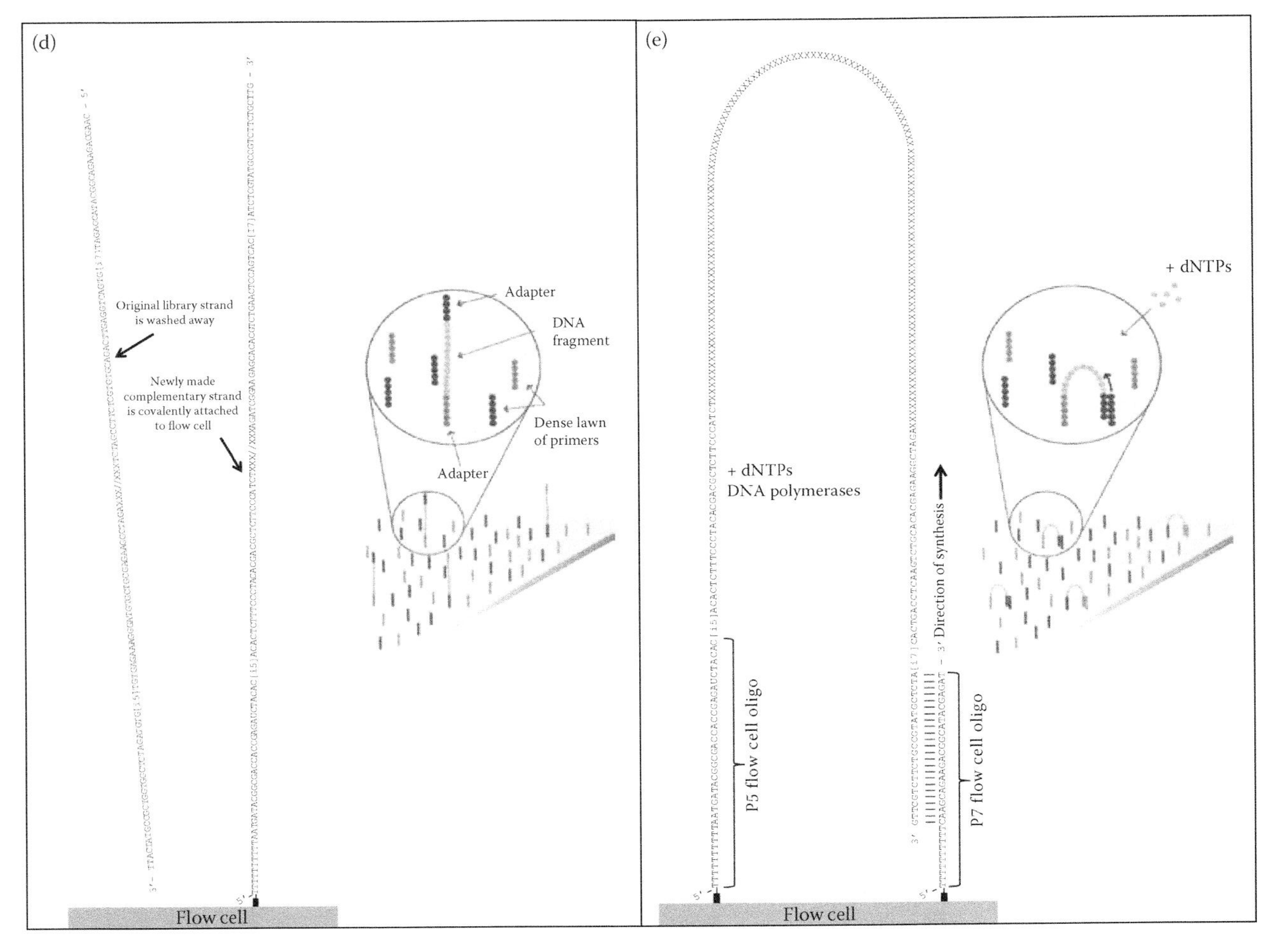

Figure 7.3. (Continued) (d) DNA duplexes are denatured using formamide and original library strands are washed away leaving complementary strands covalently attached to the flow cell. (e) Anchored library strands fold over so their adapters can hybridize or "bridge" the other flow cell oligos, PCR reagents are added, and DNA synthesis starts again. (Colored images courtesy of Illumina, Inc.)

(Continued)

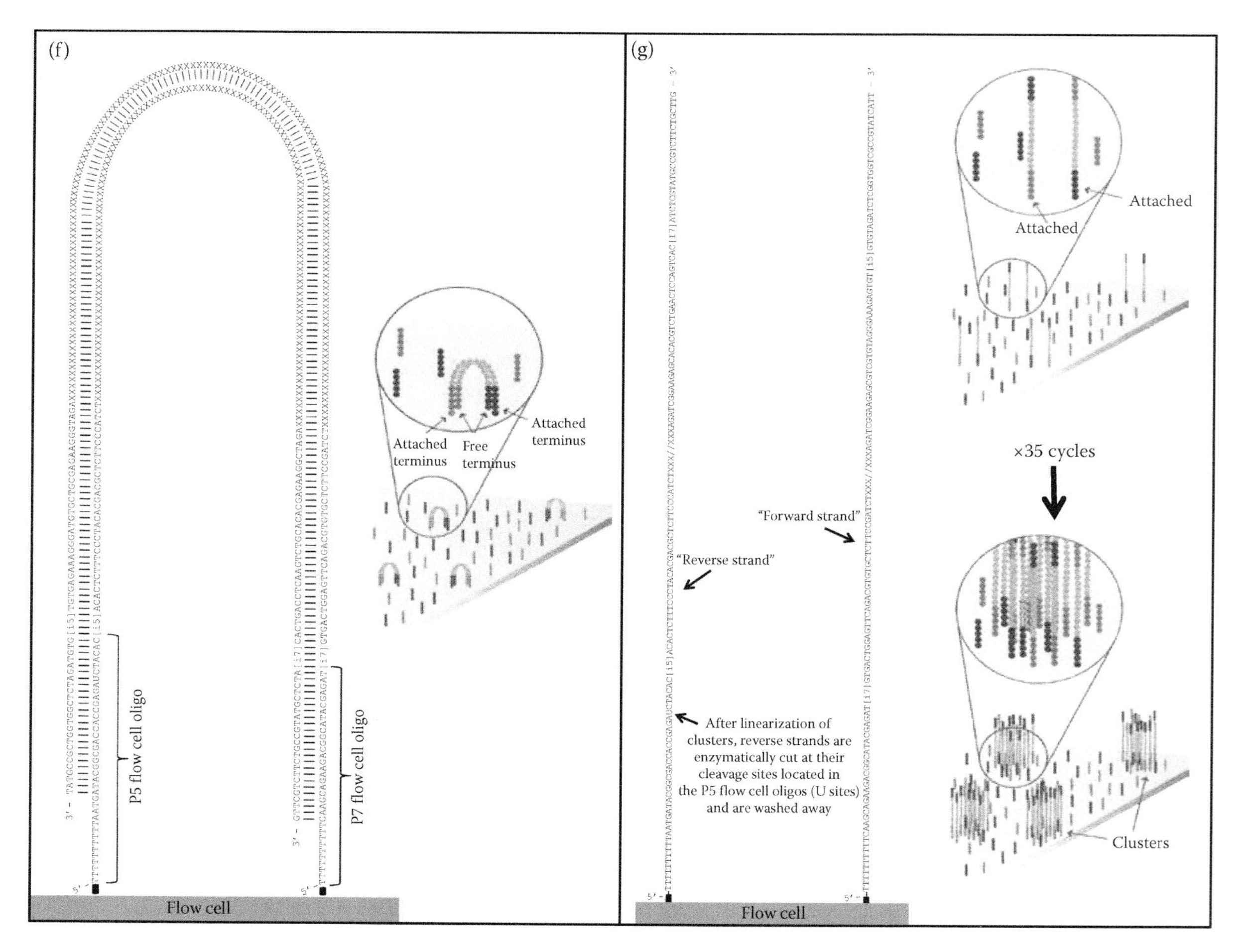

Figure 7.3. (Continued) (f) After synthesis of a new strand (gray bases), a double-stranded bridge is formed. (g) DNA duplexes are chemically denatured using 0.1 M NaOH to linearize flow-cell anchored library strands. After additional PCR cycles (35 total), clusters are generated across the flow cell surface. Following linearization of strands, the reverse strands are cleaved and washed away leaving only the forward strands attached to the flow cell. (Colored images courtesy of Illumina, Inc.)

players in the Illumina methodology. However, the design of the Illumina sequencing nucleotides differs from Sanger dye-ddNTPs in two important ways. First, the Illumina terminators can be easily "reversed" back to a normal-functioning nucleotide state after they have been incorporated into a DNA strand (i.e., 3′ hydroxyl groups can be regenerated). Secondly, their fluorophore moieties can be chemically cleaved from incorporated nucleotides after their fluorescent signals have been imaged. This modified nucleotide, which is known as a *reversible terminator*, is 3′-O-azidomethyl 2′-deoxynucleoside triphosphate (Bentley et al. 2008). The chemical structure of the reversible terminator is identical for all four bases (A, C, G, and T) except for base-specific fluors (Bentley et al. 2008). Figure 7.4a shows the chemical structure of one of these modified nucleotides, a 3′-O-azidomethyl 2′-deoxythymine triphosphate with a cleavable fluorophore. The chemical Tris(2-carboxyethyl)phosphine (TCEP) is used to cleave the fluorophore and regenerate the 3′ hydroxyl group (Figure 7.4b; Bentley et al. 2008). Thus, if the labeled reversible "T" terminator shown in Figure 7.4a becomes incorporated into a growing DNA strand, DNA synthesis will be terminated. However, following subsequent laser-detection and imaging steps, the fluorophore and 3′ blocking group can be chemically removed with regeneration of the 3′ OH group (Figure 7.4b). This now unlabeled reversible terminator can prime the addition of the next fluorescent-labeled reversible terminator.

Before the first sequencing read can be performed, all exposed 3′ ends, which include 3′ termini of the forward strands and unextended flow cell oligos, must first be blocked in order to preclude the possibility of having reversible terminators incorporated in those unwanted locations (Bentley et al. 2008). This blocking step is accomplished by using terminal transferases, which are template-independent DNA polymerases (New England Biolabs), to incorporate a single ddNTP at each available 3′ end (Figure 7.5a; Bentley et al. 2008). Next, Read 1 Sequencing Primers are annealed to the forward strand templates followed by the addition of reversible terminators and DNA polymerases (Figure 7.5a).

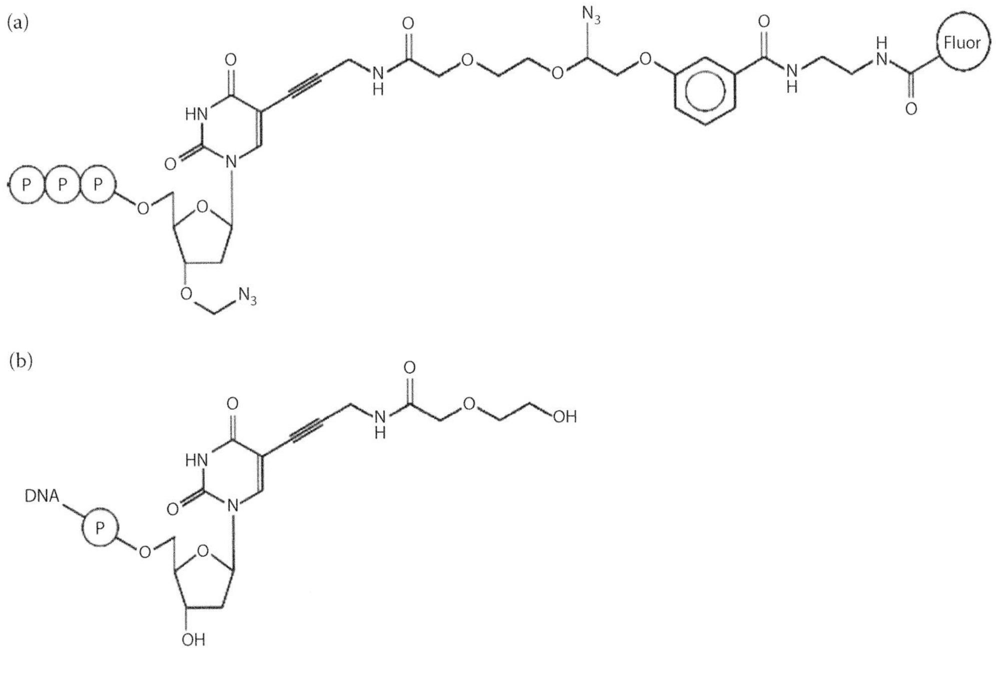

(a)

(b)

Figure 7.4. The chemical structure of a reversible terminator nucleotide. (a) Example shown is 3′-O-azidomethyl 2′-deoxythymine triphosphate with a cleavable fluorophore. The 3′ location contains a blocking group, which terminates DNA synthesis after this nucleotide has been incorporated. (b) Following cleavage of the fluorophore and 3′ blocker and regeneration of the 3′ OH group, an incorporated reversible terminator can prime the addition of a new base in the next sequencing cycle. (Reprinted from Bentley, D. R. et al. 2008. *Nature* 456:53–59. With permission.)

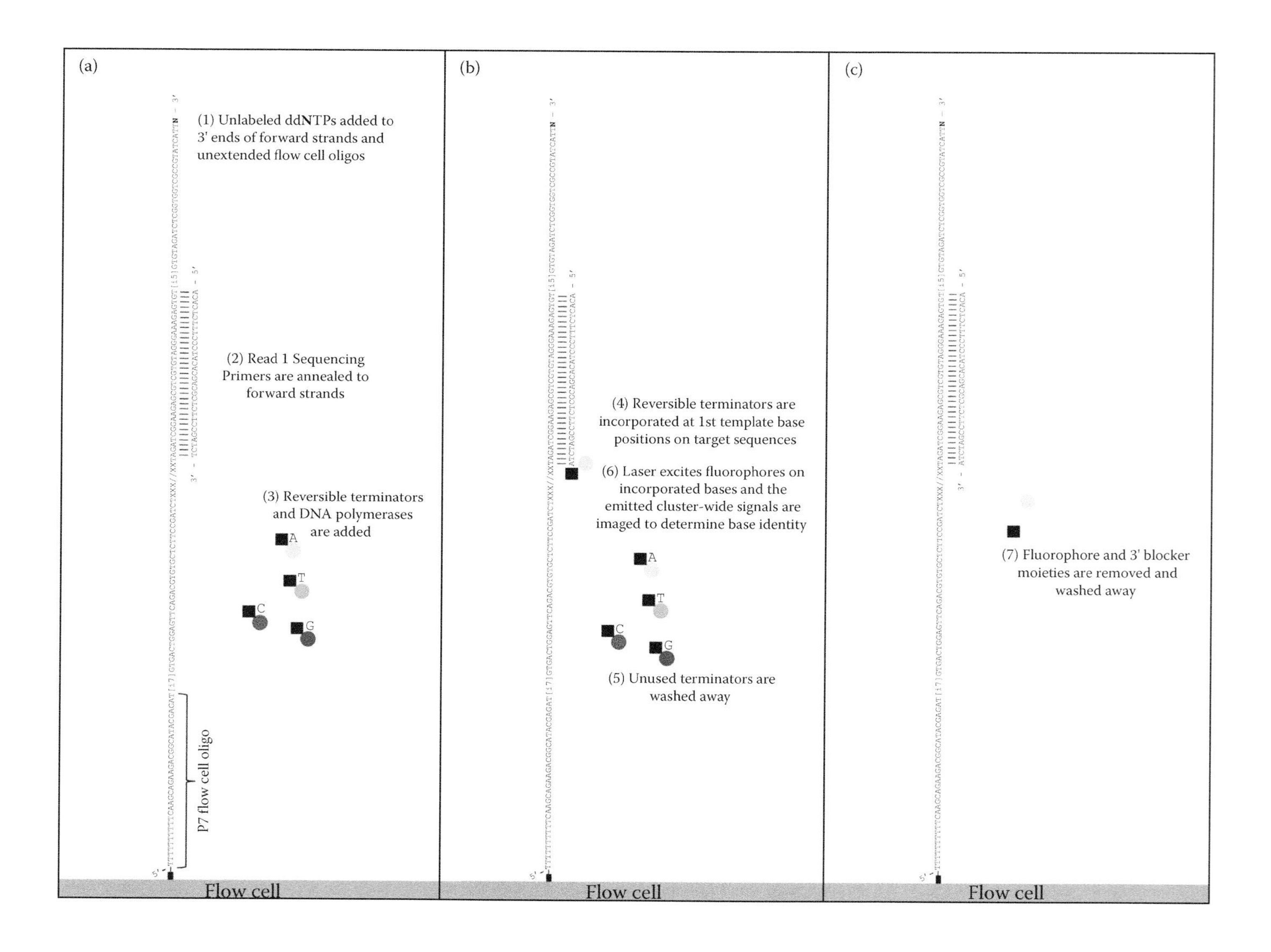

(a)
(1) Unlabeled ddNTPs added to 3' ends of forward strands and unextended flow cell oligos
(2) Read 1 Sequencing Primers are annealed to forward strands
(3) Reversible terminators and DNA polymerases are added
A
T
C
G
P7 flow cell oligo
Flow cell
(b)
(4) Reversible terminators are incorporated at 1st template base positions on target sequences
(6) Laser excites fluorophores on incorporated bases and the emitted cluster-wide signals are imaged to determine base identity
A
T
C
G
(5) Unused terminators are washed away
Flow cell
(c)
(7) Fluorophore and 3' blocker moieties are removed and washed away
Flow cell

Figure 7.5. First paired-end read of the target DNA on the forward strands. (a) First, all exposed 3′ ends must be blocked with ddNTPs. Next, Read 1 Sequencing Primers are annealed to the forward strands followed by the addition of reversible terminators and DNA polymerases. The middle section of the shown forward strand contains the target DNA (a string of Xs). The // signifies that most of the target fragment is omitted due to space limitations and the vertical dashes indicate hydrogen bonding between complementary bases. (b) A reversible terminator that is complementary to the first template base in the target sequence becomes incorporated. Unused terminators are then washed away before the fluorophore-detection and imaging steps. (c) The fluorophore and 3′ blocking group on each incorporated reversible terminator are removed and washed away. This process also regenerates 3′ OH groups on the incorporated terminators enabling them to prime the addition of a new reversible terminator in the next sequencing cycle.

After a single reverse terminator has been incorporated into the first template base position of the target library fragment, the unused terminators must be washed away. In the next step, the sequencer's laser is used to excite the fluorophores attached to the incorporated terminators (Figure 7.5b). Because all of the ~1,000 strands comprising each cluster are expected to incorporate the same base at the same position, each cluster should emit a single and strong base-specific signal, which the sequencer's imaging apparatus can detect and record. After the imaging step is done, the fluorophores and 3′ blockers on the incorporated reversible terminators are removed (Figure 7.5c). This completes one cycle of sequencing. The second cycle starts with the addition of reversible terminators and DNA polymerases and finishes with the removal of the fluorophores and 3′ blockers. The total number of cycles will equal the length of each cluster read (e.g., 100–300 bp). This step-by-step process has been termed *sequencing by synthesis* (Bentley et al. 2008; Mardis 2008). When all cycles are completed to generate the complete Read 1 sequences, the sequencing products are denatured from the template strands via treatment with 0.1 M NaOH and then washed away (Bentley et al. 2008).

As alluded to earlier, the fixed locations of the clusters plays a critical role in the Illumina sequencing process. Figure 7.6a shows an image taken of a small portion of a flow cell during the sequencing process. On this image we can see a scattering of colored dots each of which corresponds to an individual cluster that was illuminated by the sequencer's laser. Using the series of images taken—one image taken per sector on the flow cell surface per cycle, the sequencer can generate a read for each cluster. For example, in Figure 7.6b we see images for the first three cycles. Owing to the fixed location of each cluster, the sequencer can record the color of each individual cluster in a sequential manner across cycles thereby producing a read for each cluster. In our example, the cluster located at the upper left corner of the images is blue in the first cycle, green in the second, and red in the third, which gives the

(a)

(b)

Cluster location in photos	Sequence of first three bases (5′→ 3′)		
Upper left	Blue-green-red	=	GTC
Upper right	Yellow-yellow-blue	=	AAG
Lower left	Green-red-green	=	TCT
Lower right	Red-green-blue	=	CTG

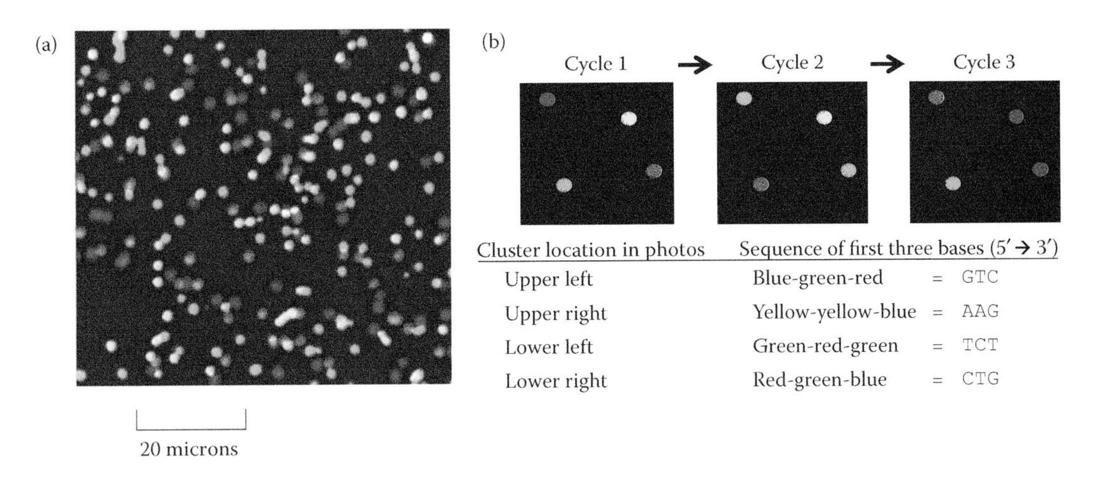

Figure 7.6. Imaging of illuminated clusters and the determination of their sequences. (a) Image showing clusters occupying a small area of a flow cell. Each color dot represents a cluster, which has a fixed location on the flow cell. (Reprinted from Bentley, D. R. et al. 2008. *Nature* 456:53–59. With permission.) (b) A simple example showing how sequences are elucidated for each cluster.

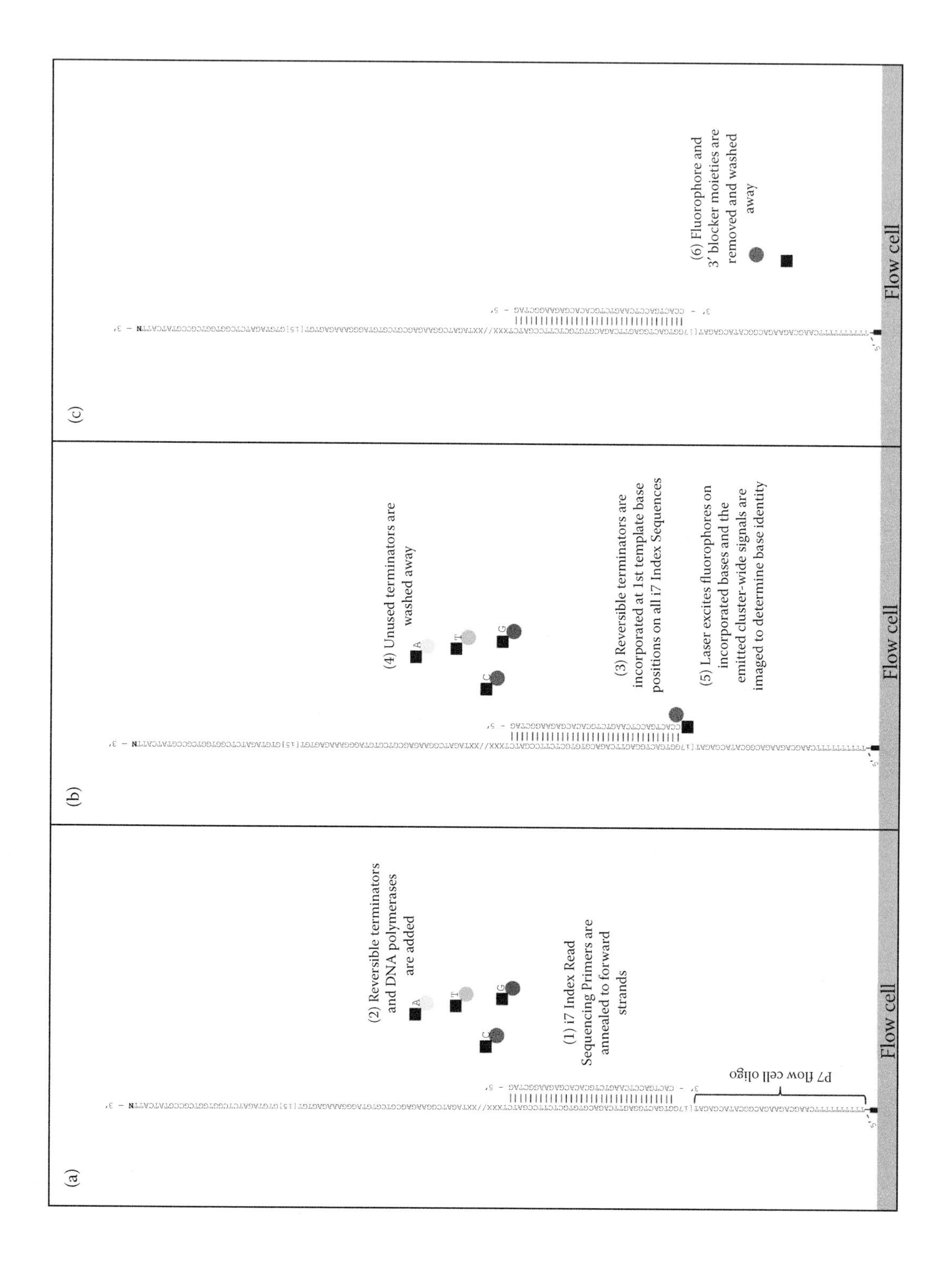

(a)
(1) i7 Index Read Sequencing Primers are annealed to forward strands
(2) Reversible terminators and DNA polymerases are added
A
T
G
C
P7 flow cell oligo
Flow cell
(b)
(3) Reversible terminators are incorporated at 1st template base positions on all i7 Index Sequences
(4) Unused terminators are washed away
(5) Laser excites fluorophores on incorporated bases and the emitted cluster-wide signals are imaged to determine base identity
A
T
G
C
Flow cell
(c)
(6) Fluorophore and 3' blocker moieties are removed and washed away
Flow cell

Figure 7.7. First index read (i7 index) on the forward strands. (a) i7 Index Read Sequencing Primers are annealed to the forward strands followed by the addition of reversible terminators and DNA polymerases. The middle section of the shown forward strand contains the target DNA (a string of Xs). The // signifies that most of the target fragment is omitted due to space limitations and the vertical dashes indicate hydrogen bonding between complementary bases. (b) A reversible terminator that is complementary to the first template base on the i7 index sequence becomes incorporated. Unused terminators are then washed away before the fluorophore-detection and imaging steps. (c) The fluorophore and 3′ blocking group on each incorporated reversible terminator are removed and washed away. This process also regenerates 3′ OH groups on the incorporated terminators enabling them to prime the addition of a new reversible terminator in the next sequencing cycle.

sequence 5′-GTC-3′ (Figure 7.6b). In reality, the number of images analyzed for this same flow cell region would number somewhere between 100 and 300 depending on the sequencer model and thus the overall read lengths would range from 100 to 300 bp for each cluster.

The second sequence to obtain from each of the clusters will be the first index sequence (i.e., i7 index). This process begins with binding of i7 Index Read Sequencing Primers to the forward strands (Figure 7.7a). The same sequencing steps used earlier to obtain all Read 1 sequences are used and thus will not be repeated (Figure 7.7a–c). Thus, eight cycles must be performed in order to obtain the complete sequence for an 8-bp index. When sequencing of the first index is completed, the sequencing products are denatured with 0.1 M NaOH and washed away.

The second index sequence represents the third sequence to be obtained from the library molecules. Unlike the Read 1 and i7 Index Read Sequencing steps, a dedicated sequencing primer is not used to obtain the i5 index sequence. Instead, the 3′ ends of the P5 flow cell oligos are used as i5 index sequencing primers (see Figure 7.1b). Thus, the process begins with restoration of 3′ OH groups on the forward strands and P5 flow cell oligos—though only the P5 flow cell oligos will participate in the i5 index sequencing process—and folding over of the forward strands so they can hybridize with the P5 flow cell oligos (Figure 7.8a). While in this single-stranded bridge configuration, the P5 flow cell oligo is used to prime the sequencing reaction; that is, instead of adding unlabeled dNTPs and DNA polymerases to generate double-stranded bridges as was done in cluster formation, reversible terminators and DNA polymerases are added to begin the sequencing of the i5 index (Figure 7.8a). The same sequencing steps are used as before (Figure 7.8a–c). However, recall that following the step in which the reverse strands were cleaved, the P5 flow cell oligos lost the last seven bases at their 3′ ends (Figure 7.3g). Thus, the first seven cycles of the i5 Index Read will be used to sequence the bases immediately upstream of the i5 index meaning a total of 15 cycles are needed to obtain a 8-bp i5 index sequence (Figure 7.8b). After all i5 Index Read cycles have been completed, the i5 index sequencing products are denatured and washed away.

To obtain the fourth and final sequence from the library molecules—the other PE read of the target fragment, the reverse strands must first be regenerated via bridge amplification (Bentley et al. 2008). This is because the binding sites for the Read 2 Sequencing Primer are located on the reverse strand (see Figure 1b). After the clusters are regenerated and then denatured to linearize the forward and reverse strands, the Goxo sites in the P7 flow cell oligos are cleaved using Fpg enzymes, which releases the forward strands allowing them to be washed away (Bentley et al. 2008). As these bridge amplification and strand cleavage steps are essentially the same as in Figure 7.3, they will not be shown again here.

With only the reverse strands now anchored to the flow cell surface, all exposed 3′ ends must again be blocked with unlabeled ddNTPs before the Read 2 Sequencing Primers are annealed (Figure 7.9a). The remaining sequencing steps are the same as for the previous reads (Figure 7.9a–c).

7.2 METHODS FOR OBTAINING MULTIPLEXED HYBRID SELECTION LIBRARIES

The goal of this section is to introduce several methods for making multiplexed Illumina libraries, which will be followed by a discussion of the procedures for performing hybrid selection with these libraries. The set of target-loci enriched libraries obtained via these procedures will then be ready for the cluster generation and sequencing-by-synthesis steps just described.

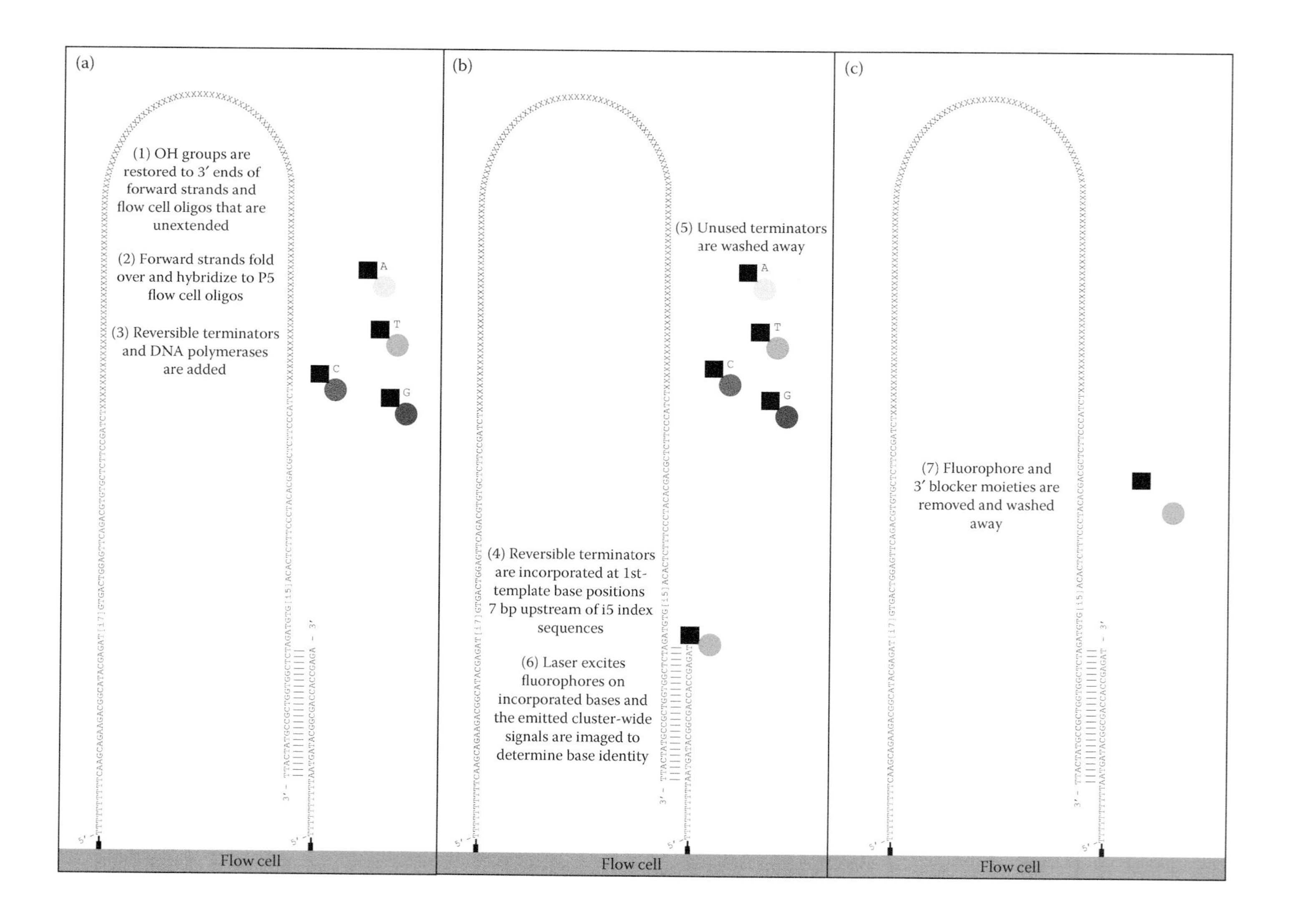

(a)
(1) OH groups are restored to 3′ ends of forward strands and flow cell oligos that are unextended
(2) Forward strands fold over and hybridize to P5 flow cell oligos
(3) Reversible terminators and DNA polymerases are added
A
T
C
G
Flow cell
(b)
(5) Unused terminators are washed away
A
T
C
G
(4) Reversible terminators are incorporated at 1st-template base positions 7 bp upstream of i5 index sequences
(6) Laser excites fluorophores on incorporated bases and the emitted cluster-wide signals are imaged to determine base identity
Flow cell
(c)
(7) Fluorophore and 3′ blocker moieties are removed and washed away
Flow cell

Figure 7.8. Second index read (i5 index) on the forward strands. (a) First, hydroxyl groups must be restored to the ends of exposed 3′ termini. Next, forward strands fold over to hybridize with truncated P5 flow cell oligos followed by the addition of reversible terminators and DNA polymerases. The middle section of the shown forward strand contains the target DNA (a string of Xs). The // signifies that most of the target fragment is omitted due to space limitations and the vertical dashes indicate hydrogen bonding between complementary bases. (b) A reversible terminator becomes incorporated adjacent to a template base 7-bp upstream of the first i5 index template base. Unused terminators are then washed away before the fluorophore-detection and imaging steps. (c) The fluorophore and 3′ blocking group on each incorporated reversible terminator are removed and washed away. This process also regenerates 3′ OH groups on the incorporated terminators enabling them to prime the addition of a new reversible terminator in the next sequencing cycle.

7.2.1 Library Preparation Approaches

Various methods for constructing Illumina libraries have been developed over the years, four of which we will examined here. The "traditional approach" reflects the methodology used in Illumina TruSeq Kits (e.g., http://www.illumina.com/products/truseq-nano-dna-library-prep-kit.html) as well as in some published non-kit protocols (e.g., Bentley et al. 2008; Bronner et al. 2014). A second method, developed by Meyer and Kircher (2010), co-opted the library-making method used for 454 sequencing (Margulies et al. 2005) for generating multiplexed sequencing libraries in a cost-effective manner. Hereafter, this will be referred to as the "Meyer and Kircher approach." A third approach, which is also based on the 454 library methods but has several important innovations that distinguish it from the Meyer and Kircher approach, was developed by Rohland and Reich (2012). Hereafter, this approach will be referred to as the "Rohland and Reich approach." A fourth approach, which uses *in vitro* transposition to construct Illumina libraries, has since become another popular method for making Illumina libraries. This method, which will be referred to as the "Nextera approach," was originally developed and sold by Epicentre Technologies (Syed et al. 2009a,b), but is now sold by Illumina (http://www.illumina.com/products/nextera_dna_library_prep_kit.html). Each of these library-making approaches can be implemented in low-throughput or high-throughput (96 sample plates) workflows.

Although these approaches to making sequencing libraries differ from each other in many ways, all are designed to convert input DNA samples into indexed libraries suitable for Illumina sequencing. In addition to commercially available library kits, a number of protocols have been published, which show, step-by-step, how these libraries can be generated. Use of these "homemade" protocols can substantially reduce the costs of making libraries—particularly if many libraries are needed. However, it must not be forgotten that the process of generating these libraries still involves an intricate series of molecular biology steps each of which must be successfully performed otherwise expensive failures can easily result if something goes wrong. Indeed, the cost of making libraries and then sequencing them on an Illumina platform is on the order of thousands of dollars. It is therefore highly recommended that researchers who do not have experience in making these libraries or do not have well-honed molecular biology skills—especially relating to genomic cloning work—should work closely with someone who is an expert and/or use commercial library kits. One advantage of kits is that they contain control reactions at different steps, which are helpful for monitoring progress and troubleshooting problems. Another advantage of these kits is that they contain ready-to-use "master mixes," which reduces the amount of pipetting and consequently speeds up the preparation process while reducing the amount of plastics consumed as well as errors (e.g., cross-contamination). Illumina provides much information on their website about making high-quality sequencing libraries (http://www.illumina.com/techniques/sequencing/ngs-library-prep.html). We will now review these four approaches to making indexed Illumina sequencing libraries.

7.2.1.1 Traditional Illumina Library Approach

In one of the earliest demonstrations of the Illumina sequencing platform, Bentley et al. (2008) used this technology to sequence the human genome. This study also provided the earliest complete and detailed description of the Illumina sequencing workflow. Since then, Illumina has offered library preparation kits, which are user-friendly compared to the non-kit methodology outlined by Bentley et al. (2008). These kits also permit the researcher to make up

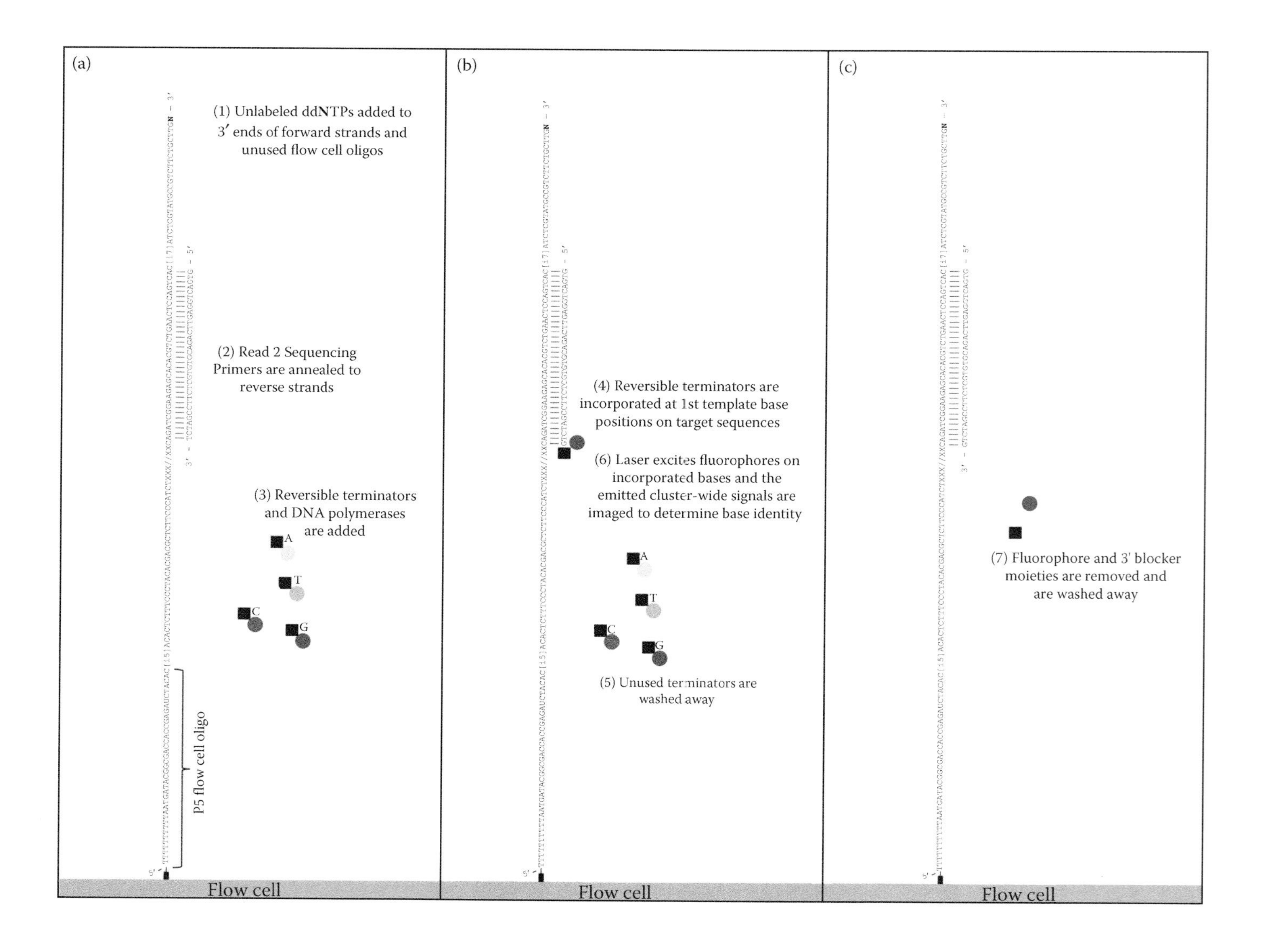

(a)
(1) Unlabeled ddNTPs added to 3′ ends of forward strands and unused flow cell oligos
(2) Read 2 Sequencing Primers are annealed to reverse strands
(3) Reversible terminators and DNA polymerases are added
A
T
C
G
P5 flow cell oligo
Flow cell
(b)
(4) Reversible terminators are incorporated at 1st template base positions on target sequences
(6) Laser excites fluorophores on incorporated bases and the emitted cluster-wide signals are imaged to determine base identity
A
T
C
G
(5) Unused terminators are washed away
Flow cell
(c)
(7) Fluorophore and 3′ blocker moieties are removed and are washed away
Flow cell

Figure 7.9. Second paired-end read of the target DNA on the reverse strands. (a) First, all exposed 3′ ends must be blocked with ddNTPs. Next, Read 2 Sequencing Primers are annealed to the reverse strands followed by the addition of reversible terminators and DNA polymerases. The middle section of the shown reverse strand contains the target DNA (a string of Xs). The // signifies that most of the target fragment is omitted due to space limitations and the vertical dashes indicate hydrogen bonding between complementary bases. (b) A reversible terminator that is complementary to the first template base in the target sequence becomes incorporated. Unused terminators are then washed away before the fluorophore-detection and imaging steps. (c) The fluorophore and 3′ blocking group on each incorporated reversible terminator are removed and washed away. This process also regenerates 3′ OH groups on the incorporated terminators enabling them to prime the addition of a new reversible terminator in the next sequencing cycle.

to 96 individually indexed libraries, which can be multiplexed before sequencing. For example, recent phylogenomic studies by Leaché et al. (2015) and McCormack et al. (2015) have used the Illumina TruSeq® Nano DNA HT Library Prep Kit and Kapa Biosystems KAPA Library Preparation Kit, respectively, in order to generate large multi-locus datasets. In addition to these kits, a number of non-kit protocols for constructing libraries have been published (e.g., Bronner et al. 2014). The traditional approach for making Illumina sequencing libraries consists of the basic steps outlined in Figure 7.10. Although both kit and non-kit approaches share most of the same steps, there are some differences between them. The main difference between the kit and non-kit based methods is that libraries are indexed during the ligation step in kits (step 4 in Figure 7.10), while the indexing step takes place during the first limited cycle PCR in non-kit protocols (step 5 in Figure 7.10). In the following discussion of the traditional approach we will focus on the Illumina kit methodology.

Step 1: Fragmentation of Input DNA Followed by SPRI For many phylogenomics applications, the starting or "input" DNA for an Illumina sequencing library usually consists of extracted genomic DNA but it can also be other forms of DNA such as short or long PCR products. Moreover, the amount of input DNA required for building an Illumina library usually varies between 100 ng and 1 μg depending on the kit or protocol being used. As the Illumina Truseq protocol points out it is extremely important to accurately quantify the input DNA prior to constructing libraries. Some researchers have reported using the UV spectrophotometers such as the Nanodrop to estimate genomic DNA concentrations for Illumina libraries when the DNA samples were free of RNA. However, other researchers have reported that UV spectrophotometric methods often yielded less reliable estimates of DNA concentrations—even when DNA samples had been treated with RNase—compared to estimates obtained using a fluorometric device such as the Qubit (F. Raposo, personal communication, 2016). Thus, when making sequencing libraries it is preferable to use a fluorometer to quantify input DNA samples. For sequencing on an Illumina platform there is an additional critical requirement: the DNA templates must be on the order of several hundred base pairs long. Accordingly, the first step in the traditional library-making process is to shear or fragment the input DNA into smaller pieces (Figure 7.10). When this procedure is performed correctly the researcher obtains a distribution of fragments that are, on average, ~200–300 bp and range ~100–1,000 bp.

Researchers have used a variety of methods for fragmenting DNA such as *nebulization*, *restriction enzymes*, *sonication*, and *acoustic shearing* (Bronner et al. 2014). Although nebulization can be used to generate a sequencing library, it suffers from one major drawback: approximately half of the input sample is lost due to vaporization, which means only a small proportion of the original sample is converted into fragments in the desired size range (Bronner et al. 2014). Another method consists of using restriction enzymes to cut up genomic DNA. The restriction enzyme method can perform well if the procedure is optimized though the G + C content of a genome may ultimately determine how well this method performs (Bronner et al. 2014). Sonication has been a very popular method for creating fragment distributions suitable for Illumina libraries. This method consists of placing a DNA sample in close proximity to a sonicator device for a specific amount of time (longer times result in shorter fragments).

However, like nebulization, use of the sonication method may result in the vast majority of a DNA sample being lost plus heat generated during the sonication procedure may damage the DNA (Bronner et al. 2014). However, one way to improve the efficiency of DNA shearing and normalize shear rates across samples is to use

Step 1: Fragmentation of input DNA followed by SPRI

Input DNA
(e.g., Genomic DNA)

Step 2: End-repair of library fragments followed by SPRI

Step 3: A-tailing of library fragments followed by SPRI*

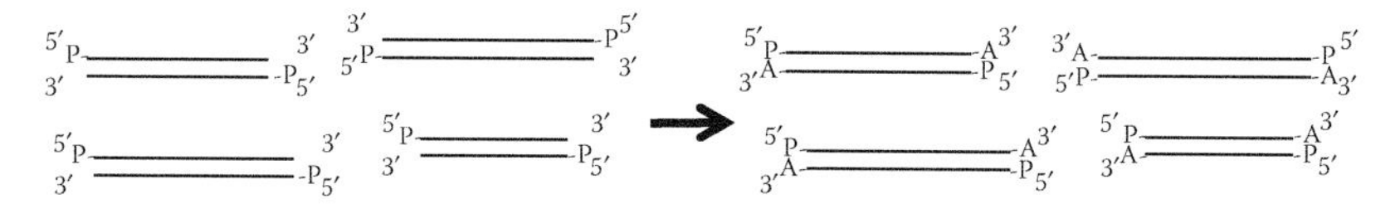

Step 4: Ligation of adapters to library fragments followed by SPRI

Adapter

Library fragment

Adapter

Step 5: First limited cycle PCR followed by SPRI

Step 6: Quantify libraries and verify fragment distributions

Step 7: Normalize and pool libraries (e.g., 4–10 libraries per pool)

In-solution hybrid selection

Figure 7.10. Basic workflow for producing a traditional Illumina sequencing library. Note that the DNA molecules shown in steps 1–4 are largely double-stranded. Following fragmentation of input DNA, the ends of these molecules become "ragged" in that they have 5′ or 3′ single-stranded overhangs. During the end-repair reaction a cocktail of enzymes are used to blunt the ends of the fragments (i.e., make double-stranded end-to-end) and to add a phosphate "P" to the 5′ end of each fragment. Following end-repair, an enzyme adds a single "A" nucleotide to the 3′ end of each fragment. Adapters have complementary 3′ "T" and 5′ "P" to enable ligation with library fragments. A limited cycle PCR is used to selectively amplify only fragments that have an adapter completely ligated to each end. Note that indexing of libraries can occur either during the adapter ligation (i.e., kit protocols) or PCR (i.e., non-kit protocols) steps. See main text for discussion of remaining steps. The asterisk means that the SPRI bead cleanup following the A-tailing reaction is not implemented in some kits, whereas it is used in other protocols (e.g., Fisher et al. 2011). This scheme reflects, with some modifications, the workflows found in the Illumina TruSeq Nano Kit, Bentley et al. (2008), Mamanova et al. (2010), Fisher et al. (2011), and Bronner et al. (2014).

special thin-walled sonication tubes (S. B. Reilly, personal communication, 2016). Focused acoustic shearing is another sound-based method for shearing DNA. In this method, a focused ultrasonicator focuses energy on the sample, which generates the desired narrow size range of fragments without losing a large portion of the original sample (Bronner et al. 2014). This is why acoustic shearing is now the preferred approach to fragmenting DNA. The Truseq kit and protocol by Bronner et al. (2014) both call for using the Covaris S220 or E220 focused ultrasonicators

(Covaris Inc.). An excellent description of Covaris use can be found in Fisher et al. (2011).

Depending on the fragmentation method used, it may be necessary to clean the fragmented DNA before proceeding with the protocol. Also, it is desirable to perform further size-selection procedures on freshly fragmented DNA samples in order to further narrow the fragment size distribution. Although fragments that are smaller or larger than the ideal sizes for Illumina sequencing can be removed in a double-size selection (see Rohland and Reich 2012 for a protocol that uses SPRI beads), in practice usually only the smaller-size fragments are eliminated. Such a size-selection step will enable a larger number of library fragments to create clusters on the flow cell, which can be sequenced. Methods used to accomplish these cleanups include using column-based cleanup kits, agarose-gel extraction followed by column-based cleanups, or SPRI beads. The preferred method at this time is to use SPRI beads owing to the method's efficacy compared to column-based methods, automation-friendliness, and availability of inexpensive generic SPRI beads (see Chapter 6).

Following size selection, it is a good idea to verify the correct fragment size distribution before proceeding with the library-construction process. Although this can be done on a simple 0.7% mini-agarose gel, the preferred approach is to check the results on a Bioanalyzer (Agilent Technologies). This is because the gel can only provide a range of sizes and a rough idea of concentration, whereas the Bioanalyzer can graphically show the actual size distribution and, importantly, indicate the average fragment size as well as the overall concentration (e.g., Mamanova et al. 2010). When performing this genome fragmentation step for the first time researchers should first practice their chosen protocol using an unimportant test DNA sample (e.g., chicken genomic DNA) before attempting the procedure using valuable genetic material and potentially wasting expensive kit reagents.

Step 2: End-Repair of Library Fragments Followed by SPRI The process of fragmenting input DNA creates a large number of smaller dsDNA fragments that have "ragged" ends. This means that many fragments have single-stranded overhangs at their 5′ or 3′ ends and may be missing phosphate groups at their 5′ ends (Figure 7.10). Left unrepaired, these fragments cannot be ligated to library sequencing adapters. Thus, before adapters can be covalently attached to both ends of each library fragment, they must first be enzymatically "end-repaired." Following the end-repair step, each fragment will have blunt or "polished" ends and phosphorylated 5′ termini (Figure 7.10).

The end-repair reaction consists of a cocktail of enzymes, which includes T4 DNA polymerase, Klenow DNA polymerase, and T4 polynucleotide kinase. These enzymes are the usual components of blunt-ending reactions: in the presence of dNTPs T4 DNA polymerase and Klenow DNA polymerase synthesize DNA strands in the 5′ to 3′ direction starting at the 3′ terminal base. Thus, used in tandem (or just T4 DNA polymerase) they are good at filling in gaps on ragged-ended fragments. In addition, both of these enzymes have 3′ to 5′ exonuclease activity and thus they can also remove 3′ overhangs, which also results in blunt-ended products. In addition to blunting fragments, it is critically important to the ligation process that all 5′ terminal bases have a phosphate group. Thus, T4 polynucleotide kinase is included in the end-repair step because this enzyme can phosphorylate the 5′ ends of each fragment lacking those phosphate groups. To complete the process of preparing library fragments for the adapter ligation step, a single "A" nucleotide must be attached to the 3′ ends of each fragment. However, before the next step can be performed, the end polished fragments must be purified otherwise the blunting enzymes will degrade the "A" overhangs soon after they are synthesized. Again, although different methods for cleaning end-repair reactions have been used in the past (e.g., column-based cleanups), SPRI bead cleanup is the preferred method for cleaning end-repair reactions.

Step 3: A-Tailing of Library Fragments Followed by SPRI The purpose of the A-tailing reaction is to add a single "A" nucleotide to the polished 3′ ends of fragments (Figure 7.10). These single 3′ A-overhangs are important because they prevent the formation of chimeric molecules, which can occur when blunt-ended fragments are ligated to each other. As we will see they are also complementary to the single 3′ T-overhangs on the adapters and thus facilitate ligation (Figure 7.10).

The key ingredients of an A-tailing reaction simply consist of the blunted fragments, dATPs, and Klenow DNA polymerase "exo-minus" (or exo-).

Klenow exo- is a mutant version of the Klenow DNA polymerase that has lost its 3′ to 5′ exonuclease activity and thus its function is to simply add a single "A" nucleotide to the 3′ ends of blunted fragments. Following the reaction, the A-tailed library fragments can be cleaned using SPRI beads (e.g., Fisher et al. 2011) though it should be noted that some kit and non-kit protocols do not involve this cleanup.

Step 4: Ligation of Adapters to Library Fragments Followed by SPRI The next step in the workflow consists of ligating adapters to A-tailed library fragments. This traditional library preparation approach uses a single adapter, whereas two distinct adapters are used in other Illumina library preparation methods (see below). If using a library preparation kit, then the adapter constructs are already in complete form; that is, they already contain PCR primer annealing sites, dual indices, and the P5 and P7 common adapters for attachment to flow cell oligos (see Table 7.1 for a list of these oligo sequences). As a side note: whereas adapters in kits are immediately ready for the ligation step (i.e., adapter sequences in Table 7.1 are pre-annealed to each other), non-kit protocols require the user to prepare the adapters for use. This initial construct is called a "forked adapter" because the two sequences comprising the adapter are perfectly complementary to each other for only a 12 bp stretch at one end, whereas the other portions of the two sequences are not complementary to each other (Figure 7.11a). This strange looking adapter design is effective for two reasons. First, the section of the adapter that contains noncomplementary sequences helps to ensure that ligation is directional in that only the correct end of adapters are ligated to A-tailed library fragments (Figure 7.11b). Secondly, the T-overhangs improve the efficiency of the ligation reaction not only because they are complementary to the single A-overhangs on the library fragments, but also by helping to prevent adapters from forming adapter–adapter dimers. In addition to having a 3′ T overhang at the blunt end of the adapter, the adjacent 5′ terminal base is phosphorylated, a feature that enables ligase enzymes to covalently link the 5′ ends of adapter strands to 3′ ends of library fragments (Figure 7.11a). Thus, when the ligation reaction is completed, the desired products will be those that have an adapter completely ligated to each end of a library fragment with the two adapters being inverted with respect to each other (i.e., no nicks will be present in the phosphate backbones of the construct (Figures 7.10 and 7.11b).

During the ligation reaction adapters lacking T-overhangs may become ligated to each other, which results in the formation of adapter dimers that are ~135 bp long. If these adapter dimers

TABLE 7.1
Oligos for Illumina sequencing based on the traditional approach

Name of oligo	Sequence (5′ → 3′)
i5 Index Adapter	AATGATACGGCGACCACCGAGATCTACAC[i5]ACACTCTTTCCCTACACGACGCTCTTCCGATCT
i7 Index Adapter	GATCGGAAGAGCACACGTCTGAACTCCAGTCAC[i7]ATCTCGTATGCCGTCTTCTGCTTG
PCR Primer 1	AATGATACGGCGACCACCGA
PCR Primer 2	CAAGCAGAAGACGGCATACGA
Read 1 Sequencing Primer	ACACTCTTTCCCTACACGACGCTCTTCCGATCT
Read 2 Sequencing Primer	CGGTCTCGGCATTCCTGCTGAACCGCTCTTCCGATCT
i7 Index Read Sequencing Primer	GATCGGAAGAGCACACGTCTGAACTCCAGTCAC

NOTE: The i5 and i7 Index Adapters each contain a unique 8 bp index sequence within the [i5] and [i7] sections of each oligo, respectively. This dual-indexing system allows for the individual identification and hence multiplexing of up to 96 different libraries. Note, oligo sequences are only shown here for purposes of understanding the library-making process using Illumina TruSeq adapters. As Illumina may upgrade its kits, some or all of these oligos may be obsolete. Always check beforehand with the sequencing service provider to make sure that your Illumina libraries are compatible with their sequencing service.

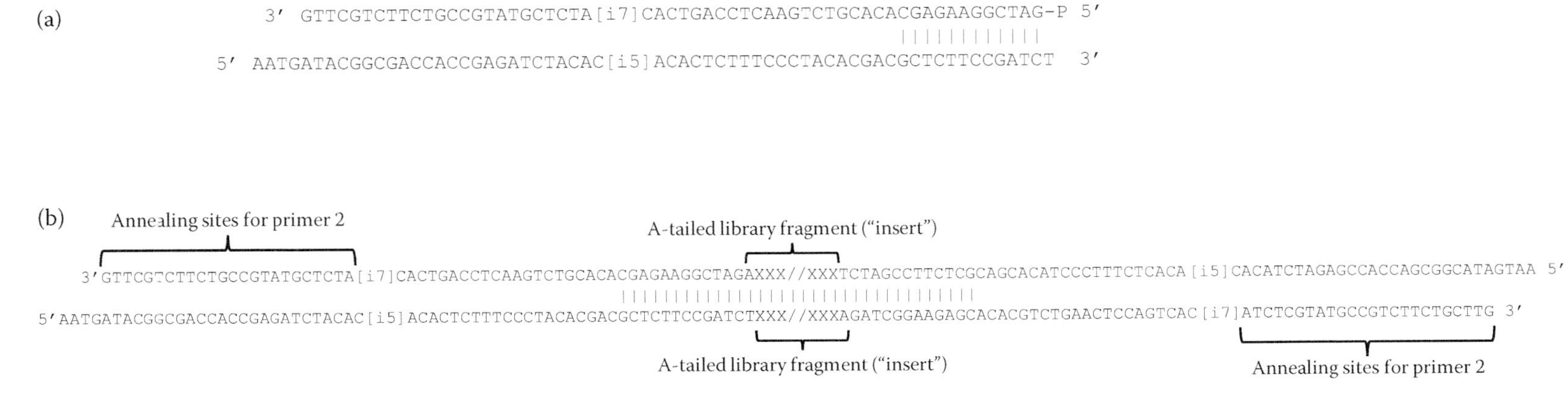

Figure 7.11. Structure of the Illumina “forked adapter” and initial adapter–fragment construct. (a) The two adapter sequences are partially hybridized to each other, which gives the molecule a “forked” appearance. Note the 3′ “T” overhang and 5′ “P” (phosphate) at the end of the adapter where the bases are complementary to each other (vertical dashes indicate hydrogen bonding between complementary bases). (b) An end-repaired and A-tailed library fragment ligated to two adapters. The middle section of the construct contains the target DNA (a string of Xs) and the // signifies that most of the target fragment length is not shown due to space limitations. For Illumina sequencing fragments are often in the 200–600 bp range depending on the protocol. Adapters contain annealing sites for only one of the two PCR primers (PE PCR Primer 2.0) used in the first limited cycle PCR. (Adapter sequences and PCR primers were obtained from Bentley, D. R. et al. 2008. *Nature* 456:53–59, and are also found in Table 7.1.) Note, these sequences are now obsolete in the Illumina product line though some non-kit protocols use them (e.g., Bronner et al. 2014).

are not removed prior to the PCR (next step in workflow), then they will be coamplified with the desired adapter-ligated library fragments. In this scenario, both adapter dimers and target DNA molecules would be capable of copopulating the flow cell thereby diminishing the quality of sequencing data. A variety of cleanup methods can be used to purify ligation products (Bronner et al. 2014). However, the SPRI bead method has been shown to outperform one column-based method in terms of eliminating adapter dimers from ligation reactions (see Figure 4 in Quail et al. 2008). The ability of SPRI beads to effectively clean ligation reactions in addition to the simplicity of the procedure makes this the best cleanup option.

Before moving on to the next step in the workflow, it is essential to verify that the library fragments were successfully ligated to the adapters. One way to assess the outcome of a ligation reaction is to compare a 1 μL sample of preligation reaction mix with 1 μL of postligation reaction side by side on a Bioanalyzer; the correctly ligated products (i.e., adapter–fragment–adapter molecules) are expected to be ~135 bp longer than unligated library fragments (Bronner et al. 2014).

Step 5: First Limited Cycle PCR (or "prehybridization PCR") Followed by SPRI These initial adapter–fragment constructs must be further modified before they can be made suitable for cluster generation and sequencing. Thus, a special type of PCR called *limited cycle PCR* is used to complete the constructs (Figure 7.10). This is called limited cycle PCR because it only involves 6–18 cycles. Note that in the entire library-making and in-solution hybrid selection workflow there are two limited cycle PCRs that must be done. The first occurs immediately after the ligation reaction (this step) while the second will be performed later in the workflow as we will see. These PCRs are commonly referred to as "enrichment PCR" or "indexing PCR" (when indices are added to the adapters via PCR). Thus to avoid ambiguity we will refer to these as the "first limited cycle PCR" and "second limited cycle PCR."

The ligation reaction purification step is primarily needed to eliminate unused adapters and especially adapter dimers. However, "purified" adapter ligation reaction will still consist of a variety of ligation products such as library fragments with an adapter ligated at each end (i.e., the correct construct; see Figure 7.11b) as well as library fragments with only one adapter or no adapters at all. Moreover, some of these products may have one or more nicks still present in their phosphate backbone due to the vagaries of a ligase reaction (e.g., missing phosphate groups on the 5′ ends of adapters or fragments). It is essential to use PCR in order to enrich the library with the correct adapter constructs. If this is not done, then library fragments containing no adapters or only a single adapter—both of which cannot generate clusters on a flow cell—will compete in the hybridization reaction thereby reducing the capture efficiency of target DNA molecules (Fisher et al. 2011; Rohland and Reich 2012). Let's now examine the mechanics of this first limited cycle PCR.

Notice that in Figure 7.11b the initial adapter construct, perhaps oddly, only shows annealing sites for one of the two PCR primers listed in Table 7.1. Where are the annealing sites for the other primer (i.e., PCR Primer 1)? The reaction starts off in a manner different from standard PCR, which is due to the ingenious design of the forked adapters. During this PCR the adapter constructs must first be resolved before they can be amplified. "Resolving the constructs" refers to the PCR-mediated process that converts initial adapter–fragment constructs into DNA molecules that are double-stranded end-to-end and exhibit the proper configuration for all downstream processes in the workflow. Understanding how this PCR works can be quite confusing owing to its unconventional nature. Thus, to appreciate exactly how the forked adapter functions in PCR we will now closely examine the first two cycles in this PCR.

The limited cycle PCR is similar to standard PCR in that each cycle consists of the familiar denaturation → primer annealing → primer extension steps. Thus, at the start of the first cycle the reaction is heated to ~95°C in order to denature the adapter–fragment constructs into single strands (Figure 7.12). The reaction is then cooled down to the primer-annealing temperature at which time only one of the two PCR primers (i.e., PCR Primer 2) anneals to each strand. Next, the samples are heated to 72°C so that new strands can be synthesized during this primer-extension step (Figure 7.12). Notice that the constructs are now completed by the end of the first cycle. The second cycle begins by denaturing the DNA into single strands once again before proceeding to the second primer-annealing step. Because DNA synthesized in the first cycle included the annealing

Figure 7.12. How the first limited cycle PCR completes adapter–fragment library constructs using the traditional Illumina approach. The denaturation and primer-extension steps in the first two cycles of the reaction are illustrated (primer-annealing steps not shown due to space limitation). The initial adapter constructs only contain annealing sites for PCR Primer 2. Annealing sites for the PCR Primer 1 become available after the first cycle and complete adapter–fragment library constructs are produced at the end of the second cycle. The 8-bp i5 and 8-bp i7 index sequences are represented by [i5] and [i7], respectively. The middle section of each strand contains the target DNA (a string of Xs). The // signifies that most of the target fragment is omitted due to space limitation and the vertical dashes indicate hydrogen bonding between complementary bases. Boldface sequences represent newly annealed primer in the same cycle and gray letters represent DNA synthesized in this reaction.

sites for the other PCR primer (i.e., PCR Primer 1) both primers can now anneal to templates and be extended (Figure 7.12). Therefore the constructs can be amplified in an exponential manner over the remaining cycles. The completed adapter–fragment construct, which was shown in Figure 7.1b, now has all sequences in their correct configuration including the P5 and P7 common adapters for the flow cell oligos, the i5 and i7 indices, Read 1 and Read 2 Sequencing Primers, and the i7 Index Read Sequencing Primer. The final procedure of this first limited PCR step is to purify the PCR products using SPRI beads.

Regarding the number of cycles used in this PCR there is an important tradeoff to consider: a larger number of cycles will generate more unique genomic fragments for sequencing but it will also duplicate many fragments and create biases in the distribution of fragments particularly if many PCR cycles are used (Fisher et al. 2011; Bronner et al. 2014). In general, it is advisable to use the fewest number of PCR cycles when enriching libraries. Bronner et al. (2014) provide some rough guidelines for specifying the number of PCR cycles depending on the amount of input DNA: 8 cycles if 500 ng is used, 12 cycles if 200–500 ng, 14 cycles if 50–200 ng, and 18 cycles if <50 ng is input.

Step 6: Quantify Libraries and Verify Fragment Distributions Typically in phylogenomic studies the libraries are pooled together before they are used in the hybrid selection phase of the workflow. To do this, however, the concentration of each library must be estimated and its fragment distribution verified. The most accurate way to estimate the concentration of a sequencing library is to perform qPCR using primers that anneal to the flow cell oligo sites on the adapters (Bronner et al. 2014). However, because qPCR is expensive, concentration estimates based on the Bioanalyzer are usually adequate for normalizing and pooling libraries (Bronner et al. 2014; A. D. Gottscho, personal communication, 2016). Many other studies have used fluorometric devices (e.g., Qubit) to estimate library concentrations. Although it may be possible to check the size range of adapter-ligated fragments on an agarose gel, this approach is not ideal because it does not show what the average adapter–fragment size is. Thus, Bioanalyzer assays represent the preferred means for examining fragment distributions of libraries. Bioanalyzer electropherograms can also indicate whether or not adapter dimers were amplified (Bronner et al. 2014). The average fragment sizes will depend largely on the input library fragment distribution at the beginning of the workflow. In any case, better cluster generation results will be obtained if the average adapter–fragment size matches the target size specified in the cluster generation kit that will be used.

Step 7: Normalize and Pool Libraries (e.g., 4–10 Libraries per Pool) In most phylogenomic studies, owing to the large number of libraries made (e.g., >25), all indexed libraries will ultimately be pooled together before being sequenced in at least one flow cell lane. However, at this point in the workflow a decision must be made as to how many indexed libraries will be combined together *for each hybridization reaction*. This is because of a tradeoff: the higher the number of libraries being "captured" in a single hybridization reaction the lower the per library hybridization efficiency. This means that pools containing a larger number of libraries will be more susceptible to "missing data" problems as different loci in different individuals will have missing sequences. Although the opposite extreme—using a single library per hybridization reaction—will yield the best capture results, this strategy is costly in terms of reagents. Another important factor to consider is genome size because larger genomes have fewer copies of each target (Lemmon et al. 2012; Portik et al. 2016; S. B. Reilly, personal communication, 2016). For example, for frogs, which have very large genomes, hybridization pools consisting of 4–6 libraries have worked well (Portik et al. 2016; S. B. Reilly, personal communication, 2016). For organisms with smaller genomes, the recent trend has been for researchers to group around 6–10 indexed libraries into individual hybridization pools (e.g., Faircloth et al. 2013, 2015; Smith et al. 2013; McCormack et al. 2015). Using the obtained concentration estimates, each library is first normalized so that all libraries will have equal representation in the resulting sequencing data. Thus, following normalization and deciding pool sizes libraries are combined in equimolar ratios to form each pool. Once this step is completed, library pools are ready to enter the hybrid selection workflow followed by cluster generation and sequencing (Figure 7.10). If problems are encountered while preparing traditional Illumina sequencing libraries, the reader can consult

Bronner et al. (2014) and Illumina Truseq protocol for troubleshooting tips.

If we look back at the entire traditional library preparation workflow in Figure 7.10, we see that there are four to five discrete SPRI bead cleanup steps. Fisher et al. (2011) developed a fantastic innovation to improve the efficiency and efficacy of this workflow, which relates to these cleanup steps. They developed a method called *with-bead SPRI*, which means that the SPRI beads added to the DNA samples following the fragmentation step *remain* with the input DNA throughout the library-making procedure. In between enzymatic reactions, only the PEG/NaCl buffer is exchanged (Figure 7.13). The library fragments are only eluted from the beads following the cleanup of PCR products made in the first limited cycle PCR. Thus, the cleanup steps are linked together rather than occurring as discrete steps in the workflow (Fisher et al. 2011).

There are many significant benefits to using this with-bead SPRI workflow. First, fewer beads are consumed, which saves a considerable amount of money. Second, this approach is automation friendly and thus large numbers of libraries can be constructed using liquid-handling robots or by humans with multichannel pipettes. Third, owing to the efficiency of the SPRI method relative to column-based purification kits as well as the fewer liquid transfer steps, 80%–90% of the DNA is recovered versus 50%–60% recovery rate for column-based cleanups (Fisher et al. 2011).

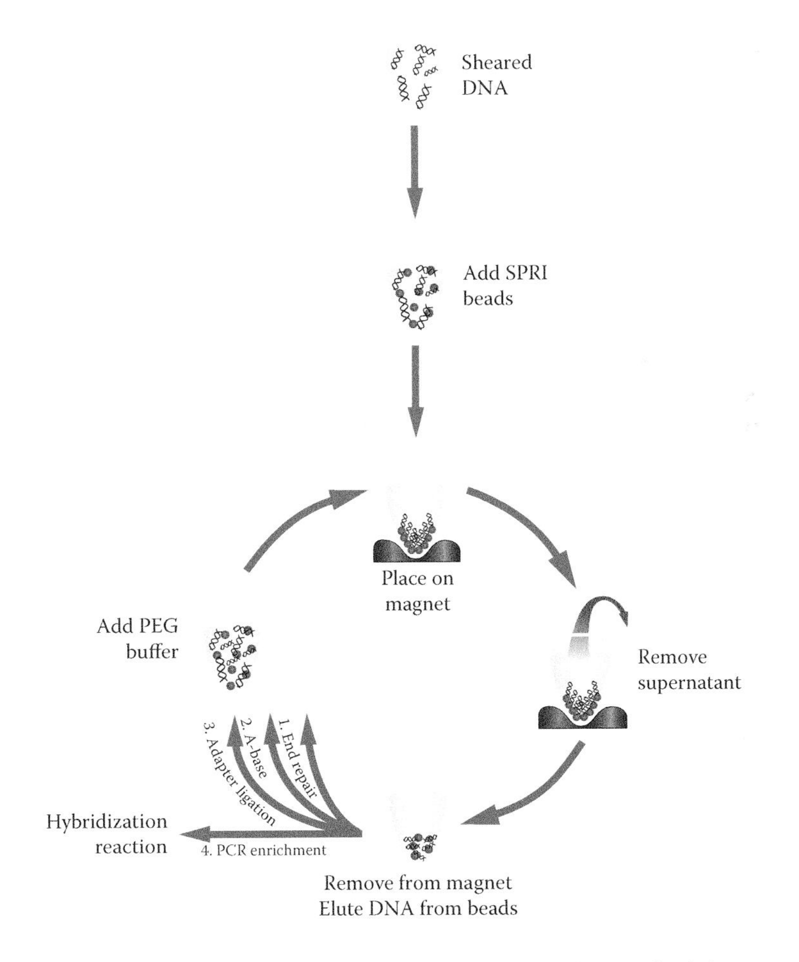

Figure 7.13. Fisher "with-bead" library preparation workflow. In contrast with earlier SPRI-bead clean-up protocols, which require the addition of new aliquots of beads at the start of each enzymatic clean-up step, the Fisher et al. method only requires the addition of fresh beads immediately after the DNA fragmentation step. In subsequent clean-up steps, the PEG buffer (and the not the beads) are exchanged. See main text and Fisher et al. (2011) for further explanation. (Reprinted from Fisher, S. et al. 2011. *Genome Biol* 12:1–15. With permission.)

Because of this improvement in the DNA recovery rate, less input DNA is needed to start a library (e.g., 100 ng can be used to obtain good hybrid selection results) and fewer PCR cycles (e.g., 6–8) are needed in the first limited cycle PCR. A fourth benefit is that the with-bead SPRI workflow can increase the diversity of a library (i.e., increase the number of unique genomic fragments) 12-fold. Moreover, the with-bead method can be used for any of the library-making approaches described in this chapter. With all these advantages it is not surprising that researchers have adopted this modification in their library-making protocols (e.g., Rohland and Reich 2012; Faircloth et al. 2015; McCormack et al. 2015).

7.2.1.2 Meyer and Kircher Library Approach

Meyer and Kircher (2010) developed a protocol for easily and inexpensively making large numbers of indexed Illumina sequencing libraries. This approach, which is based on the 454 sequencing library construction methods of Margulies et al. (2005), has been a popular method for generating enormous phylogenomic datasets (e.g., Lemmon et al. 2012; Portik et al. 2016). The basic workflow for the Meyer and Kircher approach is outlined in Figure 7.14. This approach shares many of the same steps with the traditional library preparation approach (compare with Figure 7.10) and thus we will not discuss in detail every step as we did before.

Like the traditional library preparation method, the Meyer and Kircher approach begins with the fragmentation of an input DNA sample but it does not involve an SPRI bead cleanup of the fragmentation products. However, as an SPRI bead cleanup can be used to good effect in this step (i.e., eliminate fragments too small for sequencing; Fisher et al. 2011), this can be considered an optional cleanup. Next, library fragments are end-repaired followed by an SPRI bead cleanup (Figure 7.14). Instead of A-tailing the blunt-ended and phosphorylated library fragments as is done in the traditional method, the blunt-ended fragments are ligated to nonphosphorylated adapters in a blunt-ended ligation reaction. Another difference between methods is that the Meyer and Kircher approach uses two different adapters (Figure 7.14), which are called "Adapter P5" and "Adapter P7," whereas the traditional approach uses a single forked adapter (Figures 7.10 and 7.11a). Thus, while a ligation reaction involving forked adapters will only produce one type of adapter–fragment–adapter construct, a ligation reaction with the Adapter P5 and Adapter P7 combo can generate three different adapter–fragment–adapter products: Adapter P5–library fragment–Adapter P5, Adapter P5–library fragment–Adapter P7, and Adapter P7–library fragment–Adapter P7. As you may have just realized, only the ligation products containing both Adapter P5 and Adapter P7 can create clusters on a flow cell. However, as we will see below, ligation products having two Adapter P5s and two Adapter P7s will not affect the quality of the completed library.

Let's take a closer look at the structure of Adapters P5 and P7. Table 7.2 shows the three different oligos (i.e., IS1, IS2, and IS3) that are used to construct Adapters P5 and P7. Notice that each of these three oligos is fortified with four phosphorothioate bonds on their 5′ and 3′ ends (Table 7.2). These special bonds are added to prevent 3′ to 5′ exonuclease digestion of the oligos. If these bases are not protected, then they would be vulnerable to being digested from contaminating exonucleases or from the blunting enzymes used in the end-repair reaction. Adapters P5 and P7 are made by hybridizing oligo IS1 with IS3 and IS2 with IS3, respectively (Figure 7.15a,b). The two strands of each adapter type are complementary to each other but notice that one strand on each adapter has a long 5′ overhang (Figure 7.15a,b). The existence of these overhangs ensures that each adapter can be ligated to the library fragment in only one way (Margulies et al. 2005). Unlike the Illumina forked adapter (see Figure 7.11a), Adapters P5 and P7 do not have phosphates at their 5′ ends. This means that ligase enzymes cannot seal the phosphate backbone where an unphosphorylated 5′ adapter end is positioned adjacent to the 3′ end of a library fragment. In contrast, ligase does seal the backbone where the 3′ end of an adapter comes into contact with the phosphorylated 5′ end of a library fragment. The ligation reaction thus produces an initial adapter construct with two "nicks" (or gaps) present in the phosphate backbone along with two 5′ overhangs (Figures 7.14 and 7.15c). Before this initial adapter construct can be subjected to the first limited cycle PCR, a "fill-in" reaction must be performed in which an enzyme called *Bst* DNA polymerase, Large Fragment, is used to fill-in the missing bases and seal both nicks (Margulies

Step 1: Fragmentation of input DNA followed by SPRI*

Input DNA
(e.g., Genomic DNA)

Step 2: End-repair of library fragments followed by SPRI

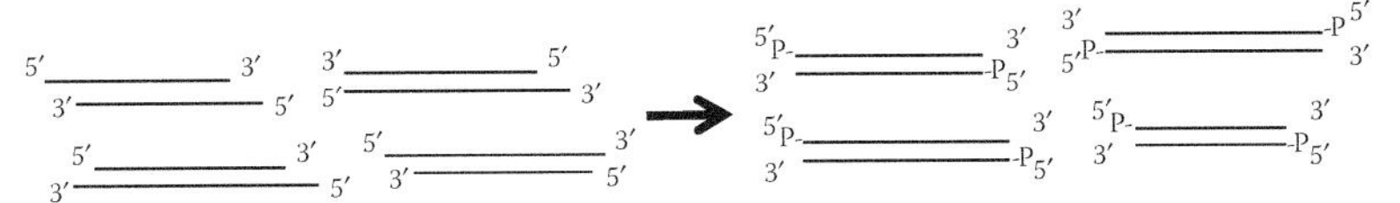

Step 3: Ligation of adapters to library fragments followed by SPRI

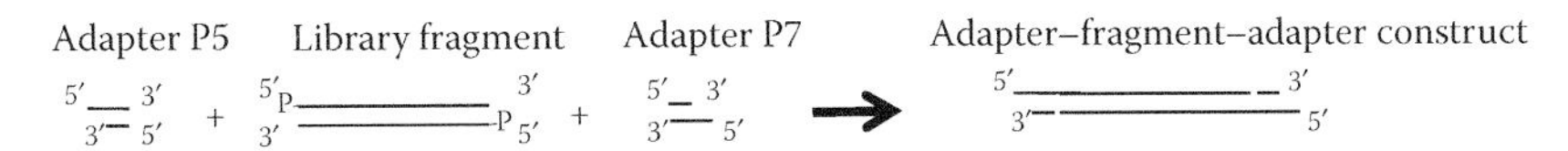

Step 4: Fill-in reaction to repair library fragments followed by SPRI

fill-in both 5′ overhangs and seal both nicks

Blunt-ended adapter–fragment–adapter construct with two continuous DNA strands

Step 5: First limited cycle PCR followed by SPRI

Step 6: Quantify libraries and verify fragment distributions

Step 7: Normalize and pool libraries (e.g., 4–10 libraries per pool)

In-solution hybrid selection

Figure 7.14. Basic workflow for producing an Illumina sequencing library using the Meyer and Kircher approach. Note that the DNA molecules shown in steps 1–4 are largely double-stranded. Following fragmentation of input DNA, the ends of these molecules become "ragged" in that they have 5′ or 3′ single-stranded overhangs. During the end-repair reaction a cocktail of enzymes are used to blunt the ends of the fragments (i.e., make double-stranded end-to-end) and to add a phosphate "P" to the 5′ end of each fragment. Following end-repair, two different adapters (Adapters P5 and P7) are ligated to the blunt-ended library fragments. In Step 4 a fill-in reaction adds bases to complement the 5′ overhangs and to seal the two nicks in the phosphate backbone (shown as gaps in the strands). Next, a limited cycle PCR is used to selectively amplify only fragments that have both types of adapters completely ligated the ends and to complete the adapter constructs by adding sample-specific indices, sequences complementary to flow cell oligos and sequencing primers. The asterisk means the SPRI bead cleanup following fragmentation is not included in the actual Meyer and Kircher (2010) protocol, but it can be used at this step to eliminate fragments too small for sequencing. See main text for descriptions of the other steps.

et al. 2005; Meyer and Kircher 2010). The resulting products consist of adapter–fragment constructs that are double-stranded end-to-end (i.e., without nicks) and contain the annealing sites for both PCR primers used in the first limited cycle PCR (Figure 7.15d; Table 7.2).

The adapter–fragment constructs are now ready to be used as templates in the first limited cycle PCR (Figure 7.14). Although this first PCR is used to "enrich" the library with the correct adapter–fragment constructs in both the traditional Illumina library preparation and Meyer and Kircher approaches, there are some differences in the PCR procedure between these two methods. While the traditional kit-based approach uses PCR to resolve the adapter constructs using a pair

TABLE 7.2
Oligos for Illumina sequencing using the Meyer and Kircher approach

Name of oligo	Sequence (5′ → 3′)
IS1_adapter.P5	A*C*A*C*TCTTTCCCTACACGACGCTCTTCCG*A*T*C*T
IS2_adapter.P7	G*T*G*A*CTGGAGTTCAGACGTGTGCTCTTCCG*A*T*C*T
IS3_adapter.P5 + P7	A*G*A*T*CGGAA*G*A*G*C
IS4_indPCR.P5	AATGATACGGCGACCACCGAGATCTACACTCTTTCCCTACACGACGCTCTT
Indexing PCR primer	CAAGCAGAAGACGGCATACGAGAT[i7]GTGACTGGAGTTCAGACGTGT
IS5_reamp.P5	AATGATACGGCGACCACCGA
IS6_reamp.P7	CAAGCAGAAGACGGCATACGA
IS7_short_amp.P5	ACACTCTTTCCCTACACGAC
IS8_short_amp.P7	GTGACTGGAGTTCAGACGTGT
BO1.P5.F	AATGATACGGCGACCACCGAGATCTACACTCTTTCCCTACACGACGCTCTTCCGATCT-Pho
BO2.P5.R	AGATCGGAAGAGCGTCGTGTAGGGAAAGAGTGTAGATCTCGGTGGTCGCCGTATCATT-Pho
BO3.P7.part1.F	AGATCGGAAGAGCACACGTCTGAACTCCAGTCAC-Pho
BO4.P7.part1.R	GTGACTGGAGTTCAGACGTGTGCTCTTCCGATCT-Pho
BO5.P7.part2.F	ATCTCGTATGCCGTCTTCTGCTTG-Pho
BO6.P7.part2.R	CAAGCAGAAGACGGCATACGAGAT-Pho
Read 1 Sequencing Primer	ACACTCTTTCCCTACACGACGCTCTTCCGATCT
Read 2 Sequencing Primer	GTGACTGGAGTTCAGACGTGTGCTCTTCCGATCT
Index Read Sequencing Primer	GATCGGAAGAGCACACGTCTGAACTCCAGTCAC

NOTE: Oligos IS1 to IS3 are used to make the initial adapters for the ligation reaction (asterisk = phosphorothioate bond); IS4 and the indexing PCR primer are used in the first limited cycle PCR; IS5 and IS6 are used to amplify a library following target capture and IS7 and IS8 can be used to amplify a library prior to the indexing PCR; BO1-06 are blocking oligos for target capture (Pho = 3′ phosphate); and Read 1, Read 2, and Index Read Sequencing Primers are used to sequence the completed library construct. Note, the [i7] sequence embedded in the indexing primer represents the 7-bp index sequence (see Supplementary File in Meyer and Kircher 2010 for a list of 228 different indexing PCR primers). (All oligos are from Meyer, M. and M. Kircher. 2010. *Cold Spring Harbor Protocols*. doi: 10.1101/pdb.prot5448.) Note, these oligos are shown here only to help illustrate this library construction approach. Readers should consult the original source by Meyer and Kircher (2010) for the actual step-by-step protocol as well as listing of all required oligos for making multiplexed Illumina libraries.

of typical nontailed PCR primers, the Meyer and Kircher approach instead uses 5′-tailed PCR primers in this PCR in order to complete each construct by adding additional sequence elements including indices. Because PCR is used to index libraries in this fashion—rather than adding them to the constructs during the ligation step—that is why this is sometimes called an indexing PCR. Note, there are traditional non-kit library protocols (e.g., Bronner et al. 2014) that also use tailed PCR primers in the first limited cycle PCR to complete the constructs similar to the Meyer and Kircher approach but we will not consider those methods here.

Let's now examine the first limited cycle PCR in the Meyer and Kircher workflow. In contrast to the traditional approach, the initial adapter construct in the Meyer and Kircher approach contains annealing sites for both PCR primers (Figure 7.15d). Thus, if we look at the first three cycles of a limited PCR using the Meyer and Kircher adapter–fragment constructs we see that both primers (IS4 and indexing PCR primer) anneal to their templates during the first cycle (Figure 7.16). By the end of the second cycle, we see that full-length adapter–fragment strands are produced and that completed double-stranded constructs only appear at the end of the third cycle. The remaining cycles are used to amplify the number of these completed constructs. Figure 7.17 shows the final sequencing-ready library construct, which

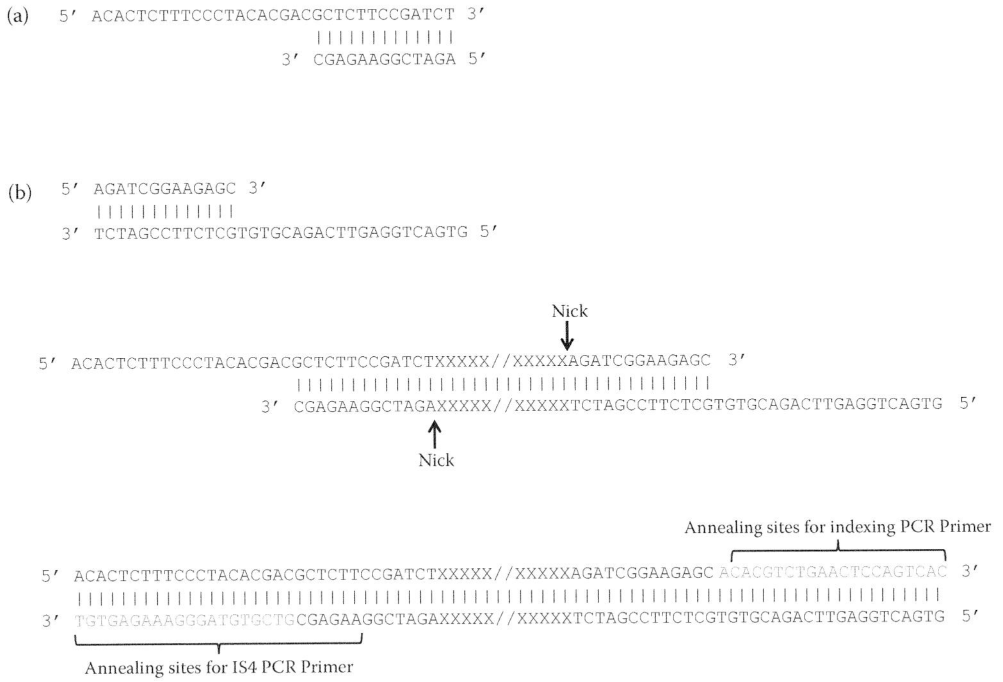

```
(a)  5′ ACACTCTTTCCCTACACGACGCTCTTCCGATCT 3′
                                |||||||||||||
                             3′ CGAGAAGGCTAGA 5′


(b)  5′ AGATCGGAAGAGC 3′
        |||||||||||||
     3′ TCTAGCCTTCTCGTGTGCAGACTTGAGGTCAGTG 5′


                                                 Nick
                                                  ↓
5′ ACACTCTTTCCCTACACGACGCTCTTCCGATCTXXXXX//XXXXXAGATCGGAAGAGC 3′
                                ||||||||||||||||||||||||||||||||
                             3′ CGAGAAGGCTAGAXXXXX//XXXXXTCTAGCCTTCTCGTGTGCAGACTTGAGGTCAGTG 5′
                                         ↑
                                        Nick

                                                                Annealing sites for indexing PCR Primer
5′ ACACTCTTTCCCTACACGACGCTCTTCCGATCTXXXXX//XXXXXAGATCGGAAGAGC ACACGTCTGAACTCCAGTCAC 3′
   ||||||||||||||||||||||||||||||||||||||||||||||||||||||||||||||||||||||||||||||||
3′ TGTGAGAAAGGGATGTGCTGCGAGAAGGCTAGAXXXXX//XXXXXTCTAGCCTTCTCGTGTGCAGACTTGAGGTCAGTG 5′
   Annealing sites for IS4 PCR Primer
```

Figure 7.15. Structure of the Meyer and Kircher Illumina adapters and initial adapter–fragment constructs. (a) Adapter P5 consists of the IS1 (above) and IS3 (below) oligos hybridized to each other (vertical dashes indicate hydrogen bonding between complementary bases). (b) Adapter P7 consists of the IS3 (above) and IS2 (below) oligos hybridized to each other. Note, phosphorothioate bonds not shown (see Table 7.2). (c) Following the ligation reaction, the initial adapter–fragment construct has long 5′ overhangs at each end as well as two nicks in the phosphate backbone at two junctions where the adapters and fragment meet. (d) A fill-in reaction repairs this initial construct by sealing the two nicks and filling in bases in a 5′ to 3′ direction (gray bases represent newly synthesized DNA). The result is a double-stranded and blunt-ended adapter–fragment construct. The middle section of the adapter–fragment constructs contains the target DNA (a string of Xs). The // signifies that most of the target fragment is omitted due to space limitation. Adapters contain annealing sites for IS4 and indexing PCR primers, which are used in the first limited cycle PCR. (Adapter sequences and PCR primers were obtained from Meyer, M. and M. Kircher. 2010. *Cold Spring Harbor Protocols*. doi: 10.1101/pdb.prot5448, and are also found in Table 7.2.)

contains the following elements: P5 and P7 flow cell adapters, 7-bp sample-specific index, annealing sites for the PCR primers in the second limited cycle PCR (IS5 and IS6 Primers), annealing sites for the Read 1 and Read 2 Sequencing Primers, and annealing sites for the Index Read Sequencing Primer.

Earlier we saw that the Meyer and Kircher ligation products consist of three types of adapter–fragment–adapter constructs. Given that only one of the three construct types can create clusters on the flow cell, should we be concerned that PCR might coamplify all three constructs and thus adversely affect sequencing data? Fortunately, this is not a problem thanks to an interesting phenomenon known as the *suppression PCR effect* (Siebert et al. 1995; Diatchenko et al. 1996; Lukyanov et al. 2007). Following the denaturation phase of PCR, the single-stranded templates cool down toward the primer-annealing temperature. Strands that have inverted terminal repeat sequences—in this case an adapter at one end and the reverse complement of this same adapter at the opposing end—are likely to form intramolecular terminal duplexes (secondary structures) that resemble "panhandles" (Siebert et al. 1995; Diatchenko et al. 1996; Lukyanov et al. 2007; Meyer and Kircher 2010). For ligation products having only a single adapter type on both ends, this means that the panhandle duplex will consist of a P5 or P7 sequence duplexed to its reverse complement sequence found at the other end of the strand. Thus, fragments having only a single adapter type will not be amplified in this reaction; instead, only the fragments having

Figure 7.16. How the first limited cycle PCR completes adapter–fragment library constructs using the Meyer and Kircher approach. The denaturation and primer-extension steps in the first three cycles of a limited PCR are illustrated (primer-annealing steps not shown due to space limitation). (*Continued*)

Figure 7.16. (Continued) The initial adapter constructs contain annealing sites for both the IS4 PCR primer and indexing PCR primer. Completed adapter–fragment constructs appear only at the end of the third cycle. The 7-bp index sequence is represented by [i7]. The middle section of the adapter–fragment constructs contains the target DNA (a string of Xs). The // signifies that most of the target fragment is omitted due to space limitation and the vertical dashes indicate hydrogen bonding between complementary bases. Boldface sequences represent newly annealed primer in the same cycle and gray letters represent DNA synthesized in this reaction.

```
                                                                                                   P7 flow cell oligo
                                                                                          3' TAGAGCATACGGCAGAAGACGAAC 5'

                                                   Read 2 Sequencing Primer                   IS6_reamp.P7
                                           3' TCTAGCCTTCTCGTGTGCAGACTTGAGGTCAGT 5'   3' AGCATACGGCAGAAGACGAAC 5'

3'TTACTATGCCGCTGGTGGCTCTAGATGTGAGAAAGGGATGTGCTGCGAGAAGGCTAGAXXX//XXXTCTAGCCTTCTCGTGTGCAGACTTGAGGTCAGTG[i7]TAGAGCATACGGCAGAAGACGAAC 5'
  ||||||||||||||||||||||||||||||||||||||||||||||||||||||||||||||||||||||||||||||||||||||||||||||||||||||||||||||||||||||||||||||
5'AATGATACGGCGACCACCGAGATCTACACTCTTTCCCTACACGACGCTCTTCCGATCTXXX//XXXAGATCGGAAGAGCACACGTCTGAACTCCAGTCAC[i7]ATCTCGTATGCCGTCTTCTGCTTG 3'

                    5' ACACTCTTTCCCTACACGACGCTCTTCCGATCT 3'   5' GATCGGAAGAGCACACGTCTGAACTCCAGTCAC 3'
                            Read 1 Sequencing Primer                Index Read Sequencing Primer
5'AATGATACGGCGACCACCGA 3'
     IS5_reamp.P5

5'AATGATACGGCGACCACCGAGAUCTACAC 3'
         P5 flow cell oligo
```

Figure 7.17. Completed Meyer and Kircher adapter–fragment construct for Illumina sequencing. The construct contains annealing sites for the P5 and P7 flow cell oligos, PCR primers for amplifying the library after target capture (i.e., primers IS5 and IS6), Read 1 and 2 Sequencing Primers, and the Index Read Sequencing Primer. The 7-bp index sequence is represented by [i7]. The middle section of the adapter–fragment construct contains the target DNA (a string of Xs). The // signifies that most of the target fragment is omitted due to space limitation and the vertical dashes indicate hydrogen bonding between complementary bases. (Oligos are from Meyer, M. and M. Kircher. 2010. *Cold Spring Harbor Protocols* doi: 10.1101/pdb.prot5448, and are also found in Table 7.2.)

both types of adapters will be amplified, which is the desired result (Meyer and Kircher 2010). When PCR is used to selectively amplify products in this manner it is referred to as *suppression* PCR (Siebert et al. 1995; Diatchenko et al. 1996; Lukyanov et al. 2007; Adey et al. 2010).

Following the limited cycle PCR, the PCR products are cleaned using SPRI beads. At this time, the libraries are quantified, normalized, and pooled in the same manner already described for the traditional approach.

7.2.1.3 Rohland and Reich Library Approach

The Rohland and Reich (2012) approach to making Illumina sequencing libraries resembles the Meyer and Kircher approach because it is also based on the 454 library preparation methods developed by Margulies et al. (2005). Indeed, their library-making workflow consists of the same basic steps found in the Meyer and Kircher workflow (see Figure 7.14) except that the former approach uses an SPRI bead cleanup step following fragmentation of the DNA, whereas the latter does not (see Figure 1 in Rohland and Reich 2012 for a detailed workflow). There are, however, a number of important differences that distinguish these approaches. First, Rohland and Reich (2012) developed a high-throughput workflow that can enable academic laboratories to generate thousands of sequencing libraries (192 indexed libraries/day) at far lower cost/library (one to two orders of magnitude) compared to commercial library kits. Another major innovation of the Rohland and Reich approach concerns the use of novel adapters, which can greatly enhance the quality of phylogenomic datasets.

Table 7.3 lists the oligos that comprise these adapters as well as other oligos required for the library and in-solution hybrid selection workflows. Unlike other Illumina adapter designs discussed in this chapter, their P5 adapter contains a 6-bp internal "barcode" sequence (Table 7.3). The barcode is used to individually mark libraries so that they can be grouped together into hybrid

TABLE 7.3

Oligos for Illumina sequencing using the Rohland and Reich approach

Name of oligo	Sequence (5′ → 3′)
Barcoded P5 adapter	CTTTCCCTACACGACGCTCTTCCGATCT [barcode]
Barcoded P5 adapter complement	[barcode] AGATCGGAA
PE P7 adapter	CTCGGCATTCCTGCTGAACCGCTCTTCCGATCT
PE P7 adapter complement	AGATCGGAAGAGC
PreHyb-PE_F	CTTTCCCTACACGACGCTCTTC
PreHyb-PE_R	CTCGGCATTCCTGCTGAACC
Sol-PE-PCR_F	AATGATACGGCGACCACCGAGATCTACACTCTTTCCCTACACGACGCTCTTC
Index PE Primer	CAAGCAGAAGACGGCATACGAGAT [index] CGGTCTCGGCATTCCTGCTGAACC
Index PE Sequencing Primer	GATCGGAAGAGCGGTTCAGCAGGAATGCCGAGACCG
PE Read 1 Sequencing Primer	ACACTCTTTCCCTACACGACGCTCTTCCGATCT
PE Read 2 Sequencing Primer	CGGTCTCGGCATTCCTGCTGAACCGCTCTTCCGATCT
Univ_Block_P5	AGATCGGAAGAGCGTCGTGTAGGGAAAG
Univ_Block_P7	AGATCGGAAGAGCGGTTCAGCAGGAATGCCGAG

SOURCE: Rohland, N. and D. Reich. 2012. *Genome Res* 22:939–946, and © 2007–2010 Illumina, Inc. All rights reserved.

NOTE: The first four oligos are used to make the two adapters for the ligation reaction; PreHyb-PE_F and PreHyb-PE_R are used to amplify the initial adapter-library fragment construct in the first limited cycle PCR; Read 1, Read 2, and Index Read Sequencing Primers are used to sequence the completed library construct, and Univ_Block P5 and P7 are blocking oligos for target capture. A 6-bp barcode sequence is used to individually mark libraries prior to making hybridization pools. A 7-bp index sequence is added during the second limited PCR to distinguish pools (see Supplementary File in Rohland and Reich 2012 for a list of 16 different indexing PCR primers). All oligos are from Rohland and Reich (2012). Note, these oligos are shown here only to help illustrate this library construction approach. Readers should consult Rohland and Reich (2012) for the actual protocol as well as listing of all required oligos for making multiplexed Illumina libraries. Always check beforehand with the sequencing service provider to make sure that your Illumina libraries are compatible with their sequencing service.

selection pools. We will return to this subject later and discuss the significance of this barcode sequence and how it can lead to better quality sequencing data.

When each pair of complementary adapter sequences is annealed to the other, they form two distinct adapters (Figure 7.18a,b). Like the Meyer and Kircher adapters, both adapters do not have phosphorylated 5′ ends, which is by design in order to prevent the formation of adapter dimers and reduce oligo costs (Rohland and Reich 2012). The single-stranded 5′ overhangs on the adapters (Figure 7.18a,b) also ensure that they can only be ligated to library fragments in the correct manner.

After the library fragments have been end-repaired/phosphorylated and SPRI-cleaned, they are ligated to the adapters. The desired ligation products are similar to those produced in a ligation reaction with the Meyer and Kircher adapters (compare Figures 7.15c and 7.18c). Accordingly, a nick-sealing and fill-in reaction using *Bst* DNA Polymerase, Large Fragment, must be performed in order to create constructs that are double-stranded from end-to-end (Figure 7.18d). Notice that this adapter construct contains the annealing sites for both PCR primers, which are used in the first limited cycle PCR (a.k.a. suppression or prehybridization PCR; Figure 7.18d) and thus both primers can begin synthesizing new strands in the first PCR cycle (not shown).

In contrast to the Meyer and Kircher approach, the first limited cycle PCR in the Rohland and Reich approach does not involve the use of 5′ tailed PCR primers. As you may recall, the Meyer and Kircher approach uses tailed primers in the first PCR to index and complete sequencing-ready constructs (see Figure 7.16). The Rohland and Reich constructs on the other hand, are not

(a)

```
5' CTTTCCCTACACGACGCTCTTCCGATCT[barcode] 3'
                          |||||||||||||||||
                       3' AAGGCTAGA[barcode] 5'
```

(b)

```
5' AGATCGGAAGAGC 3'
   |||||||||||||
3' TCTAGCCTTCTCGCCAAGTCGTCCTTACGGCTC 5'
```

(c)

```
                                                            Nick
                                                             ↓
5' CTTTCCCTACACGACGCTCTTCCGATCT[barcode]XXXXX//XXXXXAGATCGGAAGAGC 3'
                          |||||||||||||||||||||||||||||||||||||||||
                 3'       AAGGCTAGA[barcode]XXXXX//XXXXXTCTAGCCTTCTCGCCAAGTCGTCCTTACGGCTC 5'
                                           ↑
                                          Nick
```

(d)

```
                                                     Annealing sites for PreHyb-PE_R PCR primer
5' CTTTCCCTACACGACGCTCTTCCGATCT[barcode]XXXXX//XXXXXAGATCGGAAGAGCGGTTCAGCAGGAATGCCGAG 3'
   ||||||||||||||||||||||||||||||||||||||||||||||||||||||||||||||||||||||||||||||||||||
3' GAAAGGGATGTGCTGCGAGAAGGCTAGA[barcode]XXXXX//XXXXXTCTAGCCTTCTCGCCAAGTCGTCCTTACGGCTC 5'
Annealing sites for PreHyb-PE_F PCR primer
```

Figure 7.18. Structure of the Rohland and Reich Illumina adapters and initial adapter–fragment constructs. (a) One adapter contains a 6-bp "barcode" sequence that is used to mark individual libraries prior to making hybridization pools (vertical dashes indicate hydrogen bonding between complementary bases). (b) The second adapter does not contain barcode or index sequences and resembles one of the Meyer Kircher adapters (see Figure 7.15b). (c) The initial ligation product contains two nicks in the phosphate backbone and two long 5′ overhangs. Before this construct can be used in the first limited cycle PCR, the nicks must be repaired and the overhangs filled in. (d) Following the fill-in reaction, the adapter–fragment construct is double-stranded end-to-end and contains annealing sites for the PreHyb-PE_F and PreHyb-PE_R PCR Primers, which are used in the first limited cycle PCR. The middle section of the adapter–fragment construct contains the target DNA (a string of Xs). The // signifies that most of the target fragment is omitted due to space limitation. (Adapter sequences and PCR primers were obtained from Table 7.3.)

augmented during the first limited PCR; they are merely enriched with the correct adapter constructs (i.e., library fragments having both initial adapters). Thus, following the first limited cycle PCR, these constructs lack the sequences that enable them to hybridize with flow cell oligos (i.e., P5 and P7 common adapters). The rationale for making these incomplete or "truncated" adapter constructs (Rohland and Reich 2012) prior to the in-solution hybridization reaction(s) will be explained below. Only during the second limited cycle PCR will the captured truncated constructs be completed through the addition of P5 and P7 common adapters, a second index sequence, and sites for the various sequencing primers.

The remaining steps of this workflow, which are standard to all library methods, include quantifying and verifying libraries, normalizing libraries, and making prehybridization pools. As mentioned before, the current norm in phylogenomic studies is to group four to ten indexed libraries into a single pool. Each of these pools, in turn, will be used in a separate hybridization reaction (see below).

7.2.1.4 Nextera Library Approach

An alternative to the ligase-based methods for preparing Illumina sequencing libraries involves the use of the "Nextera" library preparation kit. Nextera kits for Illumina sequencing were initially developed and sold by Epicentre Biotechnologies (Madison, Wisconsin; Syed et al. 2009a,b). Illumina subsequently acquired Epicentre in 2011 and has since further refined and diversified the Nextera kits for a number of different applications (see the Illumina website: http://www.illumina.com/products/nextera_dna_library_prep_kit.html). Numerous helpful guides dealing with Nextera library preparation can be found on the Illumina Nextera website pages. At the present time, Nextera kits are available in low- or high-throughput formats and up to 96 sample-specific dual-indices can be employed. In phylogenomic studies, Nextera kits have been used to generate the initial whole-genome sequencing libraries, which were then used as input libraries for kit-based (e.g., SureSelect Custom Probe Kit, Agilent, Inc.) in-solution hybrid selection and sequencing (e.g., Faircloth et al. 2013; Meiklejohn et al. 2016).

Nextera kits have a number of advantages over other library preparation methods. First, the amount of starting material (input genomic DNA) needed for Nextera library preparation is far less than the required amount for other library approaches. For example, in an exome capture study, Nextera only requires an input amount of 50 ng of DNA, whereas a traditional library would require ~3 μg (Adey et al. 2010). Moreover, while the Nextera kits have been optimized for input of 50 ng of starting material, the Nextera XT kit has been optimized for input of only 1 ng of total genomic DNA per reaction. Even lower input amounts are possible, as Adey et al. (2010) found that inputs as low as 10 pg (0.01 ng) of genomic DNA can be used! A second advantage Nextera has is that fewer steps are involved compared to other library preparation methods. This not only conserves consumables (e.g., plastics) and labor but also reduces the risk of errors (e.g., contamination) during the library preparation procedures. The Nextera library making workflow is illustrated in Figure 7.19. We will now review the Nextera methodology.

Step 1: Tagmentation of input DNA followed by SPRI As with the previously discussed methods of library preparation a sample of purified DNA must be acquired (note that RNA can be present because it does not interfere with the tagmentation reaction). For example, in the Nextera kit approach the required amount of genomic DNA employed as starting material is only 50 ng. However, as pointed out in the Illumina Nextera protocol, the kit has been optimized for input of 50 ng of DNA. If a larger amount of DNA is used then the resulting library fragment distribution will consist of fragments that are, on average, larger than desired, whereas if <50 ng is used, then the size distribution will shift toward smaller-sized fragments. Thus, it is important to use the exact amount of input DNA specified in the protocol. Ligase-mediated library methods, on the other hand, are more robust to the input DNA amount with larger amounts generally yielding better sequencing results.

Preparation of an Illumina library using the Nextera approach is similar to the other approaches in that the starting DNA must be broken into fragments small enough for Illumina sequencing and special adapters must be ligated to the ends of each fragment. However, the Nextera approach uses a radically different means during the initial steps to accomplish this. Recall that the previous library-making methods require the

Step 1: Tagmentation of input DNA followed by SPRI

Input DNA
(e.g., genomic DNA)

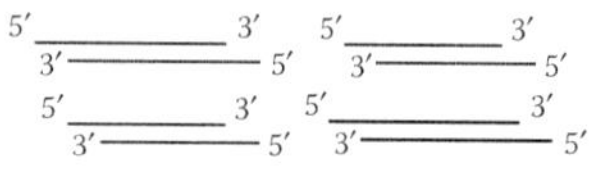

Step 2: First limited cycle PCR followed by SPRI

Step 3: Quantify libraries and verify fragment distributions

Step 4: Normalize and pool* libraries (e.g., 4–10 libraries per pool)

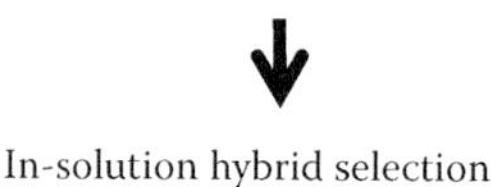

In-solution hybrid selection

Figure 7.19. Basic workflow for producing an Illumina sequencing library using the Nextera approach. This method uses transposases to simultaneously fragment input genomic DNA and attach adapters to the ends of library fragments. This process is called "tagmentation." The initial adapter–fragment constructs are then subjected to the first limited cycle PCR before they are cleaned using SPRI beads. See main text for discussion of remaining steps.

starting DNA material to be fragmented (e.g., via sonication), end-repaired, A-tailed (traditional approach only), and ligated to adapters. In contrast, the Nextera method only involves a one-step 5-minute enzymatic reaction—a process called *tagmentation*, which results in adapter-ligated fragments ready for the first limited cycle PCR step (Figure 7.19).

The tagmentation reaction uses *in vitro* DNA transposition to simultaneously fragment the input DNA into sizes appropriate for sequencing on Illumina platforms and ligating or "tagging" adapter molecules to the ends of each fragment. This is why there are no steps in this protocol that involve manual shearing of input DNA followed by separate ligation reactions. The Nextera kit is optimized so that adapters are ligated, on average, every 400–500 bp thereby creating the ideal adapter-ligated fragment distribution for the Illumina platforms. Before we continue discussing Nextera procedures, let's first review aspects of *in vitro* transposition involving the Tn5 transposase—the key player in this method.

Wildtype Tn5 transposase carries out "cut and paste" transposition in prokaryotes (Steiniger-White et al. 2004). Tn5 transposons are DNA transposons that are excised from and reinserted back into the host genome with little specificity for insertion locations (Reznikoff 2003). Briefly, the mechanism of *in vivo* transposition consists of the following steps: (1) two separate transposase subunits work together to excise a Tn5 transposon from the host DNA; (2) the two subunits form a transposase homodimer that is bound to the transposon; and (3) this newly formed "synaptic complex" then binds to a new target location in the host DNA and inserts the transposon (see Steiniger-White et al. 2004 for more details).

The wildtype form of Tn5 is not useful for *in vitro* transposition owing to its exceedingly low activity (Reznikoff 2003). However, researchers have successfully used a "hyperactive" (mutated) version of the Tn5 transposase to conduct *in vitro* transposition experiments (Reznikoff 2008). Normally, a Tn5 synaptic complex will have a single dsDNA molecule (the transposon) bound to the homodimer with each end of the DNA molecule being inserted into an active site of each transposase subunit (i.e., a loop of DNA is formed from one subunit to the other). When this complex attacks a target DNA molecule, the end result will be insertion of the entire transposon into the target DNA (see Figure 1 in Steiniger-White et al. 2004).

The ends of the Tn5 transposon contain critical sequence elements important for transposition; that is, they are essential for recognition by the transposases, synaptic complex formation, and the insertion of the transposon back into the target DNA (Steiniger-White et al. 2004). These

```
    19                 1
5′  AGATGTGTATAAGAGACAG 3′   ←  19 bp ME sequence (transferred strand)
    |||||||||||||||||||
3′  TCTACACATATTCTCTGTC 5′   ←  Complement (nontransferred strand)
```

Figure 7.20. The 19-bp ME sequence for Tn5 transposase. (Modified from Figure 1 in Bhasin, A. et al. 2000. *J Mol Biol* 302:49–63.)

19-bp sequence ends, which are called "outside end" (OE) and "inside end" (IE) sequences, are nearly identical to each other (see Figure 1 in Bhasin et al. 2000). A 19-bp "hybrid" between OE and IE elements is called the "mosaic end" (ME) sequence (Figure 7.20). As with the OE and IE sequences, this ME sequence contains the essential chemical features for enabling synaptic complex formation and subsequent transposition with target DNA. In particular, the first nucleotide position (GC base pair) of the ME (Figure 7.20), which is in contact with the active site of a transposase subunit, will be positioned closest to where a staggered double-stranded cut in the target DNA takes place during the transposition reaction. However, only one of the ME strands (i.e., transferred strand; see Figure 7.20) will be ligated to a target strand; that is, the 3′ end of this transferred strand (i.e., "G" nucleotide at position (1) will be covalently joined to the 5′ end of the newly fragmented target DNA. Researchers discovered that if transposase synaptic complexes are assembled *in vitro* using *two* free ME sequences instead of the single longer Tn5 transposon, then the ultimate result of transposition will be fragmentation of the target DNA with the ME sequences ligated to each fragment (Reznikoff 2008). Figure 7.21 shows the crystal structure of a Tn5 synaptic complex with

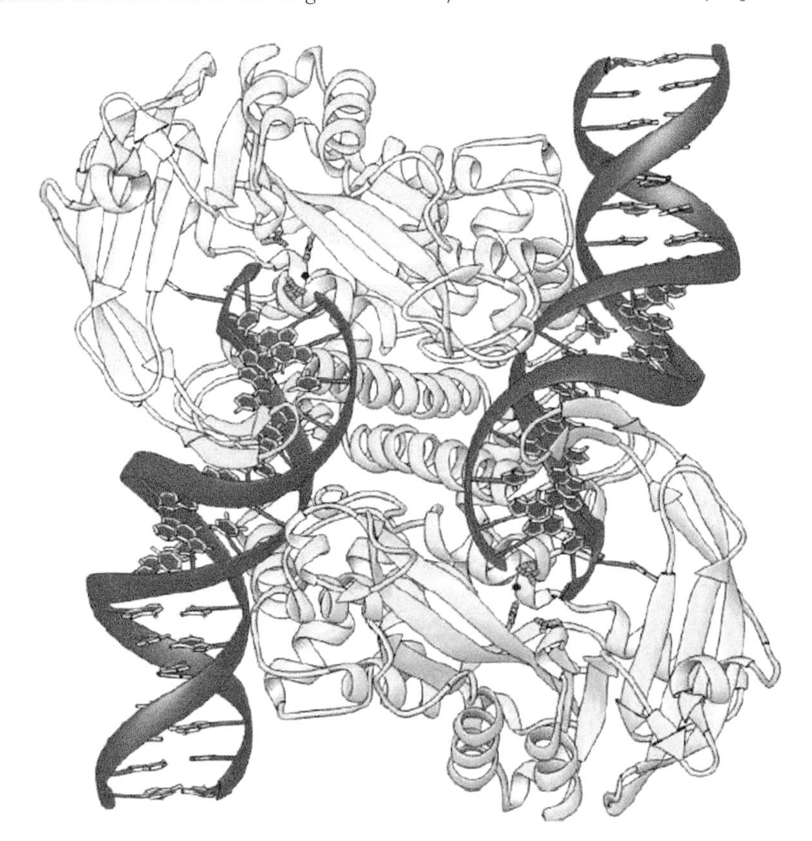

Figure 7.21. Crystal structure of Tn5 transposase/DNA complex. The transposase synaptic complex is comprised of a homodimer between two transposase subunits (yellow and blue ribbons) with two DNA transposon ends (purple) protruding from the active sites of the subunits. Catalytic residues in the active site are shown as green ball-and-stick structures and the associated Mn^{2+} ion is black. (Reprinted from Davies, D. R. et al. 2000. *Science* 289:77–85. With permission.)

TABLE 7.4
Oligos for Illumina sequencing using the Nextera approach

Name of oligo	Sequence (5′ → 3′)
Index 1 Read Primer	CAAGCAGAAGACGGCATACGAGAT [i7] GTCTCGTGGGCTCGG
Index 2 Read Primer	AATGATACGGCGACCACCGAGATCTACAC [i5] TCGTCGGCAGCGTC
Read 1 Sequencing Primer	TCGTCGGCAGCGTCAGATGTGTATAAGAGACAG
Read 2 Sequencing Primer	GTCTCGTGGGCTCGGAGATGTGTATAAGAGACAG

SOURCE: Oligonucleotide sequences © 2007–2010 Illumina, Inc. All rights reserved.

NOTE: Index 1 and 2 Read Primers are used in the first limited cycle PCR; and Read 1 and Read 2 Sequencing Primers are used to sequence the completed library construct. The i5 and i7 Indices each contain a unique 8-bp index sequence. This dual-indexing system allows for the individual identification and hence multiplexing of up to 96 different libraries. Note, oligo sequences are only shown here for purposes of understanding the library-making process using Illumina Nextera approach. As Illumina may upgrade its kits, some or all of these oligos may be obsolete. Always check beforehand with the sequencing service provider to make sure that your Illumina libraries are compatible with their sequencing service.

two transposon sequences protruding away from the homodimer. We will now see how this experimental system evolved into a novel and simple NGS library preparation method.

Researchers took advantage of the free 5′ ends of each ME sequence by tailing them with platform-specific adapter sequences (Syed et al. 2009a,b). The Illumina Nextera adapter sequences are listed in Table 7.4 along with other essential oligos for this library-making approach. We can see each of these 5′-tailed adapter or "transposon" sequences on the left side of Figure 7.22. Notice that the right half of each sequence is comprised of the critical 19-bp ME sequence, whereas the left half contains the sites necessary for the first limited cycle PCR in Step 2 (Figure 7.19). Although only the transferred strand of each adapter sequence will be ligated to the end of a library fragment via the transposition mechanism, a double-stranded adapter sequence is required for assembly of the Tn5 synaptic complex.

Figure 7.23 shows a tagmentation reaction involving ME-flanked adapter sequences. Initially, when synaptic complexes are assembled *in vitro*, there is randomness with respect to which particular ME-flanked adapters become annealed to the transposase subunits. This means that some complexes will have one of each adapter type such as the complexes having one blue and one orange adapter in Figure 7.23a, while the other two classes of complexes in the solution (not shown) will have only one adapter type (i.e., all blue or all orange adapters). In the next step of the reaction (Figure 7.23b), the two synaptic complexes that are shown attack the target DNA where they will create staggered double-stranded breaks in the target DNA and initiate strand-transfer from ME-flanked

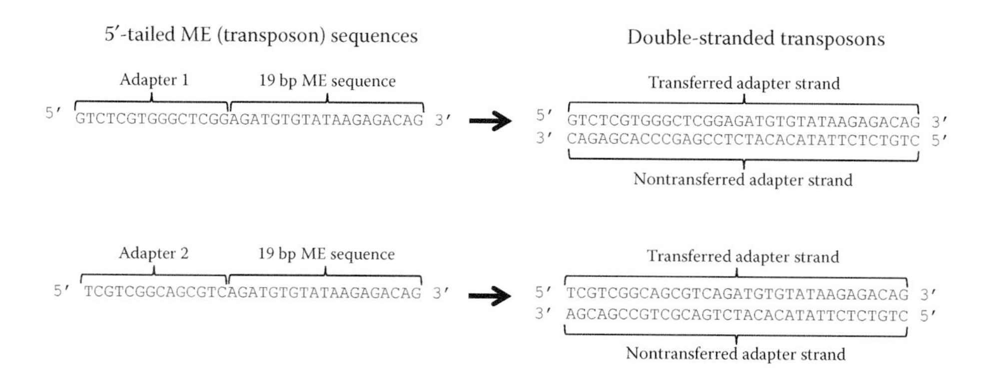

Figure 7.22. Nextera adapter sequences. Two ME sequences with 5′ tail sequences consisting of an Illumina-specific adapters 1 and 2 (left side). These tailed-ME sequences must be double-stranded "transposons" (right side) in order to form synaptic complexes with Tn5 transposases. Note that only one of the two strands of each transposon (i.e., transferred adapter strand) will be transferred to a library fragment during tagmentation. (Oligonucleotide sequences © 2007–2010 Illumina, Inc. All rights reserved.)

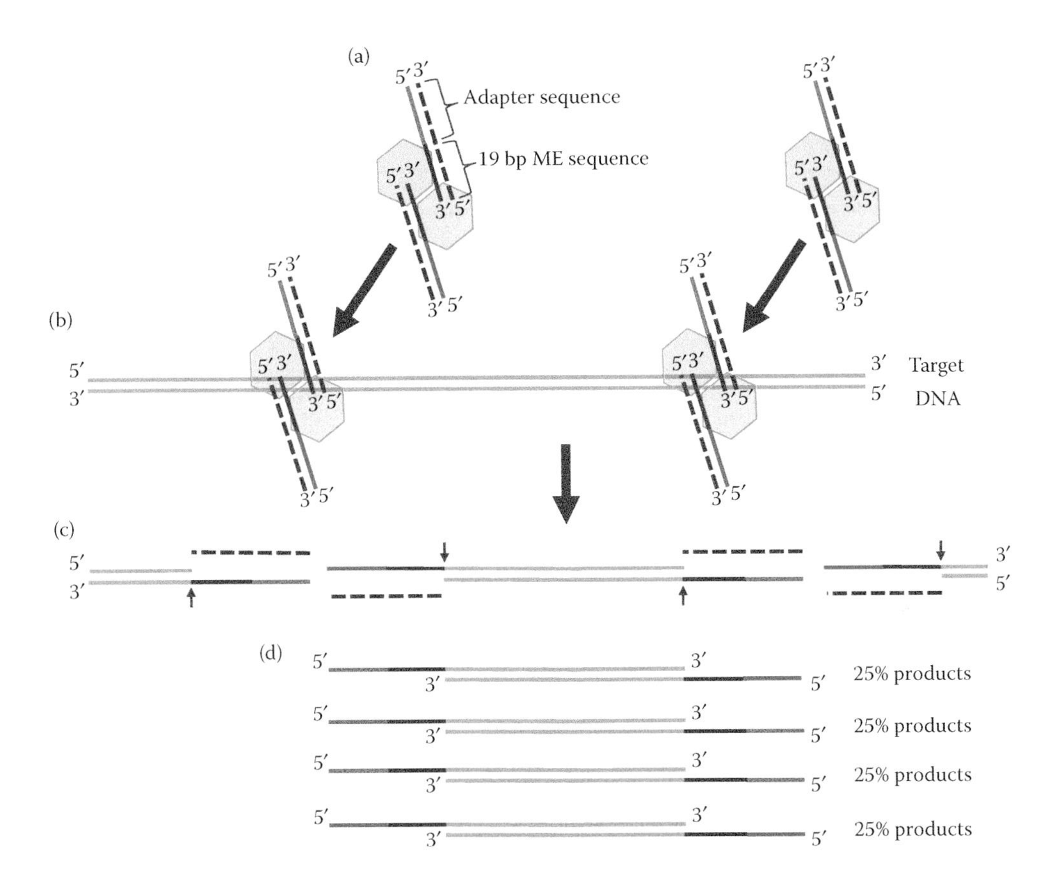

Figure 7.23. Tagmentation reaction involving two synthetic Tn5 transposase synaptic complexes and target DNA. (a) Each synaptic complex is comprised of a transposase homodimer (two hexagons) and two ME-flanked adapters. Each ME sequence terminates in the active site of a subunit (near center of each hexagon) while the adapter ends (blue and orange ends) are free and not in contact with the subunits. Solid color DNA strands are the strands that will be transferred (covalently attached) to the target DNA, while the (dashed) complementary strands are the nontransferred strands and thus will not be incorporated into target DNA. (b) Each synaptic complex independently attacks the target DNA. Upon binding, a staggered double-stranded cut is made and the 3′ ends of the ME sequences (transferred strands) subsequently become ligated to the 5′ end of the target DNA strands. (c) When the reaction is finished, the target DNA is fragmented and tagged ("tagmented") with ME-flanked adapters. Small arrows indicate the ligation points between ME-flanked adapters and target DNA. (d) Products of tagmentation. The topmost product (blue and orange ends) is the complete product shown in step (c), while the bottom three represent the other products present in the reaction. (Parts of this figure are based on Figure 2 in Steiniger-White, M. et al. 2004. *Curr Opin Struct Biol* 14:50–57 and Figure 1 in Syed, F. et al. 2009. *Nat Methods* 6:10; Syed, F. et al. 2009. *Nat Methods* 6:11.)

adapters to target DNA. Strand transfer occurs when the 3′ ends of the transferred (adapter) strands are ligated to the 5′ ends of the target DNA (Figure 7.23c). Just like the ligation reactions in the Meyer–Kircher and Rohland–Reich approaches, which also use two different adapters, there will also be an element of randomness concerning which adapter is ligated to the end of a particular fragment (Figure 7.23d). Accordingly, a limited cycle (suppression) PCR will be used to selectively amplify the library fragments having both adapters (Figure 7.19; Syed et al. 2009a,b; Adey et al. 2010).

Performing the tagmentation reaction step is simple as it consists of first adding the input DNA, enzyme buffer, and transposase enzyme/adapter complexes to a reaction tube followed by a 5-minute incubation period at 55°C in a thermocycler. At the conclusion of this *in vitro* transposition reaction, the transposases remain bound to the target DNA (Reznikoff 2003) and thus they must be removed from the tagmentation products and discarded otherwise they can interfere with downstream steps. In the past, the Illumina Nextera protocol dictated the use of column-based DNA cleanups for this purpose. However, these kits now rely on SPRI bead cleanups as they perform better than column-based cleanups (e.g., Faircloth et al. 2012).

Step 2: First Limited Cycle PCR Followed by SPRI As with the traditional and Meyer and Kircher approaches, limited cycle PCR is used to (1) complete the initial adapter constructs by adding the sample-specific indices, P5 and P7 common adapters, and annealing sites for sequencing primers; and (2) selectively amplify the correct tagmentation products (i.e., library fragments with both kinds of adapters). The Nextera kit uses a dual index system with each index being 8-bp long. Thus, one PCR primer contains the Index 1 or "i7 index" sequence while the other primer contains the Index 2 or "i5 index" sequence (Table 7.4). There are 12 and 8 different indices for the i7 and i5 index schemes, respectively, which means 96 different libraries can ultimately be pooled together into a single Illumina flow cell.

The desired tagmentation product consists of a library fragment with two 5′ overhangs each of which corresponds to a different adapter sequence (Figure 7.24a). However, inspection of the tagmentation product (Figure 7.24a) and the tailed PCR primers that will be used to amplify that product in the first limited cycle PCR (Figure 7.24b) reveals that this initial adapter–fragment construct lacks annealing sites for these primers. Instead, both adapter sequences ligated to the library fragment are actually the *same* sequence as the two primer sequences. How then, does the PCR proceed?

The trick is to begin the thermocycling program with a 3–5 minute extension step at 72°C to allow DNA polymerases a chance to fill-in the missing bases just before the normal PCR cycling begins (i.e., cycle = denaturation → annealing → extension steps). This crucial detail cannot be overlooked while performing a limited cycle PCR involving tagmentation products—to reiterate, the thermocycling program *must* begin with this 72°C step and *not* the usual 94–98°C denaturation step! Once this first extension step is completed the initial adapter constructs are double-stranded molecules from end to end and, importantly, now contain the needed annealing sites for both PCR primers (Figure 7.25). After the initial extension step has generated the PCR primer annealing sites on the adapter–fragment constructs, the thermocycler program then subjects them to at least five normal PCR cycles in order to complete these constructs and amplify their number. Similar to the limited cycle PCRs in the traditional and Meyer and Kircher approaches, fully formed constructs do not appear until after the

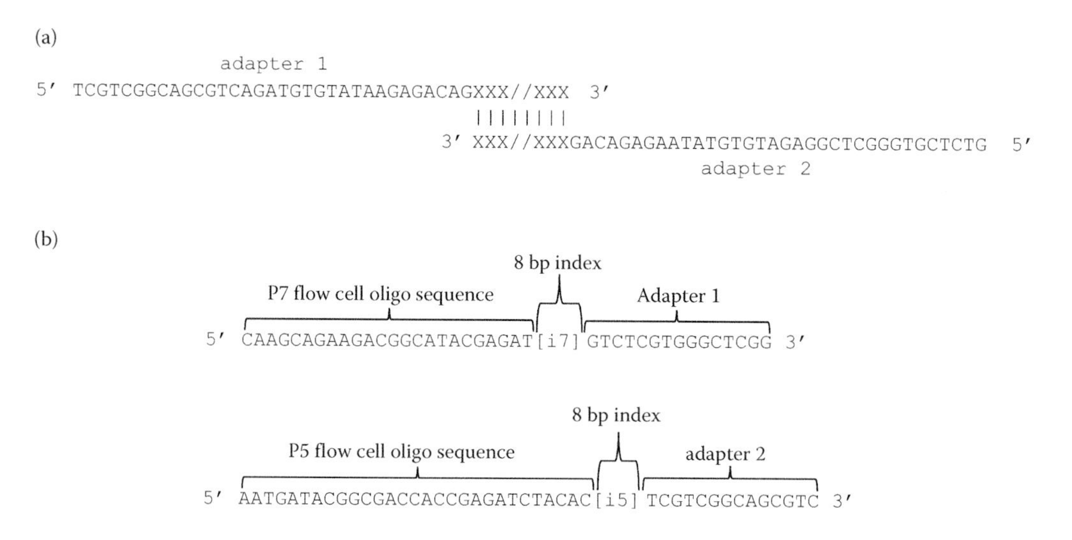

Figure 7.24. Tagmentation product and PCR primers used in the first limited cycle PCR. (a) The desired initial tagmentation product contains adapter 1 and adapter 2 sequences (vertical dashes indicate hydrogen bonding between complementary bases). The middle section of the adapter–fragment construct contains the target DNA (a string of Xs). The // signifies that most of the target fragment is omitted due to space limitation. (b) Two PCR primers, called "Index 1 Read Primer" (top) and "Index 2 Read Primer" (bottom), have 5′ tail sequences, which are used to add sequences matching the P5 and P7 flow cell oligos and to incorporate dual 8-bp indices into adapter–fragment constructs during the first limited cycle PCR. Oligos are found in Table 7.4. (Oligonucleotide sequences © 2007–2010 Illumina, Inc. All rights reserved.)

Figure 7.25. How the first limited cycle PCR completes adapter–fragment library constructs using the Nextera approach. The denaturation and primer-extension steps in the first three cycles of a limited PCR are illustrated (primer-annealing steps not shown due to space limitation). The initial adapter constructs does not contain annealing sites for the Index 1 and 2 Read Primers and thus the first step is an extension step to allow polymerases to fill in the missing bases. Completed adapter–fragment constructs appear only at the end of the third cycle. The 8-bp i5 and 8-bp i7 index sequences are represented by [i5] and [i7], respectively. The middle section of each strand contains the target DNA (a string of Xs). The // signifies that most of the target fragment is omitted due to space limitation and the vertical dashes indicate hydrogen bonding between complementary bases. Boldface sequences represent newly annealed primer in the same cycle and gray letters represent DNA synthesized in this reaction.

(Continued)

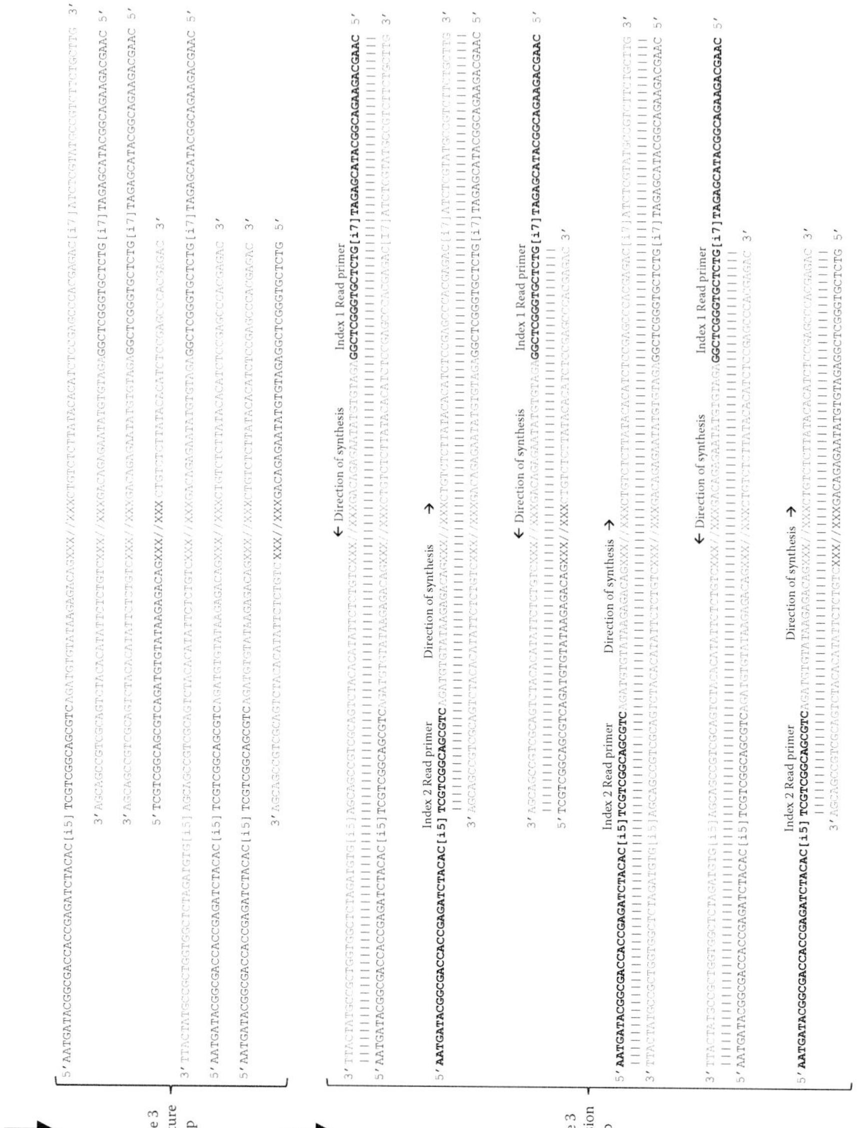

Cycle 3 denature step

5′ AATGATACGGCGACCACCGAGATCTACAC[i5] TCGTCGGCAGCGTCAGATGTGTATAAGAGACAGXXX//XXXCTGTCTCTTATACACATCTCCGAGCCCACGAGAC[i7]ATCTCGTATGCCGTCTTCTGCTTG 3′

3′ AGCAGCCGTCGCAGTCTACACATATTCTCTGTCXXX//XXXGACAGAGAATATGTGTAGAGGCTCGGGTGCTCTG[i7]TAGAGCATACGGCAGAAGACGAAC 5′

3′ AGCAGCCGTCGCAGTCTACACATATTCTCTGTCXXX//XXXGACAGAGAATATGTGTAGAGGCTCGGGTGCTCTG[i7]TAGAGCATACGGCAGAAGACGAAC 5′

5′ TCGTCGGCAGCGTCAGATGTGTATAAGAGACAGXXX//XXXCTGTCTCTTATACACATCTCCGAGCCCACGAGAC 3′

3′ TTACTATGCCGCTGGTGGCTCTAGATGTG[i5] AGCAGCCGTCGCAGTCTACACATATTCTCTGTCXXX//XXXGACAGAGAATATGTGTAGAGGCTCGGGTGCTCTG[i7]TAGAGCATACGGCAGAAGACGAAC 5′

5′ AATGATACGGCGACCACCGAGATCTACAC[i5] TCGTCGGCAGCGTCAGATGTGTATAAGAGACAGXXX//XXXCTGTCTCTTATACACATCTCCGAGCCCACGAGAC 3′

5′ AATGATACGGCGACCACCGAGATCTACAC[i5] TCGTCGGCAGCGTCAGATGTGTATAAGAGACAGXXX//XXXCTGTCTCTTATACACATCTCCGAGCCCACGAGAC 3′

3′ AGCAGCCGTCGCAGTCTACACATATTCTCTGTCXXX//XXXGACAGAGAATATGTGTAGAGGCTCGGGTGCTCTG 5′

Cycle 3 extension step

← Direction of synthesis Index 1 Read primer

3′ TTACTATGCCGCTGGTGGCTCTAGATGTG[i5]AGCAGCCGTCGCAGTCTACACATATTCTCTGTCXXX//XXXGACAGAGAATATGTGTAGA**GGCTCGGGTGCTCTG[i7] TAGAGCATACGGCAGAAGACGAAC** 5′

5′ AATGATACGGCGACCACCGAGATCTACAC[i5]TCGTCGGCAGCGTCAGATGTGTATAAGAGACAGXXX//XXXCTGTCTCTTATACACATCTCCGAGCCCACGAGAC[i7]ATCTCGTATGCCGTCTTCTGCTTG 3′

Index 2 Read primer Direction of synthesis →

5′ **AATGATACGGCGACCACCGAGATCTACAC[i5] TCGTCGGCAGCGTC**AGATGTGTATAAGAGACAGXXX//XXXCTGTCTCTTATACACATCTCCGAGCCCACGAGAC[i7]ATCTCGTATGCCGTCTTCTGCTTG 3′

3′ AGCAGCCGTCGCAGTCTACACATATTCTCTGTCXXX//XXXGACAGAGAATATGTGTAGAGGCTCGGGTGCTCTG[i7]TAGAGCATACGGCAGAAGACGAAC 5′

← Direction of synthesis Index 1 Read primer

3′ AGCAGCCGTCGCAGTCTACACATATTCTCTGTCXXX//XXXGACAGAGAATATGTGTAGA**GGCTCGGGTGCTCTG[i7] TAGAGCATACGGCAGAAGACGAAC** 5′

5′ TCGTCGGCAGCGTCAGATGTGTATAAGAGACAGXXX//XXXCTGTCTCTTATACACATCTCCGAGCCCACGAGAC 3′

Index 2 Read primer Direction of synthesis →

5′ **AATGATACGGCGACCACCGAGATCTACAC[i5] TCGTCGGCAGCGTC**AGATGTGTATAAGAGACAGXXX//XXXCTGTCTCTTATACACATCTCCGAGCCCACGAGAC[i7]ATCTCGTATGCCGTCTTCTGCTTG 3′

3′ TTACTATGCCGCTGGTGGCTCTAGATGTG[i5]AGCAGCCGTCGCAGTCTACACATATTCTCTGTCXXX//XXXGACAGAGAATATGTGTAGAGGCTCGGGTGCTCTG[i7]TAGAGCATACGGCAGAAGACGAAC 5′

← Direction of synthesis Index 1 Read primer

3′ TTACTATGCCGCTGGTGGCTCTAGATGTG[i5]AGCAGCCGTCGCAGTCTACACATATTCTCTGTCXXX//XXXGACAGAGAATATGTGTAGA**GGCTCGGGTGCTCTG[i7] TAGAGCATACGGCAGAAGACGAAC** 5′

5′ AATGATACGGCGACCACCGAGATCTACAC[i5]TCGTCGGCAGCGTCAGATGTGTATAAGAGACAGXXX//XXXCTGTCTCTTATACACATCTCCGAGCCCACGAGAC 3′

Index 2 Read primer Direction of synthesis →

5′ **AATGATACGGCGACCACCGAGATCTACAC[i5] TCGTCGGCAGCGTC**AGATGTGTATAAGAGACAGXXX//XXXCTGTCTCTTATACACATCTCCGAGCCCACGAGAC 3′

3′ AGCAGCCGTCGCAGTCTACACATATTCTCTGTCXXX//XXXGACAGAGAATATGTGTAGAGGCTCGGGTGCTCTG 5′

Figure 7.25. (Continued)

third cycle is finished (Figure 7.25). A total of five normal PCR cycles are required when using the Illumina Nextera kit. When the program stops, the adapter–fragment constructs are complete as they now contain the required P5 and P7 common adapters, annealing sites for the Read 1, Read 2, and i7 Read Sequencing Primers, and two sample-specific indices (Figure 7.26). Note, the P5 flow cell oligo not only serves to tether an adapter–fragment strand during the start of cluster generation but it also acts as the i5 Index Read Sequencing Primer. The actual sequence of the i7 Index Read Sequencing Primer is proprietary has not been disclosed by Illumina but its approximate location in the construct is shown in Figure 7.26. Illumina has also not disclosed information about the DNA polymerase(s) they include in their Nextera kits for this limited PCR step. However, Picelli et al. (2014) used the KAPA HiFi DNA polymerase (KAPA Biosystems), which synthesizes DNA in a $5' \rightarrow 3'$ direction and has $3' \rightarrow 5'$ exonuclease (i.e., proofreading) capability, to successfully amplify tagmentation products. Following the limited cycle PCR, the completed adapter–fragment constructs must be purified. The Illumina Nextera protocol uses SPRI beads to purify the PCR products.

Step 3: Quantify Libraries and Verify Fragment Distributions After PCR purification, the libraries must be quantified and verified (Figure 7.19). As with the other library preparation methods the concentration of each library can be estimated using qPCR (best method) or using a Bioanalyzer. A Bioanalyzer can also be used to verify that each library exhibits a fragment size distribution within the desired range. Libraries made using the Illumina Nextera kit without any additional size-selection steps are expected to display a size range of ~200–1,000 bp.

Step 4: Normalize and Pool Libraries (e.g., 4–10 Libraries per Pool) After libraries have been quantified and passed Bioanalyzer inspections, they can be normalized and grouped together with four to ten libraries per hybridization pool (Figure 7.19). Note, some researchers have decided to not index and pool their Nextera libraries before hybrid selection (e.g., Faircloth et al. 2012), which means that more labor and consumables must be expended in order to generate large multilocus datasets. However, as we will soon see there are advantages and disadvantages to indexing and pooling Nextera libraries prior to hybrid selection.

Although the commercial Nextera kits offer the simplest and most reliable method for constructing Illumina sequencing libraries, their expense may limit the number of libraries a given project or laboratory can produce. To help reduce the cost per library, Picelli et al. (2014) developed a "kit-free" protocol to making Tn5 transposition-mediated libraries for Illumina sequencing. This method requires additional laboratory equipment and technical expertise for molecular cloning and thus it will be more challenging than using kits. Nonetheless, because it is possible to produce vast amounts of the key reagents such as the Tn5 transposases, laboratories can make their own premade Tn5 synaptic complexes ready for producing a large number of sequencing libraries.

In their protocol, Picelli et al. (2014) detail methods for producing in-house Tn5 transposases using *E. coli* expression vectors. Once the transposase enzyme subunits are produced and purified, they can be assembled into functional synaptic complexes (i.e., Tn5 homodimers annealed to two ME-flanked adapter sequences). This approach also gives the researcher the option of making custom adapter sequences (Picelli et al. 2014).

One of the important findings in the study by Picelli et al. (2014) concerns the importance of using the proper type and concentration of a "molecular crowding agent" in the tagmentation reactions. One of the problems encountered during the early years of the transposition-mediated NGS library construction was that the researcher had little control over the fragment sizes found in the resulting libraries (Adey et al. 2010). Indeed, Adey et al. (2010) determined that without any optimized buffers or special size-selection step the observed fragment distribution averaged about 100 bp ± 47 bp. Note that the minimum fragment size from Tn5-mediated transposition is sharply delineated at ~38 bp, a limit that is likely due to steric interactions between competing synaptic complexes attacking the same stretch of target DNA in adjacent locations (Adey et al. 2010). Although these fragment sizes are smaller than the desired 400–500 bp range for Illumina sequencers, Adey et al. (2010) correctly predicted that further optimization of reaction buffers could increase the sizes of the fragments and produce a more desirable size range.

As we discussed in Chapter 6, a solution of PEG can be used in a simple and inexpensive method

```
                                                                                                            P7 flow cell oligo
                                                                                                  3'TAGAGCATACGGCAGAAGACGAAC 5'
                                                                          Read 2 Sequencing Primer
                                                                  3'GACAGAGAATATGTGTAGAGGCTCGGGTGCTCTG 5'

3' TTACTATGCCGCTGGTGGCTCTAGATGTG[i5]AGCAGCCGTCGCAGTCTACACATATTCTCTGTCXXX//XXXGACAGAGAATATGTGTAGAGGCTCGGGTGCTCTG[i7]TAGAGCATACGGCAGAAGACGAAC 5'
   ||||||||||||||||||||||||||||||||||||||||||||||||||||||||||||||||||||||||||||||||||||||||||||||||||||||||||||||||||||||||||||||||||||||||
5' AATGATACGGCGACCACCGAGATCTACAC[i5]TCGTCGGCAGCGTCAGATGTGTATAAGAGACAGXXX//XXXCTGTCTCTTATACACATCTCCGAGCCCACGAGAC[i7]ATCTCGTATGCCGTCTTCTGCTTG 3'

                                                                   5' i7 Index Read Sequencing Primer 3'
                              5'TCGTCGGCAGCGTCAGATGTGTATAAGAGACAG 3'
                                        Read 1 Sequencing Primer
 5'AATGATACGGCGACCACCGAGAUCTACAC 3'
 P5 flow cell oligo and i5 Index Read Sequencing Primer
```

Figure 7.26. Completed Nextera adapter–fragment construct for Illumina sequencing. The construct contains annealing sites for the P5 and P7 flow cell oligos, Read 1 and 2 Sequencing Primers, and an i7 Read Sequencing Primer (sequence not shown). The 8-bp i5 and 8-bp i7 index sequences are represented by [i5] and [i7], respectively. The middle section of the construct contains the target DNA (a string of Xs). The // signifies that most of the target fragment length is not shown due to space limitations and the vertical dashes indicate hydrogen bonding between complementary bases. (Oligonucleotide sequences © 2007–2010 Illumina, Inc. All rights reserved.)

for purifying and size-selecting DNA fragments such as PCR products and sequencing library fragments. Picelli et al. (2014) experimented with various types and concentrations of PEG in tagmentation reactions. Importantly, they showed how PEG in particular can modulate fragment size in libraries depending on the molecular weight of each PEG polymer type and its concentration. Their results showed that the Illumina Nextera XT kit produced the best library results as the fragment distribution ranged from 200 to 1,000 bp and was centered over the desired sizes of 400–500 bp (Figure 7.27). Interestingly, the authors nearly duplicated these results by using their in-house Tn5 transposase enzymes and a buffer containing a 5% solution of PEG 8000 (Figure 7.27). Other concentrations and crowding agents not only yielded inferior library results, but also the exclusion of any crowding agent caused a complete failure (Figure 7.27). These results are important because achieving the optimal size range of adapter-ligated fragments in a tagmentation reaction will lead to more efficient library construction, even with picogram-scale input DNA (Picelli et al. 2014).

As pointed out earlier, Nextera kits now use SPRI beads to clean tagmentation products and thus this method may also perform well with the Picelli et al. protocol. An alternative way to accomplish the task of stripping the transposases away from the tagmentation products was discovered by Picelli et al. (2014). These authors found that a 0.1% SDS "stripping buffer" was effective at purifying tagmentation products as evidenced by their good quality PCR results. Further experimentation by these authors revealed that a buffer with a concentration of 0.2% SDS gave the highest PCR yields while a 0.3% SDS concentration caused a complete PCR failure (Picelli et al. 2014). Besides its simplicity, the stripping buffer approach has the added advantage of being well suited for a high-throughput sequencing library workflow (Picelli et al. 2014). It remains to be seen which of the two methods—SPRI beads or the stripping buffer—should be the preferred approach to cleaning tagmentation products.

7.2.2 In-Solution Hybrid Selection

The attainment of whole-genome or "shotgun" sequencing libraries is a major achievement toward the goal of acquiring large multilocus datasets for phylogenomic studies. However, before these libraries can be sequenced they must be reduced down to the subset of library fragments representing the set of target loci (e.g., exons, anchored loci, etc.). Once these target DNA fragments have been isolated they can be prepared for sequencing on an Illumina platform.

Gnirke et al. (2009) developed a method called *in-solution hybrid selection*, which has since become the standard approach for obtaining phylogenomic datasets consisting of hundreds to thousands of loci for dozens to hundreds of individuals. In-solution hybrid selection typically employs a complex mixture of biotinylated RNA probes, which are 60–120 bp long

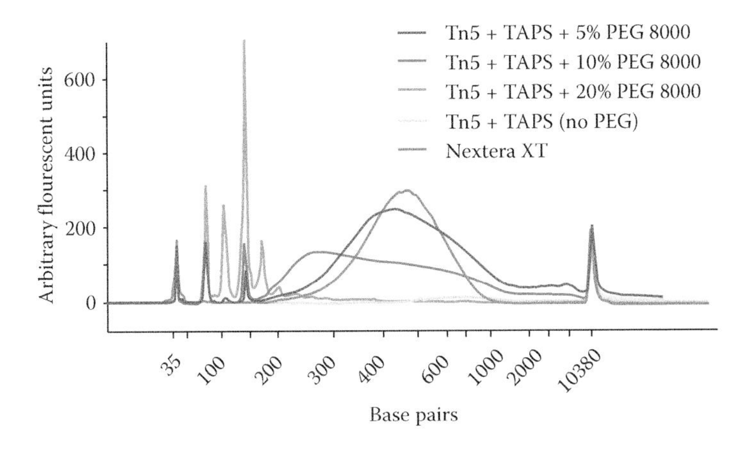

Figure 7.27. Effects of different crowding agents on tagmentation reactions. The Bioanalyzer electropherogram shows the results of tagmentation reactions involving different concentrations of PEG 8000, no PEG, or using an Illumina Nextera XT kit. (Reprinted from Picelli, S. et al. 2014. *Genome Res* 24:2033–2040. With permission.)

(Lemmon and Lemmon 2013) and are complementary to a set of target genomic loci. During an in-solution hybridization reaction, which is performed under stringent thermal conditions, RNA probes hybridize with complementary single-stranded (target) library molecules. The biotin moieties, which are incorporated into each probe at the time of oligonucleotide synthesis via transcription, play a key role together with streptavidin-coated magnetic beads in capturing RNA–DNA hybrids while in solution. Following a "pull down" step in which a magnetic separator is used to immobilize the RNA–DNA hybrids, the nontarget DNA fragments are washed away, which leaves behind target DNA for sequencing. The in-solution hybrid selection method has been analogized to the practice of fishing whereby RNA "baits" are used to catch target DNA fragments (i.e., "the catch") from a "pond" of whole-genome library fragments (Gnirke et al. 2009; Lemmon and Lemmon 2013).

Although there are many similarities between in-solution and "on-array" (i.e., microarray) approaches, one big difference between the two is that the former uses a vast excess of oligonucleotide probes relative to library templates, while the latter employs a reverse strategy (Gnirke et al. 2009; Mamanova et al. 2010). The high probe-to-template ratios that characterize in-solution hybridization reactions has several advantages over the on-array approach including: (1) improved probe–template hybridization efficiency; (2) much less library template is required for hybrid selection reactions; and (3) hybridization reactions can be conducted in 96-well microplates on thermocyclers and do not require specialized equipment (Gnirke et al. 2009; Mamanova et al. 2010). Thus, library inputs for in-solution hybrid selection experiments only need to be in the 50–500 ng range rather than necessitating microgram quantities of library DNA.

Figure 7.28 illustrates the basic workflow of an in-solution hybrid selection experiment, which requires at least 2 days to complete (Blumenstiel et al. 2010). Note that Figure 7.28 is simplistic in that it only illustrates the major steps involved in hybrid selection. Other crucial steps such as DNA purification steps (e.g., using SPRI beads) immediately follow some of the steps shown in Figure 7.28 and thus it is critically important to follow actual hybrid selection protocols unless certain modifications are clearly warranted and are well understood by the researcher. Although hybrid selection experiments are now routinely performed using commercially available kits, additional helpful references containing protocols, information about key step-modifications, and troubleshooting tips can be found (e.g., Blumenstiel et al. 2010; Mamanova et al. 2010; Fisher et al. 2011; Rohland and Reich 2012). We will now review the major steps in the in-solution hybrid selection workflow.

Step 1: Acquire set of Biotinylated RNA "Baits" The initial step of an in-solution hybrid selection experiment is to obtain a set of biotinylated RNA "baits" (Figure 7.28). These probe sets are acquired in two different ways. First, many phylogenomic studies especially those involving vertebrates can likely use an existing probe set. For example, a number of probe sets for phylogenomic studies have been designed for anchored loci such as UCE-anchored loci (http://ultraconserved.org) and AE loci (http://anchoredphylogeny.com). The total number of different genomic loci targeted by these probe sets varies but at this time it ranges from ~500 (Lemmon et al. 2012) to more than 5,000 loci (McCormack et al. 2015). Thus, if a probe set that is suitable for a planned study already exists, then an RNA bait kit can simply be custom-ordered from a company that produces these kits. The SureSelect Kit (Agilent) and MYbaits kit (MYcroarray, Inc.) represent two of the more popular RNA bait kits at the present time. The alternative way to acquire a probe set is to design it yourself and then order a custom-made kit from one of the aforementioned suppliers. In principle, a set of RNA baits can target any portion of a genome (e.g., anonymous loci) and thus a researcher is free to design a probe set comprised of any aggregate of genome-wide loci. In Chapter 8, we will visit the issue of how these probe sets are designed.

Step 2: Heat-Denature the "Pond" and "Blocking" DNA Each pool of indexed libraries made during the library preparation workflow is used as input whole-genome library DNA (i.e., "pond DNA") in a hybridization reaction. However, before the RNA baits can be hybridized to the target DNA, the pond DNA and special DNA additives called "blocking DNA" must first be heat denatured (in the absence of the RNA baits) in a thermocycler, then allowed to cool down to the hybridization temperature (Figure 7.28). This is a necessary procedure because all blocking DNAs must be

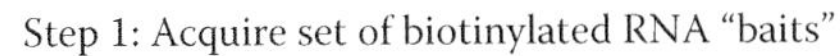

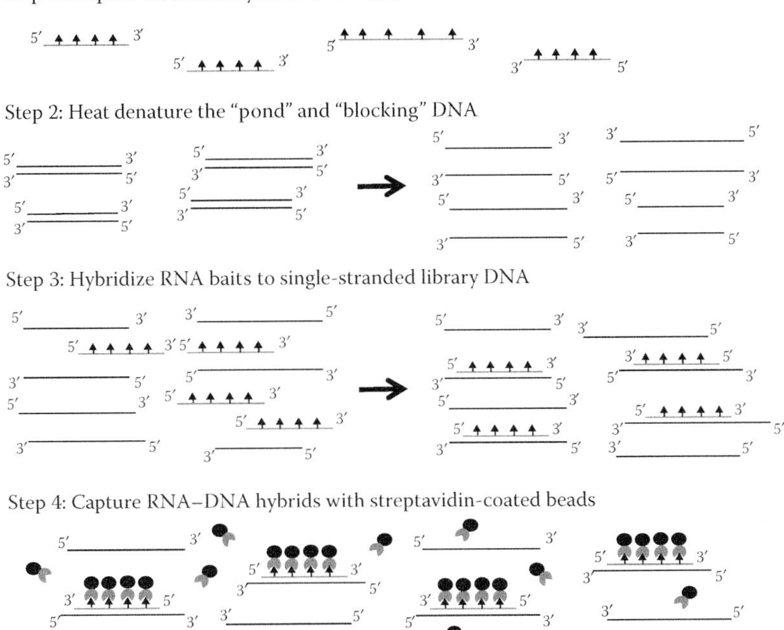

Step 5: "Pull down" captured RNA–DNA hybrids onto magnet

Step 6: Eliminate nontarget DNA via series of washes

Step 7: Second limited cycle PCR followed by SPRI

Step 8: Quantify libraries and verify fragment distributions

Step 9: Normalize and pool the pooled libraries

Cluster generation and sequencing

Figure 7.28. Basic workflow for in-solution hybrid selection. See main text for descriptions of each step.

single-stranded and free of secondary structures in order to function properly before the RNA baits are placed in the pond DNA mixture to initiate the hybridization reaction. The rationale and use of blocking DNA is discussed below. Thus, the first thermocycler step is a "denaturation step" of 95–98°C for 5 minutes followed by a "hybridization step" of 65°C for about 24 hours.

Pond DNA is a heterogeneous mixture of genomic elements some of which are the target DNA molecules. Because the hybridization reaction is performed at a high level of stringency for DNA duplex formation (i.e., at high temperatures), in principle RNA baits should only anneal to their complementary library DNA strands suspended in the solution. These target DNA molecules can then be harvested and sequenced. However, even at these high temperatures, other unwanted molecular hybrids between "off-target" and "on-target" library strands can also form. If off-target library strands are harvested along with the target library strands, then the output sequence data will consist of many "by-catch" (to continue the fishing analogy) sequences that do not map to any of the desired loci.

Hodges et al. (2009) described two scenarios whereby off-target library strands can be captured during a hybridization reaction. In the first, if a library DNA strand contains a stretch of sites that are complementary to an RNA bait *and* an adjacent

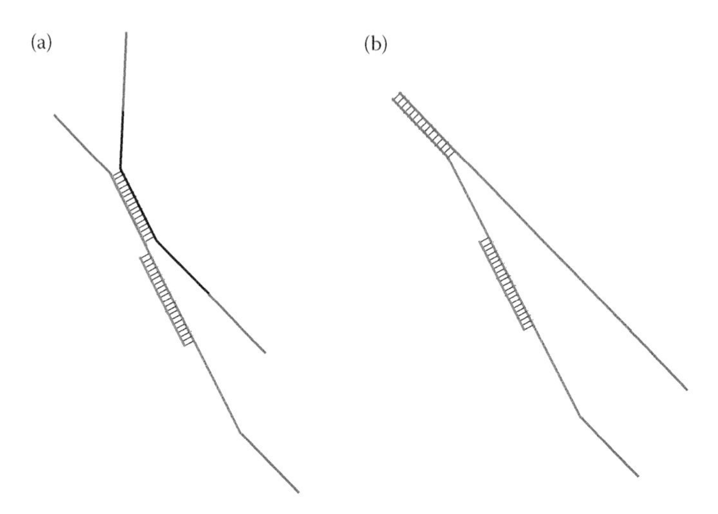

Figure 7.29. Two "by-catch" scenarios whereby nontarget library fragments may be captured during an in-solution hybridization reaction. (a) In one scenario, nonspecific hybridization can occur between two different library fragments (orange and black) when stretches of repetitive DNA are found in both fragments. (b) In the other scenario, complementary adapter sequences (blue) can anneal to each other regardless of their respective fragment DNA sequences (orange and green). In both scenarios, if the target DNA fragment is also hybridized to an RNA probe (gray), then both target and nontarget fragments will be captured and sequenced. (Modified after Figure 2 of Hodges, E. et al. 2009. *Nat Protoc* 4:960–974. With permission.)

stretch of bases consisting of repetitive DNA, then a target strand can simultaneously anneal to its correct RNA bait and to the repetitive DNA found in an off-target library strand (Figure 7.29a). In this scenario, both the target and off-target library strands will be captured and likely sequenced. A second type of unfavorable intermolecular interaction that can occur during a hybridization reaction consists of having the adapter sequence on a target DNA strand anneal to the complementary adapter sequence found on a nontarget DNA strand (Figure 7.29b). Again, if a RNA bait is annealed to a target strand whose adapter sequence is duplexed with the adapter sequence found on an off-target strand, then both target and off-target strands will be captured and probably sequenced. This phenomenon whereby off-target library strands are captured and sequenced because they had annealed to adapters on target library strands during the hybridization reaction is called "daisy chaining" (Mamanova et al. 2010; see Figure 2 in Rohland and Reich 2012). As we will see below, daisy chaining can drastically reduce the efficiency of a hybridization reaction.

There are two main strategies aimed at minimizing the by-catch problem in hybridization reactions. One strategy is to use a "blocking DNA mix" in the hybridization reaction (Hodges et al. 2009) while the other is to construct libraries with shorter adapter sequences before hybrid selection (Faircloth et al. 2012; Rohland and Reich 2012). We will first consider the use of blocking DNA additives and later in this chapter we will discuss the effects of adapter length on in-solution hybridization efficiency.

"Blocking DNA mix" consists of different types of DNA that are used to competitively anneal to stretches of repetitive DNA and adapters on pond DNA strands in order to block these sites from catching unwanted library strands. A type of DNA called Cot-1 DNA, which is a repeat-rich DNA that has been selected for its high affinity for repetitive DNA, is used to neutralize stretches of repetitive DNA found in pond library strands (Hodges et al. 2009). Although it was originally thought that Cot-1 DNA should be obtained from the same (or similar) species as species represented by the libraries (Hodges et al. 2009), in practice it may not matter which Cot-1 DNA is used as blocking DNA because repetitive DNA can evidently anneal to a variety of nonhomologous sequences. For example, Faircloth et al. (2015) found that hybrid selection of hymenopteran libraries was more efficient when chicken Cot-1 DNA was used as the repetitive DNA blocker compared to actual hymenopteran Cot-1 DNA. Thus, human or chicken Cot-1 DNA has been used as the repetitive DNA blocker in many vertebrate phylogenomic studies. Salmon sperm is another additive that is sometime used to help minimize unfavorable molecular interactions during the hybridization reaction. Another strategy that has worked well is to mix different Cot-1 DNAs. For example, good hybridization results were obtained for several reptile and amphibian species by using a mix consisting of 1/3 human Cot-1 DNA, 1/3 chicken Cot-1, and 1/3 salmon sperm Cot-1 (S. B. Reilly, personal communication, 2016). Similarly Portik et al. (2016) used a mix comprised of 1/3 human Cot-1, 1/3 mouse Cot-1, and 1/3 chicken Cot-1 DNA in their hybrid selection of pooled frog libraries. A third and critical component to the blocking mix is comprised of forward and reverse complements to the adapter sequences. It is essential to match the adapter blocking oligos as closely as possible with the adapters used to make the library constructs. It is therefore common practice to replace the kit-supplied adapter blocking mix with a custom-made adapter blocking mix that better matches the adapters used in the libraries. For example, Lemmon et al. (2012) replaced the adapter blocking mix contained in the SureSelect kit (i.e., Block #3) with a custom-made indexed blocking mix (see Geller 2011 for oligo sequences and recipe) appropriate for indexed Illumina libraries. Custom-designed adapter blocking oligos can also be found in Meyer and Kircher (2010) and Rohland and Reich (2012). Faircloth et al. (2015) designed blocking oligos for their custom-made adapters, each of which included a string of 10 inosines to block the 10-bp index sequences. Similarly, Leaché et al. (2015) used a custom blocking mix matched to the Illumina Truseq Nano adapters, which also used inosines in place of the index bases. The recent study by Portik et al. (2016) evaluated the performance of three different types of adapter blocking oligos on hybridization results. These included the adapter blocking oligos included in the MYbaits kit, short blocking oligos that do not protect the index bases, and the *xGen* blocking oligos made by IDT. Based on their results, the authors concluded that the *xGen* blocking oligos performed much better than the other oligos included in their test. Accordingly, given that hybridization reactions are typically carried out with commercial baits kits, careful consideration should be given to which adapter blocking mix should be used.

How much pond DNA is input into a hybrid selection experiment? Typically, the total amount of input pond DNA used for the hybrid selection procedure usually ranges between 100 and 500 ng (e.g., Blumenstiel et al. 2010) and protocols usually presume that the pond DNA itself is at a concentration of 100 ng/μL. The original in-solution hybrid selection study of Gnirke et al. (2009) used 500 ng of pond DNA together with 500 ng of RNA baits. However, these authors noted that they obtained comparable results regardless of whether their inputs were 100 ng or 500 ng each of pond and baits (see also Faircloth et al. 2015). Given the high cost of baits and occasions when pond concentrations are low, it may be desirable to use the minimum amount of baits allowed by the protocol (e.g., 100 ng). However, for organisms with very large genomes (e.g., amphibians), it is preferable to use at least 250 ng of each library in a hybridization reaction; thus if four frog libraries are pooled together, then 1 μg total of library DNA would comprise a hybridization pool (S. B. Reilly, personal communication, 2016).

Step 3: Hybridize RNA Baits to Single-Stranded Library DNA After the pond DNA and blocking DNAs have been denatured, the thermocycler lowers the temperature down to the standard hybridization temperature of 65°C. For the next 16–48 hours the hybridization reaction will occur at this temperature, which is high enough to ensure that RNA baits largely only anneal to complementary stretches of DNA within the library DNA strands (Figure 7.28). The process of setting up hybridization reactions should be done exactly according to the baits protocol in order to minimize the occurrence of by-catch and nonspecific RNA–DNA hybrids during the reaction. Note, to achieve the best possible capture results, it is critically important to also do the following: (1) carefully mix, by pipetting, your library pool and hybridization mix and avoid creating bubbles or splashing liquid on the sides of the reaction tubes and (2) be sure to heat the thermocycler lid during the hybridization reaction (S. B. Reilly, personal communication, 2016).

To complete the reaction setup, the following procedures must be performed in the correct sequence. When the thermocycler's heat block arrives at 65°C, the machine must be temporarily paused. This is to allow some time for the Cot-1 DNA and adapter blocking oligos to hybridize with repetitive DNA and adapters on the library DNA strands, respectively, before the

RNA baits are mixed with the pond DNA. At this time, the hybridization buffer mix (i.e., hybridization buffer, RNA baits, and RNase block) must be prewarmed at 65°C thus they are placed into the thermocycler's heat block near to the tubes containing the libraries. After the preheating period is completed, the hybridization buffer mix is transferred into the pond DNA tubes while all tubes are still sitting in the heat block. For safety reasons it should be kept in mind that the thermocycler's heated block is hot enough to cause severe burns and thus extreme caution should be exercised. After all reagents have been mixed together the lid of the thermocycler is closed and program is resumed to allow the hybridization reaction to proceed.

Step 4: Capture RNA–DNA Hybrids with Streptavidin-Coated Beads When the hybridization reaction is completed the next step is to "capture" each of the RNA–DNA hybrids while they are still in solution. This is accomplished by adding streptavidin-coated paramagnetic beads to the hybridization reactions. The streptavidin-coated beads are able to capture the RNA–DNA hybrids because streptavidin (a bacterial protein) has an exceptionally high affinity for the biotin moieties (a form of vitamin B) that are incorporated into the RNA baits (Figure 7.28). The streptavidin–biotin interaction is evidently one of the strongest noncovalent biological interactions (Holmberg et al. 2005).

Step 5: "Pull Down" Captured RNA–DNA Hybrids onto Magnet Next, a magnetic separator is used to "pull down" the beads attached to the RNA–DNA hybrids, which immobilizes them against the tube wells near the magnet. The supernatant contains nonhybridized DNA and is therefore removed and discarded (Figure 7.28).

Step 6: Eliminate Nontarget DNA via Series of Washes Although a hybridization reaction may produce a sufficient number of the target RNA–DNA hybrids, these reactions may also generate a number of hybrids involving nontarget DNA. If nontarget library fragments having both the P5 and P7 common adapters are not somehow eliminated, then they will populate the flow cell along with the target DNA resulting in lower quality sequence data. Fortunately, these nonspecific hybrids are weakly bound together owing to their nonhomologous sequence matches and thus they can be disassociated via a series of stringency washes (Figure 7.28).

The first wash step is a "low stringency" wash in which a wash buffer consisting of nuclease-free water, saline sodium citrate (SSC), and sodium dodecyl sulfate (SDS) is incubated with the hybridization reactions at room temperature. This first wash buffer helps to strip away nontarget DNAs that are weakly bound to RNA baits or are sticking to the inside of the tube. At the end of the brief incubation period the supernatant containing these unwanted DNAs is removed using a pipette and discarded. The second wash is a "high stringency" wash, which is needed in order to dissociate the more stable nonspecific RNA–DNA hybrids. This wash is more stringent than the first for two reasons. First, the second wash buffer contains a 10-fold lower concentration of SSC (salt buffer) than is found in the first wash buffer. Secondly, the second wash is conducted at a much higher temperature (65°C) than was used during the first wash (room temperature). Following this second wash step, the collection of captured DNA fragments, which at this time are still bound to the biotinylated RNA baits on the magnet, must be amplified in a second limited cycle PCR.

Step 7: Second Limited Cycle PCR Followed by SPRI Following the hybrid selection and wash steps, a second (or "post hybridization") limited cycle PCR must be performed (Figure 7.28). If the captured library constructs are already sequencing-ready (i.e., are indexed and contain P5 and P7 common adapters), then this PCR is merely used to enrich the library with the captured library strands so that a sufficient number of sequencing templates can be added to the flow cell. However, if the captured constructs do not contain all the required elements for sequencing, then this PCR is used to (1) complete the constructs using 5′ tailed PCR primers in order to add one or two indices as well as the P5 and P7 common adapters and (2) amplify these newly completed constructs.

Before these PCRs can be set up and run, captured DNA templates must be prepared. The original method for accomplishing this involved using a solution of sodium hydroxide (NaOH) to chemically break the hydrogen bonds linking together the RNA and DNA strands (e.g., Blumenstiel et al. 2010). Thus, the tubes (or plate) containing the hybridization reactions are removed from the magnet before a solution of NaOH is added to the reactions. Next, a neutralization buffer consisting of Tris–HCL is added, and, after a brief incubation period, the magnetic separator is used to immobilize the bead–RNA bait complexes. Supernatants containing the captured ssDNAs are then transferred to clean tubes (or plate) where they are cleaned using SPRI beads followed by elution with nuclease-free water (Blumenstiel et al. 2010). The captured DNA samples are then ready to serve as PCR templates.

Fisher et al. (2011) developed a far simpler alternative method for preparing the captured DNA. Instead of employing the standard chemical-denaturing and SPRI bead cleanup approach, the RNA–DNA hybrids are simply allowed to air dry while they are on the magnet. This drying procedure is initiated immediately after the removal and disposal of the second wash buffer (i.e., high stringency buffer) while the tubes (or plate) are on magnet. Next, the magnet is removed and nuclease-free water is pipetted into the same hybridization tubes (or plate) in order to resuspend the dried RNA–DNA hybrids. The other PCR reagents are then added before the samples are placed into a thermocycler to conduct the PCR. Fisher et al. (2011) termed this procedure *off-bead PCR* because target DNA strands are released from the streptavidin bead-RNA bait complexes during the first heat-denaturation step of the PCR. After the off-bead PCR is completed, the magnet is used to pull down and immobilize the streptavidin beads while the supernatant containing the PCR products is saved to other clean tubes (or plate). The PCR products are cleaned using a standard SPRI bead cleanup.

The off-bead PCR method has significant advantages over the NaOH-denaturing approach. Besides the simplicity and automation friendly aspects of the method, Fisher et al. (2011) observed an approximately 3-fold increase in captured product yield following off-bead PCR compared to the NaOH-based approach. This method represents a major improvement in the hybrid selection methodology and thus it has been used in some phylogenomic studies (e.g., Faircloth et al. 2015) and is now implemented in some RNA bait kits (e.g., MYbaits and SureSelect kits).

This second limited cycle PCR typically consists of 15–20 normal cycles. Like the first limited cycle PCR, only PCR proofreading (high fidelity) DNA polymerases should be used. One interesting difference between this PCR and other PCRs already discussed is that the templates are single-stranded at the start because the RNA baits only capture one library strand at a time. However, this does not present a problem because one of the two PCR primers can readily synthesize complete complementary strands during the first cycle. Thus, starting in the second cycle there will be two complementary strands that can be copied because both will have annealing sites for each primer. Thenceforth, target templates can be copied at an exponential rate over the course of the remaining PCR cycles.

As you will recall, the library constructs generated by the Rohland and Reich (2012) approach contained a 6-bp barcode sequence, which allows for the pooling of multiple libraries prior to hybridization. Because these library constructs have truncated adapters and thus are not yet in complete form, 5′ tailed primers must be used in the second limited cycle PCR in order to add a 7-bp index sequence to the P7 side of the construct as well as the P5 and P7 common adapters. Figure 7.30 shows the first three cycles of a second limited cycle PCR involving a Rohland and Reich truncated-adapter construct. As the reaction mixture is heated toward the first denaturation temperature, the RNA baits become dissociated from the library DNA strands and thus the first extension step will effectively only involve one of the two complementary strands. However, at the conclusion of the first cycle, library strands containing the binding sites for both PCR primers will be ready to participate in DNA synthesis starting in the second cycle (Figure 7.30). By the end of third cycle, fully formed constructs start to appear among the reaction products (Figure 7.30). When the PCR is finished, the products will be largely comprised of fully-formed adapter constructs derived from the original captured library DNAs (Figure 7.31). Notice that the PE Read 1 Sequencing Primer is used to sequence the 6-bp barcode *and* one end of the target fragment (Figure 7.31). Thus, the first six bases to be sequenced using the PE Read 1 Sequencing Primer will actually be the barcode followed by the target fragment. The 7-bp index is, however, sequenced using its own sequencing primer (i.e., Index PE Sequencing Primer; Figure 7.31).

Step 8: Verify and Quantify Libraries Following the second limited cycle PCR, each posthybridization library pool must be verified and quantified. Verification of libraries is generally done two ways. First, a Bioanalyzer is used to confirm the size range distribution for each pool. The electropherogram should show a distribution of

Cycle 1 denature step
Cycle 1 extension step
Cycle 2 denature step
Cycle 2 extension step
Cycle 3 denature step
Cycle 3 extension step
Sol-PE-PCR_F primer
Direction of synthesis
Index PE primer

Figure 7.30. How the second limited cycle PCR completes adapter–fragment library constructs using the Rohland and Reich approach. The denaturation and primer-extension steps in the first three cycles of a limited PCR are illustrated (primer-annealing steps not shown due to space limitation). Because this PCR is begun with a single adapter–fragment strand that was captured in the hybridization reaction, only one of the two primers can bind to the template in the first cycle. However, in the second cycle binding sites are available for both PCR primers. Completed adapter–fragment constructs appear only at the end of the third cycle. The 7-bp index sequence is represented by (index). The middle section of the adapter–fragment constructs contains the target DNA (a string of Xs). The // signifies that most of the target fragment is omitted due to space limitation and the vertical dashes indicate hydrogen bonding between complementary bases. Boldface sequences represent newly annealed primer in the same cycle and gray letters represent DNA synthesized in this reaction.

fragments with the peak centered between 250 and 500 bp depending on initial size distribution generated at the start of the library making process and taking into account adapter lengths. The other critical verification assay that should be performed is to evaluate hybrid selection performance via qPCR; in other words, verify that target molecules were selectively captured and nontarget molecules depleted following each hybridization reaction (Hodges et al. 2009; Bi et al. 2012; Peñalba et al. 2014; Faircloth et al. 2015).

In this "relative qPCR" procedure (Faircloth et al. 2015) a small subset of target and nontarget loci from random genomic locations are subjected to qPCR in order to compare the amounts of specific target and nontarget templates present in pre- versus posthybridization pools. To perform this procedure, custom locus-specific PCR primers for 2–3 target loci and 2–3 nontarget loci are first designed (Chapter 8 discusses primer design) to amplify a 100–150 bp section of each target and nontarget locus (Hodges et al. 2009; Peñalba et al. 2014; S. B. Reilly, personal communication, 2016). These primers should be 20 bp long, have melting temperatures (Tm) between 58 and 60°C, and must be tested beforehand using regular PCR procedures to ensure that they robustly amplify the correct-sized products and do not produce other artifacts such as primer dimers (Hodges et al. 2009; Peñalba et al. 2014). Next, each test locus is qPCR-amplified in duplicate or triplicate with templates supplied from pre- and posthybridization library pools (see Hodges et al. 2009 and Peñalba et al. 2014 for protocols). Thus, if duplicate reactions are set up, then each locus will be amplified in two replicates using prehybridization pool template and in two replicate reactions using posthybridization template. After qPCR is completed, within-locus C_T values output from qPCR analyses are compared to each other. The C_T value (cycle threshold) is defined as the cycle number in which the detected amount of fluorescently labeled products is first distinguishable from a horizontal threshold value (i.e., background fluorescence "noise"). The C_T value is useful because it is correlated with the amount of starting template. Thus, if hybrid selection was successful, then C_T values for posthybridization library templates should be shifted 6–10 cycles to the left of C_T values for prehybridization library templates; in contrast, C_T values for nontarget loci should show the opposite with posthybridization C_T values shifted 4–10 cycles to the right of prehybridization C_T values (Hodges et al. 2009; Peñalba et al. 2014; S. B. Reilly, personal communication, 2016). We can easily visualize differences between C_T values by examining plots of qPCR amplification curves for each locus (Figure 7.32). Notice that the C_T values in the top plot of Figure 7.32 are located near the 20 and 30 cycle marks, while the values in the bottom plot are approximately at 18 and 28 cycles. Because the within-locus results show that posthybridization C_T values are shifted to the left of prehybridization C_T values by at least five cycles, this hybrid selection reaction can be considered successful (Hodges et al. 2009; Peñalba et al. 2014; S. B. Reilly, personal communication, 2016). Though not shown here, amplification curves for nontarget loci should show evidence that they were depleted in posthybridization pools.

Once each posthybridization pool has passed the relative qPCR test, they must be quantified. Although a Bioanalyzer or Qubit fluorometer can be used to quantify each library pool, the most accurate way to quantify library pools is to use qPCR. However, unlike the relative qPCR we just discussed, an Illumina qPCR quantification kit should be used, which includes primers that anneal to the adapters.

Step 9: Normalize and Pool the Pooled Libraries After quantification, library pools are ready to be pooled together in equimolar ratios (i.e., "pooling of pools"). The number of final pools will depend on the number of Illumina lanes that will be used

```
                                                                                                                          P7 flow cell oligo
                                                                                                                  3'TAGAGCATACGGCAGAAGACGAAC 5'
                                                                         PE read 2 Sequencing Primer
                                                                 3' TCTAGCCTTCTCGCCAAGTCGTCCTTACGGCTCTGGC 5'

3' TTACTATGCCGCTGGTGGCTCTAGATGTGAGAAAGGGATGTGCTGCGAGAAGGCTAGA [barcode]XXXXX//XXXXXTCTAGCCTTCTCGCCAAGTCGTCCTTACGGCTCTGGC [index]TAGAGCATACGGCAGAAGACGAAC 5'
   ||||||||||||||||||||||||||||||||||||||||||||||||||||||||||||||||||||||||||||||||||||||||||||||||||||||||||||||||||||||||||||||||||| ||||||||||||||||||||||||
5' AATGATACGGCGACCACCGAGATCTACACTCTTTCCCTACACGACGCTCTTCCGATCT[barcode] XXXXX//XXXXXAGATCGGAAGAGCGGTTCAGCAGGAATGCCGAGACCG [index]ATCTCGTATGCCGTCTTCTGCTTG 3'

                                                                  5' GATCGGAAGAGCGGTTCAGCAGGAATGCCGAGACCG 3'
                                 5'ACACTCTTTCCCTACACGACGCTCTTCCGATCT 3'     Index PE Sequencing Primer
                                          PE Read 1 Sequencing Primer
5' AATGATACGGCGACCACCGAGAUCTACAC 3'
          P5 flow cell oligo
```

Figure 7.31. Completed Rohland and Reich adapter–fragment construct for Illumina sequencing. The construct contains annealing sites for the P5 and P7 flow cell oligos, PE Read 1 and 2 Sequencing Primers, and the Index PE Sequencing Primer. Notice that the 6-bp barcode is sequenced together with the Read 1 sequence. The 7-bp index sequence is represented by (index). The middle section of the adapter–fragment construct contains the target DNA (a string of Xs). The // signifies that most of the target fragment is omitted due to space limitation and the vertical dashes indicate hydrogen bonding between complementary bases. (Adapter sequences and primers were obtained from Rohland, N. and D. Reich. 2012. *Genome Res* 22:939–946, and are also found in Table 7.3.)

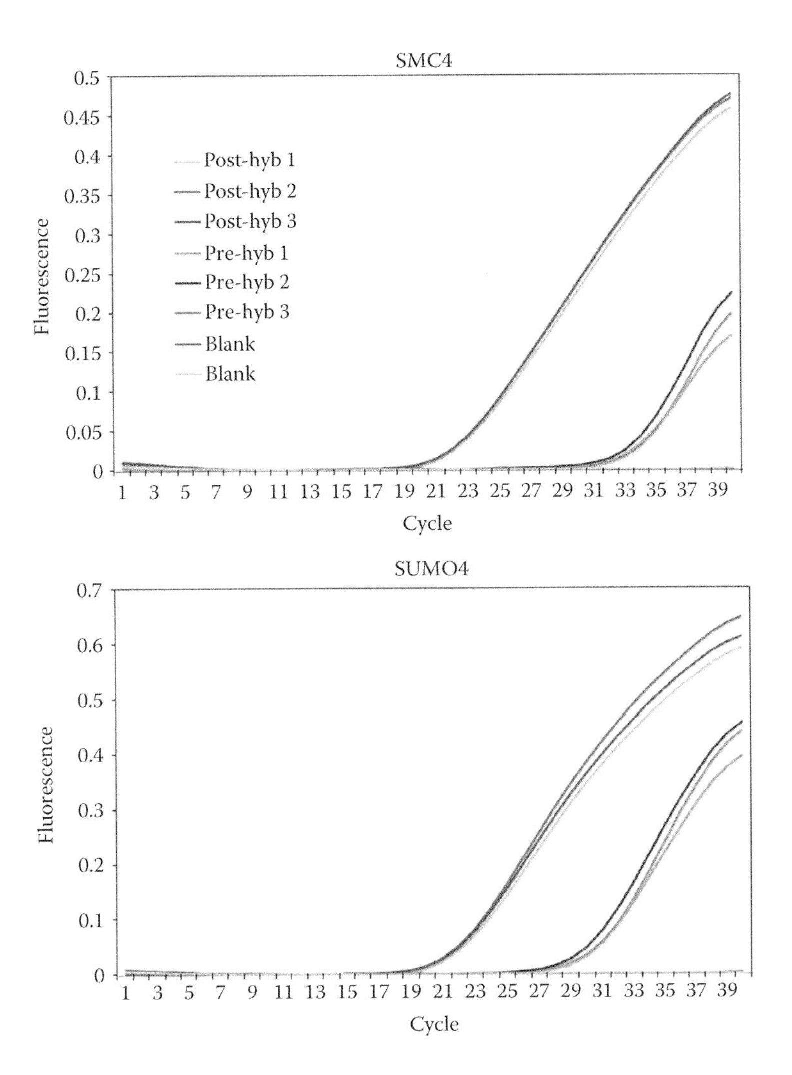

Figure 7.32. qPCR analyses showing levels of hybrid selection efficiency for two targeted loci. The top amplification plot shows qPCR results for the first target locus, which was amplified using the pre- and posthybridization pools as templates (curves pre-hyb 1–3 and post-hyb 1–3, respectively). Three replicates each of pre- and posthybridization qPCRs and two replicates of negative controls (blanks) were performed. The bottom amplification plot shows the results for the other targeted locus. The ~6–10 cycle shifts of the post-hyb curves to the left of the pre-hyb curves indicates that the hybrid selection reactions were successful. The x-axis shows the qPCR cycle number and y-axis corresponds to the accumulation of products (amount of detected fluorescence). This procedure can also be performed using 2–3 nontarget loci in order to verify their depletion following the hybridization reactions (results not shown). (Reprinted from Hodges, E. et al. 2009. *Nat Protoc* 4:960–974. With permission.)

though for phylogenomic studies this will usually mean that a single pool will be needed.

7.2.3 Indexing, Pooling, and Hybrid Selection Efficiency Revisited

Because phylogenomic studies often include genetic samples from dozens to hundreds of individuals, performing pooled hybridization reactions can greatly economize RNA baits (which are very expensive) as well as save labor in the lab. Thus, libraries must be indexed prior to the hybridization step if they are to be grouped together into hybridization pools. Although most of the steps in the library-making and in-solution workflows are comparable across studies, one aspect of this work that does vary concerns the indexing and pooling steps. Table 7.5 shows eight different indexing and pooling schemes that could be done using the methods described in this chapter. Schemes

TABLE 7.5
Eight possible indexing and pooling schemes for the in-solution hybrid selection workflow

Scheme	Adapter ligation	1st PCR	Pooling?	Hybrid capture	2nd PCR	Pooling?	Sequencing	References
A	Indexing		Yes	Capture each pool		No		x
B	Indexing		Yes	Capture each pool		Yes		1
C	Indexing		No	Capture each library		Yes		x
D		Indexing	Yes	Capture each pool		No		x
E		Indexing	Yes	Capture each pool		Yes		2, 3, 4
F		Indexing	No	Capture each library		Yes		x
G			No	Capture each library	Indexing	Yes		5, 6
H	Indexing		Yes	Capture each pool	Indexing	Yes		7

NOTE: The top row with the arrows shows the sequence of major steps in the in-solution hybrid selection workflow (not all steps shown). The indexing step can be performed during adapter ligation (A–C), during the first limited cycle PCR (D–F), or during the second limited cycle PCR (G and H). The only exception is scheme H in which indexes are added at two separate times (see main text). Pooling of libraries can only take place after the individual libraries have been indexed and following the first or second PCR. Below, the term "adapter ligation" is used to mean traditional ligase-mediated or transposase-mediated adapter attachment (i.e., tagmentation) and "hybrid capture" refers to whether an individual hybridization reaction includes a single library (i.e., "capture each library") or a pool of libraries (i.e., "capture each pool"). References: 1 = Faircloth et al. (2015); 2 = Lemmon et al. (2012); 3 = Smith et al. (2013); 4 = Meiklejohn et al. (2016); 5 = Meyer and Kircher (2010); 6 = Faircloth et al. (2012); 7 = Rohland and Reich (2012). x = unknown if studies have used this scheme.

A–C, which tend to be implemented in library kits, do the indexing step at the time of adapter ligation. The next three (D–F), which are usually done in non-kit protocols, index libraries during the first limited cycle PCR. Scheme G delays the indexing step until the second limited cycle PCR and H introduces index steps during the ligation and second PCR steps. Interestingly, at least half of these schemes have been used in studies showing that none of them is evidently dominant over the others in practice. What are the advantages and disadvantages to each of these schemes?

The schemes in which pooling is done in two stages—first to make hybrid capture pools followed by "pooling the pools" (B, E, H)—are certainly better for studies that have large numbers of samples (e.g., >30 libraries). However, for studies with a smaller number of libraries, making only the hybrid pools (A, D) will still be beneficial from the standpoint of conserving RNA baits. If hybrid selection is performed on single libraries, which are indexed before or after hybrid selection (C, F, G), then hybridization efficiencies are expected to be better than the results obtained from pooled samples. A drawback to the C, F, G approaches is that they are not practical for studies with large numbers of libraries. Scheme H may, at first glance, appear peculiar—after all, why does this approach, which was developed by Rohland and Reich (2012), introduce an index into the adapter–fragment construct during the ligation and second limited cycle PCR steps?

During the ligation step in the Rohland and Reich library-making approach, one adapter consisting of a 6-bp barcode sequence and binding sites for a nontailed PCR primer is ligated to one end of the library fragment while the other adapter only contains sites for the other nontailed PCR primer (Figure 7.18). A second (7-bp) index is added along with the missing sequence elements during the second limited cycle PCR (Figures 7.30 and 7.31). Thus, the initial barcoding of libraries permits the pooling of samples for hybridization using minimal-length adapters, whereas the index added later is used to distinguish pools. The combination of barcode and index sequences makes this approach amenable to high-throughput hybrid selection and sequencing projects (Rohland and Reich 2012). However, what is the significance of these short adapters?

Recall our discussion from earlier in this chapter concerning the potentially adverse effects of daisy chaining on hybridization reactions. Hodges et al. (2009) surmised that long adapter sequences—those containing the P5 and P7 common adapters, index sequences, and binding sites for PCR and sequencing primers—are at high risk of hybridizing to complementary adapter sequences, which could result in the capture and sequencing of unwanted library molecules

TABLE 7.6
Comparison of Illumina adapter lengths for adapter constructs used in pooled hybridization reactions

Types of adapters	Length of P5 adapter	Length of P7 adapter
Illumina TruSeq Nano HT	70 (8)	66 (8)
Meyer and Kircher (2010)	58	65 (7)
Rohland and Reich (2012)	34 (6)	33
Illumina Nextera	70 (8)	66 (8)

NOTE: Shown are the lengths (in bp) of adapters for several different adapter types made for Illumina library construction. Number within parentheses represents the number of bases for the index (when present). Adapter lengths are for adapters used in hybrid selection reactions and they are based on Figures 7.1b, 7.17, 7.18d, and 7.26.

(Figure 7.29b). Rohland and Reich (2012) tested this hypothesis and found that libraries with their "truncated" adapters, which are approximately half the length of full-length adapters (Table 7.6), had 2-fold higher hybridization efficiencies compared to libraries with long adapters. Hybrid efficiency here refers to the percentage of reads that mapped to target loci. Thus, daisy chaining appears to have a severe effect on the efficiency of hybrid selection reactions when long adapters are used.

One strategy that can be used with library-making approaches that produce long indexed adapters prior to the hybridization step is to simply not add the P5 and P7 common adapters and indices until the second limited cycle PCR (scheme G in Table 7.5; Meyer and Kircher 2010; Faircloth et al. 2012). However, this approach is not practical for studies with large numbers of libraries. As pointed out earlier, careful selection of adapter blocking oligos may also help ameliorate the daisy chaining problem but additional studies are needed to determine if an optimal indexing and pooling scheme exists. On reflection, perhaps this is not a big problem after all, as recent studies (e.g., McCormack et al. 2015) show that it is still possible to collect thousands of loci despite the use of long adapters in hybridization reactions.

7.3 COST-EFFECTIVE METHODS FOR OBTAINING MULTIPLEXED TARGETED-LOCI LIBRARIES

In-solution hybrid selection represents the best approach for obtaining large phylogenomic datasets consisting of hundreds to thousands of DNA sequence loci. However, a major drawback is that this approach is likely to be prohibitively expensive for many smaller phylogenomic laboratories owing to the high cost of obtaining commercially made probe kits. Fortunately, there are a number of other more cost-effective approaches for utilizing Illumina sequencing to generate phylogenomic datasets. One of these methods is a form of in-solution hybrid selection, which uses biotinylated PCR amplicons as baits instead of RNA (Maricic et al. 2010; Peñalba et al. 2014). In this method, which has been termed *sequence capture using PCR-generated probes* or "SCPP" (Peñalba et al. 2014), PCR baits can be easily generated in any laboratory at low cost (i.e., mainly the cost of the primers) thereby obviating the need to purchase expensive bait kits (Maricic et al. 2010; Peñalba et al. 2014). Another approach, which is known as *parallel tagged amplicon sequencing* or "PTS," involves the sequencing of pooled adapter-tagged PCR products (O'Neill et al. 2013). Although these methods could potentially be used to produce datasets comprised of hundreds of loci, they are practical for obtaining datasets of relatively modest sizes (e.g., 20–100 loci). The cost-effectiveness of these methods, however, means that smaller labs can obtain phylogenomic datasets consisting of (at least) dozens of loci for dozens to hundreds of individuals. We will now review these two approaches.

7.3.1 Sequence Capture Using PCR-Generated Probes (SCPP)

Maricic et al. (2010) described a method for using PCR product baits to capture target loci fragments from indexed Illumina libraries that had been pooled together. The authors illustrated the utility of this method, which can be applied to any set of target loci for which PCR products can be produced, by generating a sufficient number of reads from a single flow cell lane to construct 46 complete mitochondrial genomes. As can be seen in Figure 7.33, their workflow largely resembles the in-solution workflow we examined earlier in this chapter. This overview of the workflow shows that the PCR baits and indexed sequencing libraries (i.e., the "pond") must first be prepared before the hybrid selection reaction can take place (Figure 7.33).

To begin preparation of the PCR baits, PCR products for the target sequences in at least one exemplar individual must be generated. In the Maricic et al.

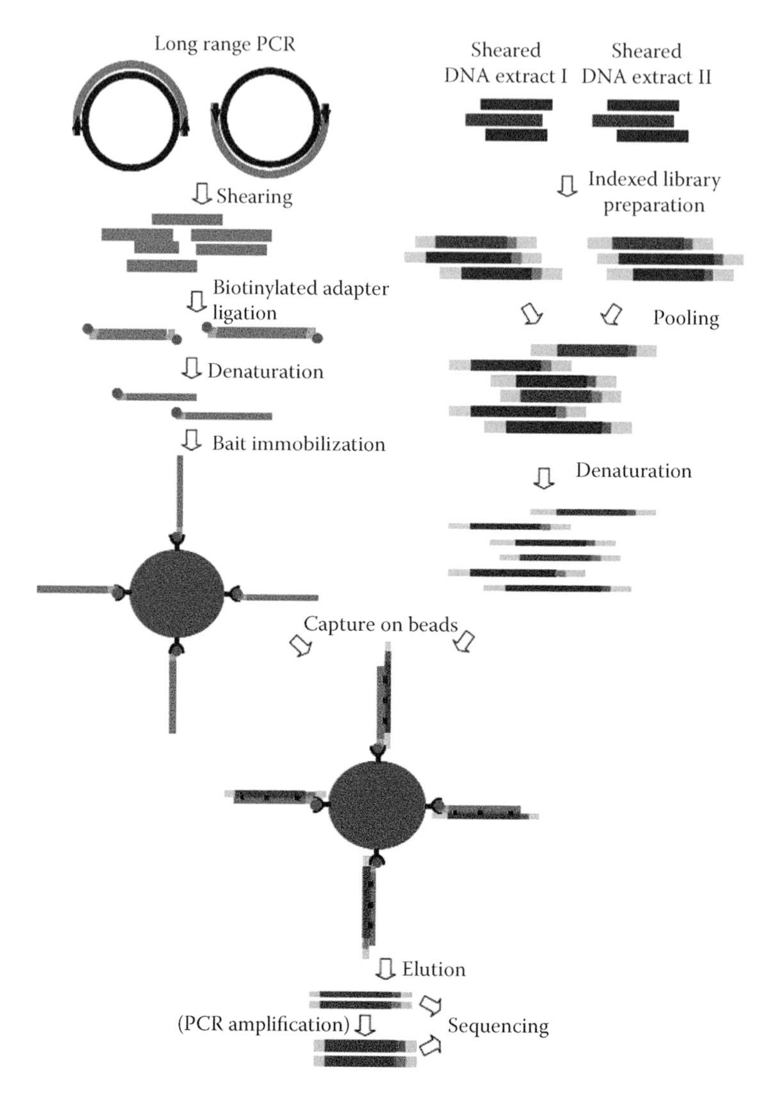

Figure 7.33. Overview of the Maricic et al. SSCP workflow applied to the sequencing of entire human mitochondrial genomes. The preparation of baits (top left) begins with the generation of two long PCR products that cover the entire mitochondrial genome. The PCR products are cleaned then pooled before they are sheared into fragments. Next, biotinylated adapters are ligated to the fragmented PCR products followed by a denaturation step to generate single-stranded bait DNA (light red) for immobilization on streptavidin-coated beads. Indexed libraries are constructed (top right) then pooled before being denatured into ssDNA. The hybridization reaction (bottom) proceeds after the bead-immobilized PCR baits are combined with the pooled library DNA. Captured library fragments (dark red) are eluted and directly sequenced or they are PCR-amplified before the sequencing step. Color codes for library DNAs: Indices are green and pink while the adapters are light gray. Thick lines indicate dsDNA while thin lines show ssDNA. (Reprinted from Maricic, T. et al. 2010. *PLoS One* 5:e14004. With permission.)

(2010) study, the authors used long PCR to produce two long mitochondrial DNA fragments that covered the entire mitochondrial genome for one of the genetic samples. Next, they SPRI-cleaned the fragments and then quantified their concentrations using a Nanodrop. In the following step, the two PCR products were combined together in equimolar ratios before subjecting the pool to a sonication treatment in order to generate a fragment size range of 150–850 bp. The fragments were biotinylated by ligating biotin-labeled adapters to each fragment. Following a column-based purification step, the biotinylated baits were immobilized on streptavidin-coated beads (Figure 7.33).

The library making process began with DNA extraction from the 46 samples (Figure 7.33).

Next, all DNA extracts were subjected to a sonication treatment to create fragment sizes between 150 and 850 bp. A portion of each fragmented sample was then used as input DNA to generate indexed Illumina libraries via the Meyer and Kircher (2010) approach discussed earlier. Following the first limited cycle PCR, the concentration of each library was measured using a Nanodrop before all libraries were grouped together, in equimolar amounts, to form a hybridization pool.

The hybridization reaction procedure began with the preparation of a hybridization mixture consisting of pond DNA, blocking oligo mix, Agilent blocking mix, and Agilent hybridization buffer. The mixture was then heat-denatured at 95°C to generate single-stranded adapter-ligated library fragments (Figure 7.33). The mixture was incubated at a temperature that allowed the blocking oligos and DNAs to bind to their targets (65°C) before bead-bound PCR baits were added to complete the hybridization reaction mixture. The hybridization reaction was incubated at 65°C for 48 hours. Later, nonhybridized DNA was discarded and a series of wash steps, which increased in stringency with each step, were used to further eliminate nonspecific hybrids. Captured DNA fragments were released from baits (i.e., eluted) via a treatment with NaOH. Next, the captured DNA was quantified using qPCR. If a sufficient amount of template was obtained from the hybridization reaction, then the next step is cluster generation followed by sequencing. However, if insufficient template is found, then a second limited cycle PCR must be used to further enrich the library pool. If a second limited cycle PCR is used, then it is essential to use the fewest number of cycles possible in order to minimize the number of duplicated fragments as well as to lower the risk of recombination between amplicons representing different individuals (Maricic et al. 2010). In addition to obtaining enough sequence reads to assemble all 46 mitochondrial genomes, the authors noted that they found no evidence of numts contamination in their sequence data.

Peñalba et al. (2014) modified the Maricic et al. (2010) method to make it more suitable for phylogenomic studies involving groups of organisms with varying levels of intraclade divergences. These modifications were primarily for improving capture efficiencies of target loci particularly when libraries with varying phylogenetic distances to the PCR baits are being captured. Peñalba et al. (2014) also used standard library-making approaches (i.e., Illumina Truseq kit and Meyer-Kircher [2010] approach).

The first key modification involved diversifying intralocus sequences within PCR bait sets. Maricic et al. (2010) had generated their PCR bait set from long PCR products obtained from a single individual. While this approach is expected to work well for studies operating at intraspecific levels or recently diverged species, hybridization efficiencies would likely suffer in studies involving species with moderate to high divergences. Thus, Peñalba et al. (2014) broadened the hybridization capabilities of their PCR bait sets by including amplicons obtained from species that bracketed levels of sequence divergences found within each study group. For each locus, good quality PCR products obtained from different species are pooled together and then electrophoresed in an agarose gel. Next, each target band is excised from the gel and purified using a gel-purification kit. Though this procedure is much more laborious than the SPRI-bead based purification used in the Maricic et al. protocol (Figure 7.34), it should yield pure target amplicons by eliminating all smaller and larger PCR artifacts.

After all loci have been extracted from the gel and purified, they are pooled together in equimolar ratios. Another difference between the Maricic et al. and Peñalba et al. methods is that the former used sonication to shear the long PCR products into smaller fragments, while the latter leaves the PCR amplicons intact (Figure 7.34). The pooled amplicons are biotinylated the same way in both protocols (i.e., via ligating biotinylated adapters to the ends of amplicons), but the Peñalba et al. approach involves making two aliquots of PCR baits for each project (Figure 7.34). In both protocols, the next step is to denature PCR baits to make them single-stranded and then immobilize them on streptavidin-coated magnetic beads.

Peñalba et al. (2014) made several modifications to the hybridization reaction procedures in an effort to enhance capture efficiencies of target sequences. The first of these changes was to add Cot-1 DNA to the hybridization mix in order to reduce the chances of capturing nontarget library DNA containing repetitive DNA (Figure 7.34). The second change was to use a longer hybridization time (72 hours) and a touch-down PCR program that begins at the standard hybridization

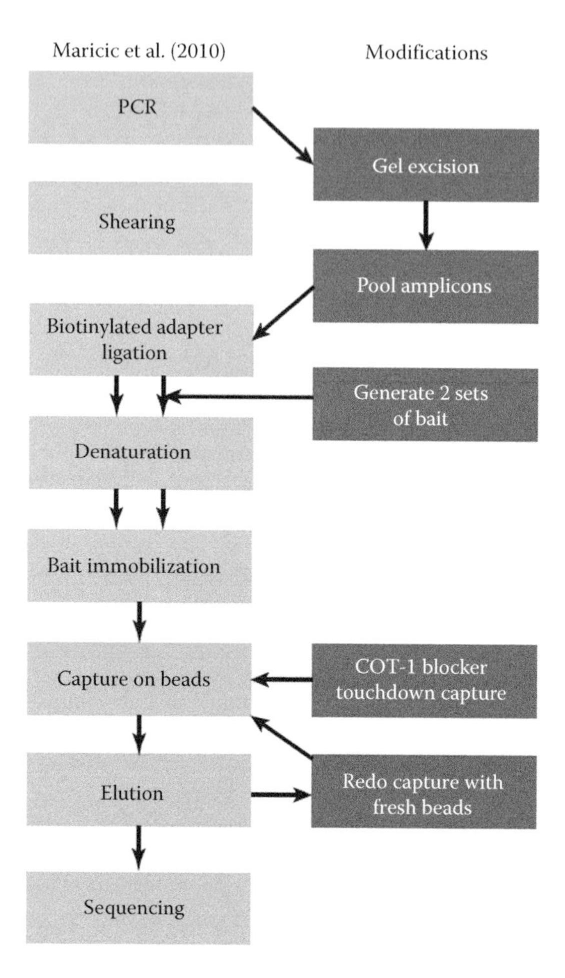

Figure 7.34. Overview of the Peñalba et al. modified SSCP workflow. The original SSCP workflow developed by Maricic et al. (2010) is shown on the left in the light gray rectangles while the modified steps used by Peñalba et al. (2014) are shown in the dark gray rectangles to the right. Note, this workflow only shows the PCR bait preparation and hybridization procedures and not the library-making methods, which are standard. (Reprinted from Peñalba, J. V. et al. 2014. *Mol Ecol Resour* 14:1000–1010. With permission.)

temperature of 65°C and slowly decreases in temperature down to a final temperature of 60°C. By gradually lowering the stringency of the hybridization temperature, it is believed that capture efficiency of more divergent sequences (relative to the bait sequences) can be increased. The third change was to redo the capture with a set of fresh PCR baits. In this procedure, the eluate from the first hybridization reaction is subjected to a second hybridization reaction using the second (fresh) aliquot of PCR baits (Figure 7.34). As discussed earlier, it is also important to perform a relative qPCR assay in order to evaluate the success of SCPP reactions. Peñalba et al. (2014) provide a detailed explanation of this procedure involving SCPP reactions.

The modified SCPP method allows for capitalization of primers in the literature or new primers can be designed and used (Chapter 8). This method is therefore well suited for small labs that already have many primer sets optimized for their study systems (Peñalba et al. 2014). In practical terms, this method can be used to generate phylogenomic datasets of modest sizes (25–100 target loci) for nonmodel organisms involving shallow-level divergences (e.g., phylogeography) or deeper level divergences (Peñalba et al. 2014).

7.3.2 Parallel Tagged Amplicon Sequencing

The study by O'Neill et al. (2013) showed how PTS can be used to generate phylogenomic datasets consisting of nearly 100 nuclear loci. These authors used 454 sequencing to obtain haplotype sequences from 95 independent loci representing 93 individual tiger salamanders. Although it is possible to obtain datasets with a larger number of loci, notice that their study required a minimum of 8,835 PCRs! In another study on amphibians, Barrow et al. (2014) used PTS with Illumina sequencing to generate haplotype sequences for 27 nuclear and three mitochondrial loci from 44 individual chorus frogs. While PTS may not be practical for harvesting hundreds or thousands of loci from genomes, this method does have at least one advantage over in-solution hybrid selection methods when anonymous loci are used.

As you will recall, one of the desirable properties of anonymous loci is that they are believed to often meet the neutrality assumption. Moreover, in Chapters 2 and 3, we learned that a large fraction of all transposable elements in genomes are no longer functional and thus these sequences—provided they are not linked to functional sites—can represent ideal anonymous loci. However, researchers have long expressed worries about designing PCR-based loci containing repetitive DNA because of concerns about amplifying and sequencing paralogous copies. Thus, repetitive DNA is usually masked or not given consideration for designing new anonymous loci. For reasons that will be made clear in Chapter 8, PCR primers, when properly designed, can faithfully obtain the same "orthologous" sequences even when these sequences contain one or more transposable elements. This high level of specificity means that special measures for avoiding repetitive DNA such as designing PCR primers in regions of the genome not containing repetitive DNA need not be used for PTS. In contrast, RNA or PCR product probes used in in-solution hybrid selection studies will, if they contain repetitive DNA sequences, be more prone to obtaining multiple copies of loci thereby potentially leading to the creation of datasets that are contaminated with paralogous sequences. Even if the paralogy problem didn't exist or if you were to use anonymous loci free of repetitive DNA, there is another reason why it would not be practical to use in-solution hybridization methods that use RNA probes to generate anonymous loci datasets. Unless a given anonymous loci bait set were to be applied to closely related species groups, given the lack of sequence conservation typical of anonymous loci, a new bait set would need to be developed and purchased for each study group, a practice that would likely be cost-prohibitive for most labs. Thus, PTS used with Illumina sequencing is a cost-effective method for generating datasets consisting of dozens of single-copy, neutral, and independent loci for different organismal groups.

REFERENCES

Adessi, C., G. Matton, G. Ayala et al. 2000. Solid phase DNA amplification: Characterisation of primer attachment and amplification mechanisms. *Nucleic Acids Res* 28:e87–e87.

Adey, A., H. G. Morrison, X. Xun et al. 2010. Rapid, low-input, low-bias construction of shotgun fragment libraries by high-density *in vitro* transposition. *Genome Biol* 11:R119.

Barrow, L. N., H. F. Ralicki, S. A. Emme, and E. M. Lemmon. 2014. Species tree estimation of North American chorus frogs (Hylidae: *Pseudacris*) with parallel tagged amplicon sequencing. *Mol Phylogenet Evol* 75:78–90.

Bentley, D. R., S. Balasubramanian, H. P. Swerdlow et al. 2008. Accurate whole human genome sequencing using reversible terminator chemistry. *Nature* 456:53–59.

Bhasin, A., I. Y. Goryshin, M. Steiniger-White, D. York, and W. S. Reznikoff. 2000. Characterization of a Tn5 pre-cleavage synaptic complex. *J Mol Biol* 302:49–63.

Bi, K., D. Vanderpool, S. Singhal, T. Linderoth, C. Moritz, and J. M. Good. 2012. Transcriptome-based exon capture enables highly cost-effective comparative genomic data collection at moderate evolutionary scales. *BMC Genomics* 13:403.

Blumenstiel, B., K. Cibulskis, S. Fisher et al. 2010. Unit 18.4 Targeted exon sequencing by in-solution hybrid selection. *Curr Protoc Hum Genet* 18:1–18.

Brito, P. H. and S. V. Edwards. 2009. Multilocus phylogeography and phylogenetics using sequence-based markers. *Genetica* 135:439–455.

Bronner, I. F., M. A. Quail, D. J. Turner, and H. Swerdlow. 2014. Improved protocols for Illumina sequencing. *Curr Protoc Hum Genet* 79:18.2.1–18.2.42.

Brown, S. M. 2013. *Next-Generation DNA Sequencing Informatics*. Cold Spring Harbor: Cold Spring Harbor Laboratory Press.

Cronn, R., A. Liston, M. Parks, D. S. Gernandt, R. Shen, and T. Mockler. 2008. Multiplex sequencing of plant chloroplast genomes using Solexa sequencing-by-synthesis technology. *Nucleic Acids Res* 36:e122–e122.

Diatchenko, L., Y. F. Lau, A. P. Campbell et al. 1996. Suppression subtractive hybridization: A method for generating differentially regulated or tissue-specific cDNA probes and libraries. *Proc Natl Acad Sci USA* 93:6025–6030.

Edwards, S. and S. Bensch. 2009. Looking forwards or looking backwards in avian phylogeography? A comment on Zink and Barrowclough 2008. *Mol Ecol* 18:2930–2933.

Faircloth, B. C., M. G. Branstetter, N. D. White, and S. G. Brady. 2015. Target enrichment of ultraconserved elements from arthropods provides a genomic perspective on relationships among Hymenoptera. *Mol Ecol Resour* 15:489–501.

Faircloth, B. C., J. E. McCormack, N. G. Crawford, M. G. Harvey, R. T. Brumfield, and T. C. Glenn. 2012. Ultraconserved elements anchor thousands of genetic markers spanning multiple evolutionary timescales. *Syst Biol* 61:717–726.

Faircloth, B. C., L. Sorenson, F. Santini, and M. E. Alfaro. 2013. A phylogenomic perspective on the radiation of ray-finned fishes based upon targeted sequencing of ultraconserved elements (UCEs). *PLoS One* 8:e65923.

Fedurco, M., A. Romieu, S. Williams, I. Lawrence, and G. Turcatti. 2006. BTA: A novel reagent for DNA attachment on glass and efficient generation of solid-phase amplified DNA colonies. *Nucleic Acids Res* 34:e22–e22.

Fisher, S., A. Barry, J. Abreu et al. 2011. A scalable, fully automated process for construction of sequence-ready human exome targeted capture libraries. *Genome Biol* 12:1–15.

Geller, E. 2011. High-throughput indexed library preparation and pooled Agilent exome enrichment for Illumina sequencing platform. http://openwetware.org/images/1/10/Geller_exome.pdf (accessed May 9, 2016).

Gnirke, A., A. Melnikov, J. Maguire et al. 2009. Solution hybrid selection with ultra-long oligonucleotides for massively parallel targeted sequencing. *Nature Biotechnol* 27:182–189.

Hodges, E., M. Rooks, Z. Xuan et al. 2009. Hybrid selection of discrete genomic intervals on custom-designed microarrays for massively parallel sequencing. *Nat Protoc* 4:960–974.

Holmberg, A., A. Blomstergren, O. Nord, M. Lukacs, J. Lundeberg, and M. Uhlen. 2005. The biotin–streptavidin interaction can be reversibly broken using water at elevated temperatures. *Electrophoresis* 26:501–510.

Leaché, A. D., A. S. Chavez, L. N. Jones, J. A. Grummer, A. D. Gottscho, and C. W. Linkem. 2015. Phylogenomics of phrynosomatid lizards: Conflicting signals from sequence capture versus restriction site associated DNA sequencing. *Genome Biol Evol* 7:706–719.

Lemmon, A. R., S. A. Emme, and E. M. Lemmon. 2012. Anchored hybrid enrichment for massively high-throughput phylogenomics. *Syst Biol* 61:727–744.

Lemmon, E. M. and A. R. Lemmon. 2013. High-throughput genomic data in systematics and phylogenetics. *Annu Rev Ecol Evol Syst* 44:99–121.

Lukyanov, S. A., K. A. Lukyanov, N. G. Gurskaya, E. A. Bogdanova, and A. A. Buzdin. 2007. Selective suppression of polymerase chain reaction and its most popular applications. In *Nucleic Acids Hybridization Modern Applications*, 29–51. Springer: Netherlands.

Mamanova, L., A. J. Coffey, C. E. Scott et al. 2010. Target-enrichment strategies for next-generation sequencing. *Nat Methods* 7:111–118.

Mardis, E. R. 2008. Next-generation DNA sequencing methods. *Annu Rev Genomics Hum Genet* 9:387–402.

Margulies, M., M. Egholm, W. E. Altman et al. 2005. Genome sequencing in microfabricated high-density picolitre reactors. *Nature* 437:376–380.

Maricic, T., M. Whitten, and S. Pääbo. 2010. Multiplexed DNA sequence capture of mitochondrial genomes using PCR products. *PLoS One* 5:e14004.

Masoudi-Nejad, A., Z. Narimani, and N. Hosseinkhan. 2013. *Next Generation Sequencing and Sequence Assembly: Methodologies and Algorithms* (Vol. 4). New York, Heidelberg, Dordrecht, London: Springer Science & Business Media.

McCormack, J. E., S. M. Hird, A. J. Zellmer, B. C. Carstens, and R. T. Brumfield. 2013. Applications of next-generation sequencing to phylogeography and phylogenetics. *Mol Phylogenet Evol* 66:526–538.

McCormack, J. E., W. L. Tsai, and B. C. Faircloth. 2015. Sequence capture of ultraconserved elements from bird museum specimens. *Mol Ecol Resour* doi: 10.1111/1755-0998.12466.

Meiklejohn, K. A., B. C. Faircloth, T. C. Glenn, R. T. Kimball, and E. L. Braun. 2016. Analysis of a rapid evolutionary radiation using ultraconserved elements: Evidence for a bias in some multispecies coalescent methods. *Syst Biol.* doi: 10.1093/sysbio/syw014.

Meyer, M. and M. Kircher. 2010. Illumina sequencing library preparation for highly multiplexed target capture and sequencing. *Cold Spring Harbor Protocols.* doi: 10.1101/pdb.prot5448.

O'Neill, E. M., R. Schwartz, C. T. Bullock et al. 2013. Parallel tagged amplicon sequencing reveals major lineages and phylogenetic structure in the North American tiger salamander (*Ambystoma tigrinum*) species complex. *Mol Ecol* 22:111–129.

Peñalba, J. V., L. L. Smith, M. A. Tonione et al. 2014. Sequence capture using PCR-generated probes: A cost-effective method of targeted high-throughput sequencing for nonmodel organisms. *Mol Ecol Resour* 14:1000–1010.

Picelli, S., Å. K. Björklund, B. Reinius, S. Sagasser, G. Winberg, and R. Sandberg. 2014. Tn5 transposase and tagmentation procedures for massively scaled sequencing projects. *Genome Res* 24:2033–2040.

Portik, D. M., L. L. Smith, and K. Bi. 2016. An evaluation of transcriptome-based exon capture for frog phylogenomics across multiple scales of divergence (Class: Amphibia, Order: Anura). *Mol Ecol Resour* doi: 10.1111/1755-0998.12541.

Quail, M. A., I. Kozarewa, F. Smith et al. 2008. A large genome center's improvements to the Illumina sequencing system. *Nat Methods* 5:1005–1010.

Reznikoff, W. S. 2003. Tn5 as a model for understanding DNA transposition. *Mol Microbiol* 47:1199–1206.

Reznikoff, W. S. 2008. Transposon Tn 5. *Annu Rev Genet* 42:269–286.

Rohland, N. and D. Reich. 2012. Cost-effective, high-throughput DNA sequencing libraries for multiplexed target capture. *Genome Res* 22:939–946.

Shendure, J. and H. Ji. 2008. Next-generation DNA sequencing. *Nat Biotechnol* 26:1135–1145.

Shendure, J., R. D. Mitra, C. Varma, and G. M. Church. 2004. Advanced sequencing technologies: Methods and goals. *Nat Rev Genet* 5:335–344.

Siebert, P. D., A. Chenchik, D. E. Kellogg, K. A. Lukyanov, and S. A. Lukyanov. 1995. An improved PCR method for walking in uncloned genomic DNA. *Nucleic Acids Res* 23:1087.

Smith, B. T., M. G. Harvey, B. C. Faircloth, T. C. Glenn, and R. T. Brumfield. 2013. Target capture and massively parallel sequencing of ultraconserved elements (UCEs) for comparative studies at shallow evolutionary time scales. *Syst Biol*. doi: 10.1093/sysbio/syt061.

Steiniger-White, M., I. Rayment, and W. S. Reznikoff. 2004. Structure/function insights into Tn5 transposition. *Curr Opin Struct Biol* 14:50–57.

Syed, F., H. Grunenwald, and N. Caruccio. 2009a. Optimized library preparation method for next-generation sequencing. *Nat Methods* 6(10).

Syed, F., H. Grunenwald, and N. Caruccio. 2009b. Next-generation sequencing library preparation: Simultaneous fragmentation and tagging using *in vitro* transposition. *Nat Methods* 6(11).

Toews, D. P., L. Campagna, S. A. Taylor et al. 2015. Genomic approaches to understanding population divergence and speciation in birds. *Auk* 133:13–30.

Turner, E. H., S. B. Ng, D. A. Nickerson, and J. Shendure. 2009. Methods for genomic partitioning. *Annu Rev Genomics Hum Genet* 10:263–284.

CHAPTER EIGHT

Developing DNA Sequence Loci

Once a researcher decides which type of locus or loci will be needed for a phylogenomic study, the next problem to solve is how to begin the process of generating a DNA sequence dataset. Many types of DNA sequence loci are PCR-based, which of course means they require pairs of DNA primers to target particular genomic regions for sequencing. As we saw in Chapters 6 and 7, PCR products can be used as input templates for Sanger sequencing or NGS (e.g., Illumina sequencing). However, we have not considered how PCR primers are designed in the first place.

During the early years of molecular phylogenetics and phylogenomics a researcher could acquire PCR primers for DNA sequence loci via three methods. In the first method, which was the only way to develop novel genomic loci, a researcher had to use genomic cloning methods. This method, however, had major drawbacks because it required researchers to have considerable expertise in molecular biology skills and access to a sophisticated molecular biology laboratory. This cloning-based technique was therefore not a practical option for many researchers. The second method consisted of using existing *universal primers* borrowed from published studies. The original universal primers developed by Kocher et al. (1989) enabled large numbers of researchers to sequence the same mitochondrial genes in many different metazoan species. Unfortunately, the numbers of universal primers available at that time in the literature—and indeed for many years thereafter—was limited to a small number of mitochondrial and nuclear genes. The successful use of these primer sets also depended on a trial and error process of repeated PCR testing of different primers in the laboratory. Such *cross-species PCR* experiments did not always produce successful amplifications owing to mismatches between primers and templates. "Cross-species PCR" is a PCR experiment in which a pair of primers developed from the genome of one species is used (hopefully) to amplify the same locus in the genome of a different species. The third method for acquiring loci consisted of custom-designing universal primers. As this approach depended on the availability of existing DNA sequences obtained from other organisms, it largely limited researchers to obtain sequence data for preexisting loci. Despite the limitations of the universal primer approach, these special primers represented one of the key innovations that helped spur the early growth of molecular systematics (Palumbi 1996) and they are still vital today. A striking example is the pair of mitochondrial universal primers originally developed for invertebrates more than 20 years ago (Folmer et al. 1994), which later launched the DNA barcoding program (Hebert et al. 2003). Many DNA barcoding studies—even on vertebrates—are still using these original universal primers.

Two decades later, the situation concerning primer availability has dramatically improved due to the existence of online primer databases and advent of genome-enabled methods for developing new loci (Thomson et al. 2010). Later in this chapter, we will be covering the topic of designing primers for new loci including how to make universal primers. For many types of studies, however, it may be unnecessary to make new primers because the requisite primers can often be obtained from published sources. It is not uncommon for researchers to locate suitable primers from previously published work performed on shared or closely related study organisms. In recent years, journals such as *Molecular Ecology Resources* and *Conservation Genetics Resources* have published articles offering sets of newly designed primers. Researchers may also find primers in public online databases such the *Molecular Ecology Resources* (http://tomato.bio.trinity.edu/) and *Barcode of Life Database* or "BOLD" (http://www.boldsystems.org/) websites. Thus, at the start of a study it may be worth your time to search the literature or online databases for existing primers, an approach that can save you time and money in attempting to design your own primers. Literature sources not only contain the actual primer sequences, but other useful information can also be found such as expected PCR product sizes, optimal annealing temperatures, and thermocycling profiles. It is important to understand that success of a cross-species PCR experiment is dependent upon how the primers were designed as well as the number and nature of base differences between the primers and target templates—even a single mismatch, in one of the two primers, can be sufficient to cause PCR failure. Although obtaining primers from published sources is the simplest method of primer acquisition, this approach may not be effective for many studies.

Researchers equipped with knowledge on how to design new loci are not constrained by the availability of primers in the literature. Thus they are free to design their own custom primers thereby greatly expanding the number of research possibilities. Also, the capability to design custom primers empowers researchers to obtain higher quality PCR results when only poorly functioning primers are available. In order to design new PCR-based loci a researcher must have access to one or more genomic template sequences for the target locus or loci and then design primers according to well-established rules of primer design. The existence of vast amounts of genomic sequences in online databases as well as NGS sequencing are providing researchers with templates for primer design while the primer design step can be satisfied by using primer design software or by manual design methods. As we will see in this chapter, primer design software can perform remarkably well for making some types of primers but not for all types of primers. Accordingly, because some types of primers must still be manually constructed, it is essential for a researcher to obtain a solid knowledge of primer design rules, which we will now review before discussing the various methods for developing phylogenomic loci.

8.1 PRIMER DESIGN THEORY

As we already know the forward and reverse primers are located in the regions immediately flanking the target sequence. Thus, if we know the flanking sequences, then we might be able to design a pair of primers that can amplify the target in a PCR experiment. Although it is possible that a researcher could design a good-functioning set of primers without consideration of the molecular properties of candidate primers except for observing the 18 bp minimum length rule we learned in Chapter 5, it is more likely that such primers would fail to some degree in PCR. This is because there are a number of potential design flaws that can seriously impair the proper functioning of a primer. What are the potential flaws that could afflict a primer? In general, these flaws consist of primers interacting with themselves (i.e., forming secondary structures or "hairpins"), interacting with other primers (e.g., forming primer dimers), or amplifying nontarget stretches of the genome. To address this problem, primer design rules have been developed. These well-established rules enable a researcher (or primer design software) to narrow down a list of candidate primers to those that are most likely to function well in PCR. Although there is no guarantee that a newly designed primer will work well in PCR, you can greatly increase your chances of making good primers by practicing careful primer design.

8.1.1 Rules of Primer Design

As PCR began to take off in the late 1980s, the need for an effective means of designing new PCR primers arose. Researchers realized that primers that effectively amplify their target loci could be developed if certain criteria were followed during the design process. These criteria were largely based on the chemistry of duplex formation between two primers, between primer and DNA template, and also whether or not primers could form secondary or "hairpin" structures. To meet this need, Rychlik and Rhoads (1989) developed the first software (http://oligo.net/) for primer design, which implemented these selection criteria. A short time later, Innis and Gelfand (1990) formalized a set of primer design "rules," which, together with the work by Rychlik and Rhoads (1989), has formed the foundation of primer design methodology. Another popular primer design program called *Primer3* (Rozen and Skaletsky 1999; Koressaar and Remm 2007; Untergasser et al. 2012; http://bio-info.ut.ee/primer3/), which was based on the program *Primer 0.5* (Lincoln et al. 1991), has also greatly facilitated the task of choosing new primers and many other similar programs are available on the internet (Abd-Elsalam 2003). Although the "default" settings in primer design software often produces excellent results, many of these programs allow the user to alter the default settings to decrease or increase the stringency of searches. Changing any of the options usually affects the number of possible primers output from searches. We will now examine the most commonly applied rules. These rules are not given in order of importance though a brief consideration of which rules can be relaxed is given at the end of this discussion.

Rule 1: Primers Should Be 18–28 bp Long As was shown in Chapter 5, primers for routine PCR should be at least 18 bases long in order to achieve sufficient target template specificity for most PCR applications. Primers shorter than this minimum size may be susceptible to the problem of *mis-priming*. Mis-priming occurs when a primer anneals to nontarget genomic locations long enough for DNA polymerase to begin synthesis of a new DNA strand. The severity of mis-priming ranges from relatively minor effects such as consumption of reagents (e.g., dNTPs) to the deleterious consequences of coamplification of double-stranded PCR artifacts. The presence of multiple bands in an agarose gel is evidence of a mis-priming problem. As we will see below, other design defects can cause mis-priming problems in PCR.

What about the upper limit to the length of primers? Although the upper length limit of a primer is less constrained, primers should be less than 29 bases to allow for rapid annealing to the template (Innis and Gelfand 1990). In most studies, primers range from 18 to 25 bases long. Unless you have reason to do otherwise, for routine PCR applications it is best to choose primers that are close to 20 bases long with a search range of 19–22 bases.

Rule 2: Primers Should Have 40%–60% G + C Content Primer that have >60% G + C content may cause a number of PCR problems including an increased incidence of mis-priming, the formation of secondary structures called "hairpins," and stable primer–primer duplexes called "dimers." For this reason the G + C content of a primer should be kept as low as possible. Innis and Gelfand (1990) recommended an upper limit of 60% for the G + C content of a primer. The lower limit to G + C content seems to be less important. Suggested lower limits range from 50% (Innis and Gelfand 1990) down to 40% or even lower (Rychlik 1995). Note that for some types of A + T-rich templates (e.g., noncoding nuclear DNA) the primers will likely have a low G + C content therefore primers having a G + C content in the 30%–40% range may represent the only options. As we will see below under other primer design rules, the base sequence and total number of bases determine the optimal temperature at which a primer will anneal to its target template.

Rule 3: Avoid Complementarity at the 3′ End of the Primer In Chapters 5 and 6, we briefly considered what primer dimers are and how they can impact the quality of a DNA sequence. We will now look at the nature of primer dimers in more detail and see why it is critically important to ensure that the 3′ end of the primer does not contain a sequence of bases that could either make it self-complementary or complementary to the 3′ end of the other primer in the reaction. When a primer is self-complementary at the 3′ end it can lead to the formation of a type of primer dimer called a *homodimer*, whereas a 3′ end duplex formed between forward and reverse primers is called a *heterodimer*. Figure 8.1a shows an example of a homodimer, which formed because the 3′ end of one primer contained a 5′-CCGG-3′ sequence. There are two important consequences of this duplex formation. First, notice that *two* primer–template junctions are created, which could allow DNA polymerase to start extending DNA strands at both ends of the duplexed molecule. This can result in the production of double-stranded PCR products that are slightly shorter than the two primers joined end to end (Figure 8.1b). The second consequence of this dimer product formation is that binding sites for forward and reverse primers are synthesized, meaning each strand of this product can serve as a template in all future cycles (Figure 8.1b). Thus, as the reaction mixture is heated toward the denaturation temperature in the first cycle, the genomic DNA and newly formed dimer products denature into single-stranded templates and thus become available for primers during the initial annealing

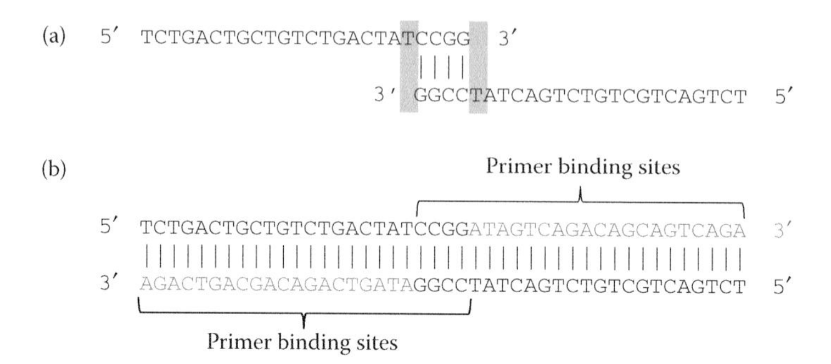

Figure 8.1. Formation of a homodimer in PCR. (a) Self-complementarity at the 3′ end of a primer causes formation of homodimers. Two primer–template junctions are formed in the homodimer (gray rectangles). (b) DNA polymerase (not shown) synthesizes DNA (gray bases) resulting in the amplification of a 44 bp PCR product. In subsequent cycles each of the strands can serve as a template for both primers ("primer binding sites"). Vertical dashes represent hydrogen bonding between complementary bases.

step. Given the high concentration of unincorporated primer in the reaction, the dimer products will proliferate exponentially and compete with the target products for reagents. When the reaction is completed, there is a great abundance of at least two PCR products: the target product and the primer dimers. Unless the primer dimers are somehow removed from the completed reaction and discarded before sequencing, they too will generate extension products in a cycle sequencing reaction with the end result being the diminished quality of DNA sequence chromatograms.

However, as the 4-bp duplex in Figure 8.1a cannot form during the annealing temperatures of PCR experiments (i.e., this duplex is only stable at temperatures less than 10°C), how then is it possible for the primer dimer to become replicated throughout the PCR process? The answer to this question highlights the critical nature of the first step of the first PCR cycle—specifically the time from when all the reagents are combined in the PCR tube until the first denaturation step is reached. Because the PCR tubes are initially at a cooler temperature (4–25°C) from the moment all the reagents are combined until the start of the first cycle, these conditions could enable dimer formation between primers because *Taq* polymerase can still function at these lower temperatures. At these lower temperatures *Taq* can begin extending off the primers and produce short double-stranded products prior to the first denaturation step. When the first denaturation step is reached, both the genomic DNA and newly created primer dimer templates become single stranded. Thus by the time the first annealing step is reached, there will be *two* sets of templates available for the primers, which ultimately results in coamplification of target and primer dimer templates.

Not all primer–primer duplexes are necessarily detrimental to PCR. When checking a particular primer or primer pair using primer design software (or by visual inspection), it is possible that the analysis will suggest possible primer–primer duplex formation involving internal bases or the 5′ ends (Figure 8.2). Although these duplexes may temporarily form before the first denaturation step, at no time can DNA polymerase synthesize DNA on these dimers simply because neither of these dimers contains primer–template junctions. Therefore, these particular dimers cannot lead to the mass production of double-stranded PCR artifacts. It is conceivable that the formation of these dimers can potentially lower the efficiency of a PCR reaction in other ways. For example, if the primer–primer duplex is stable enough—even at typical PCR annealing temperatures, then these primer–primer interactions might decrease the target product yield by depriving the target templates of available primers. However, as we will see below, this is unlikely to occur on the basis of predicted primer–primer duplex stabilities.

Thus far, we have been considering a vague notion of "stability" of primer–primer or primer–DNA template duplexes, but what factors determine the stability of a DNA double helix? While hydrogen bonding between complementary bases on adjacent strands contributes to helix

```
(a)          5′   ATACACTGCCGGTGACTATCATG   3′
                              ||||
          3′   GTACTATCAGTGGCCGTCACATA   5′

(b)                          5′ GGCCACTGCTGTCTGACTATCATG   3′
                                     ||||
        3′  CATGATAGTCAGACAGCAGTCCGG   5′
```

Figure 8.2. Two primer–primer duplexes that have little or no effect on PCR efficiency. (a) Internal self-complementarity may lead to duplex formation between two primers but not to homodimer formation. (b) Self-complementarity at the 5′ end of a primer may also lead to duplex formation between two primers but not to homodimer formation. Note in both cases a primer–template junction cannot be formed therefore DNA synthesis cannot occur. Vertical dashes represent hydrogen bonding between complementary bases.

stability, other factors play significant roles in determining the molecular stability of duplexes; these include entropic forces caused by water molecules that are displaced to the outside of the helix and especially by base stacking interactions of adjacent bases (Watson et al. 2014). Consider two dsDNA molecules of equal length. Given that each base pair follows Watson–Crick pairing rules and each duplex experiences comparable entropic forces, which duplex is expected to be the more stable one? It is often thought that base composition of a sequence plays a large role in the stability of a duplex, as sequences with a large number of G + C bases tend to be more stable than sequences containing fewer of these bases. Thus, one might conclude that the duplex with the higher G + C content might be the more stable duplex of the two. Although base composition correlates with duplex stability, it is not base composition *per se* that determines duplex stability. Rather, stability is largely determined by the thermodynamic interactions of adjacent or nearest-neighbor bases on the same strand—in other words, stability is determined by the primary nucleotide sequence (Breslauer et al. 1986). Thus, stability of a dsDNA molecule is primarily due to the sum of its nearest-neighbor interactions. This implies that longer DNA duplexes will be more stable than shorter duplexes all else equal. Recognition of this fundamental idea, which has since been formalized as *nearest-neighbor thermodynamic theory*, has provided researchers with an effective means for predicting the stability of DNA duplexes and has proved useful to applications such as PCR and DNA sequencing. An important extrinsic factor that determines duplex formation between two complementary ssDNA molecules is temperature. We saw in Chapters 5 and 6 that a thermocycler is used to modulate the thermal environment of the PCR reaction between denaturing DNA duplexes at high temperatures and facilitating proper primer–template duplexes at lower temperatures. Thus, it is the length and sequence of a primer together that ultimately determine its optimal annealing temperature to target genomic sequences in PCR.

The relevant metric for quantifying the stability of short DNA duplexes is the Gibb's Free Energy or "ΔG" (Breslauer et al. 1986; Rychlik 1995), which was named in honor of Josiah Willard Gibbs, a scientist who made seminal contributions to physics, chemistry, and mathematics (Josiah Willard Gibbs, Wikipedia). Predicted ΔG values, which can be positive or negative, are in units of kcal/mol. Duplex formation is energetically favored when ΔG is negative and the strength of stability for this duplex increases as ΔG becomes more negative. Because the stability of a DNA duplex is determined by the identity of its nearest-neighbor bases, the stability of a particular duplex can simply be calculated by summing the free energies of each pair of bases in a sequence. Breslauer et al. (1986) provided a thermodynamic library for all ten unique nearest-neighbors (Table 8.1). This library consists of the free energies for each possible dinucleotide that can be found in a Watson–Crick DNA duplex. Inspection of the free energy parameters for each nearest-neighbor pair reveals that the most stable dinucleotide groups are CG/GC, CC/GG, GC/CG, and GG/CC (i.e., their respective ΔG values are −3.6, −3.1, −3.1, and −3.1), while in contrast all the pairings involving A's and T's have less negative ΔG's and are therefore relatively less stable (Table 8.1). Other

TABLE 8.1
Nearest-neighbor free energy (ΔG) parameters for all 10 dinucleotide pairs at 25°C

Interaction	ΔG
AA/TT	1.9
AT/TA	1.5
TA/AT	0.9(1.0)
CA/GT	1.9
GT/CA	1.3
CT/GA	1.6
GA/CT	1.6
CG/GC	3.6
GC/CG	3.1
GG/CC	3.1

SOURCE: Data from Table 8.2 of Breslauer, K. J. et al. 1986. *Proc Natl Acad Sci USA* 83:3746–3750. With permission.

NOTE: ΔG values are in units of kcal/mol. For example, the nearest-neighbor group AT/TA is interpreted as a 5′–AT–3′ dinucleotide that is hydrogen bonded to its complementary sequence 3′–TA–5′.

workers subsequently refined these parameters (see SantaLucia 1998), but the values in Table 8.1 are good approximations for evaluating stabilities of short DNA duplexes.

Breslauer et al. (1986) and SantaLucia (1998) provided equations for estimating the total ΔG of a duplex. Their equations compute the total ΔG by summing the free energies of each nearest-neighbor dinucleotide then correcting this estimate using additional terms in the equations (see Breslauer et al. 1986 and SantaLucia 1998 for details). For purposes of primer design, we will follow Rychlik (1995) and use a modified equation that omits the correction terms. This modified equation predicts the total uncorrected free energy ΔG for a given dimer duplex at 25°C:

$$\Delta G = \sum \Delta G_i, \quad (8.1)$$

where i is each nearest-neighbor group.

Using Equation 8.1 together with Table 8.1 we can estimate the stability of the 3′ end duplex shown in Figure 8.1a. If we momentarily ignore the part of each primer not involved in the primer–primer interaction and focus only on the bases involved in the duplex, then the duplex consists of double-stranded 4-bp stretch of DNA (Figure 8.1a). Notice there are three different nearest-neighbor groups each of which will contribute to the total uncorrected ΔG value for the duplex. Using Table 8.1 we can estimate ΔG for this duplex:

$$\Delta G = \Delta G(CC) + \Delta G(CG) + \Delta G(GG)$$
$$\Delta G = -3.1 + (-3.6) + (-3.1)$$
$$\Delta G = -9.8 \text{ kcal/mol}$$

The above calculation follows the example in Rychlik (1995) and can easily be done without a calculator. Now that we have estimated the ΔG for the 3′ end homodimer duplex in Figure 8.1a, we need to address the following question: *if we used a primer that forms homodimers with a duplex stability of ΔG = −9.8 kcal/mol, then what is the expected impact on PCR product yield?* Rychlik (1995) showed the dependence of PCR product yield on 3′ duplex stability of primer dimers, which is shown in Figure 8.3. The curve in Figure 8.3 suggests that our ΔG of −9.8 kcal/mol may severely diminish PCR product yield. Rychlik's results show that once the ΔG for a 3′ end duplex exhibits higher stability (i.e., more negative) than ΔG = −4 kcal/mol, we can expect PCR product yields to be sharply reduced. Primers with negative ΔG 3′ stability values should be avoided because, even if the formation of the unwanted duplexes does not lead to the formation and accumulation of primer dimers, these partially stable duplexes may interfere with the synthesis of target products (Rychlik 1995). Thus, it is best to design primers that are free of 3′ end dimer issues.

There are occasions when, for lack of other primer options, a researcher must use a pair of primers that produce primer dimers along with

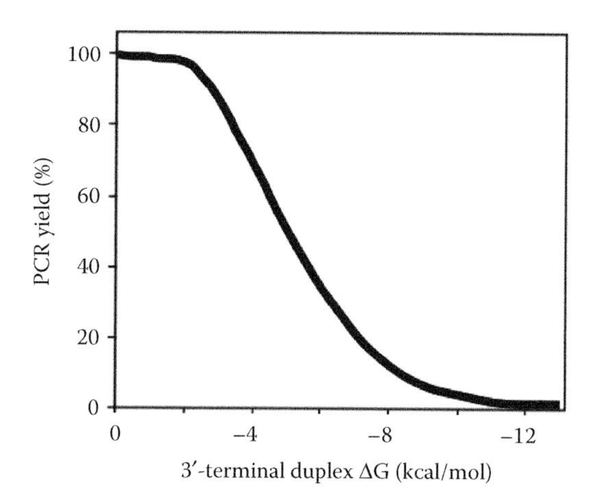

Figure 8.3. Dependence of PCR yield on the ΔG of 3′-terminal duplexes. (Reprinted from Rychlik, W. 1995. *Mol Biotechnol* 3:129–134. With permission.)

a well-amplified target product. In this situation, there are three strategies that can be used to obtain high-quality target sequences. First, if the primer dimers are homodimers and the product is short enough (<700 bp), then simply sequencing with the primer not involved in the dimer formation will likely yield a high-quality sequence. If the PCR reaction produces heterodimers, then sequencing the amplicons in both directions can solve the problem. Although each sequence will have low-quality base calls in the first ~40–50 bases, they will have higher quality base calls (≥Q20 Phred scores) for the remainder of the sequences. This will allow the researcher to construct a high-quality contig sequence using both chromatograms. A second strategy is to use methods of PCR product cleanup (Chapter 6) that are effective at removing primer dimers from target products (e.g., PEG or some commercial spin-column kits). These methods can be effective but they can be costly in terms of sample processing time and cost of kits. A third method attempts to prevent the formation of primer dimers during PCR via the "hot start" method (Chou et al. 1992). It should now be clearer exactly *how* the hot start method can prevent the formation of primer dimers. These harmful DNA templates can only form during the brief time period prior to the first denaturation step in PCR because the remainder of the PCR steps will be carried out at temperatures that prohibit the initial formation of primer–primer duplexes. Thus, if *Taq* is added to the reaction at a temperature well above the permissible temperatures for the formation of short DNA duplexes, then it is likely that the 3′ end duplexes will not form and thereby preclude the mass production of alternative priming sites for *Taq*.

Rule 4: Avoid Runs of Three or More C's and G's at the 3′ End of Primer When designing primers it is critically important to ensure that the bases at the 3′ end of a primer should not contain runs of 3 or more "G" or "C" bases otherwise mis-priming will likely occur (Innis and Gelfand 1990). For example, in Figure 8.4a we see a primer with a "sticky" or relatively stable 3′ end (consisting of a 4-bp CGCG sequence) hybridizing to a nontarget template location. Mis-priming occurs when the 3′ terminal end of the primer duplexes with a nontarget strand long enough to allow the DNA polymerase to synthesize a new strand of DNA (Figure 8.4b). In contrast, primers with relatively less stable 3′ ends, but highly stable internal sections or 5′ ends, are not likely to mis-prime (Rychlik 1995).

How sticky must a 3′ end of a primer be for mis-priming to occur? Rychlik (1995) suggested that the 3′ end stability of a primer could be quantified by calculating free energy of the 3′ end *pentamer*, which we will refer to as $\Delta G_{pentameter}$. Note, that here we are specifically referring to the stability of the last five bases located at the 3′ end of the primer. Based on empirical data, Rychlik (1995) showed that primers with relatively high stabilities at the 3′ pentameter ($\Delta G_{pentameter} = -9$ to -12 kcal/mol or lower) performed poorly in PCR owing to mis-priming problems, whereas primers with low to medium 3′ end stabilities ($\Delta G_{pentameter} = -5$ to -9 kcal/mol) performed better. In Figure 8.5, we see 20 primers of the same

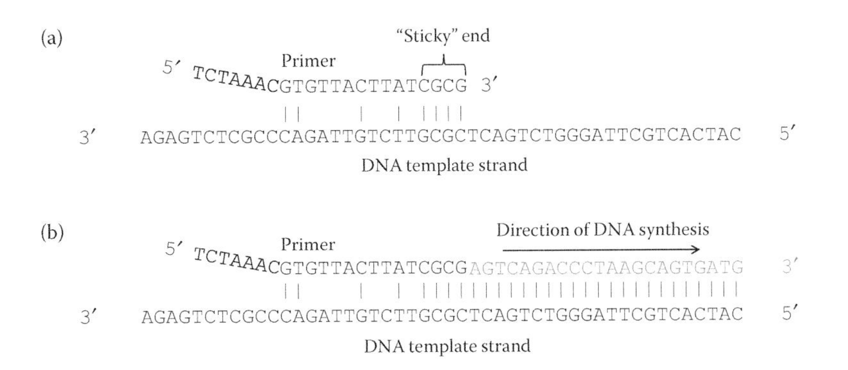

Figure 8.4. Mis-priming caused by primers with "sticky" 3′ ends. (a) Partial annealing of primer to nontarget location (i.e., genomic DNA or PCR amplicon). The resulting partial duplex contains a primer–template junction, which can be extended by DNA polymerase. (b) DNA polymerase (not shown) begins synthesizing new strand of DNA (gray bases). Vertical dashes represent hydrogen bonding between complementary bases.

length but with different 3′ end pentamers. On the right side of Figure 8.5 are shown the ΔG values for each pentamer. As you will notice, when the G + C content increases in the pentamers, the end stabilities also tend to increase (Figure 8.5). Accordingly, if we designate a critical threshold for primer end stability of $\Delta G_{pentameter} = -9$ kcal/mol for specific (ideal) primers, then we can see that primers with a total G + C count of two or less at their 3′ ends are within this range. Some primers with a G + C count of three located within the 3′ pentamer also have acceptable levels of end stability but it depends on their nearest-neighbors (e.g., GACTC and TAGGG both have 3′ ends that are less stable than $\Delta G = -9$ kcal/mol threshold; Figure 8.5). Rychlik (1995) suggested that a primer with a low stability 3′ end but higher stability middle section or 5′ end is a more specific primer and less susceptible to mis-priming problems.

Although incidences of mis-priming can be reduced if the 3′ primer end does not contain too many G's and C's, some researchers have suggested that having a terminal 3′ base consisting of a G or C—called a "GC clamp"—may help ensure proper priming (e.g., Kwok et al. 1990). Provided that the 3′ end pentamer of a primer exhibits low to medium stability, then such a GC clamp may improve primer performance (Rychlik 1995). However, empirical evidence provided by Haas et al. (1998) in addition to the great abundance non-GC clamp primers published in the literature suggests that a primer could have any of the four possible bases at the 3′ terminus and still function properly.

Rule 5: Avoid Palindromic Sequences within the Primer There is another form of self-complementary sequence that, if present in a primer, can cause serious PCR problems. If a primer contains a *palindromic sequence*, then it is self-complementary to itself and can therefore form a secondary structure or *hairpin* structure (Rychlik and Rhoads 1989; Innis and Gelfand 1990). A hairpin can form when a primer contains a palindromic sequence with a string of least three loop-forming bases located in the middle of the palindromic sequence (Figure 8.6a). Note, it is sterically impossible for a loop to form with less than three bases (Freier et al. 1986; Rychlik and Rhoads 1989). By forming a single-stranded loop structure (the bases in the loop do not hydrogen bond to other bases), the primer can fold over on itself and form a base-paired duplex—called the *stem*—with the other half of the palindromic sequence (Figure 8.6b). If the hairpin involves the 3′ end of the primer then a primer–template junction can form leading to self-priming and hence primer extension (Figure 8.6c). Just

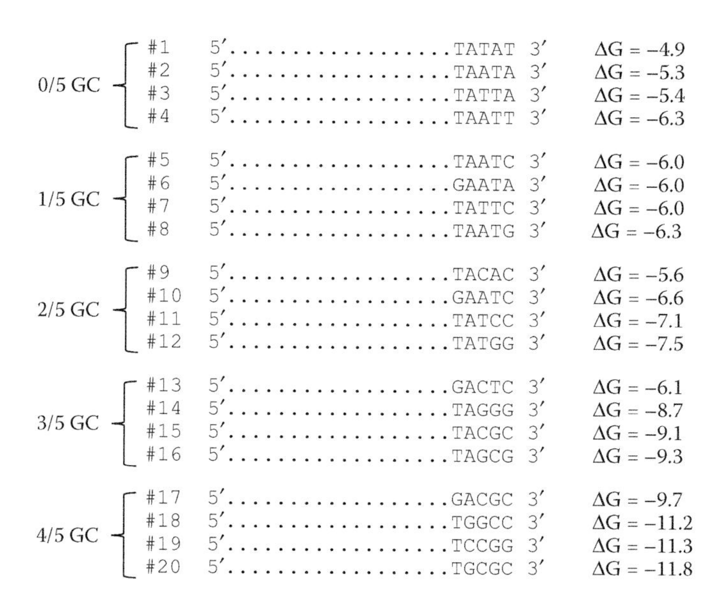

Figure 8.5. Stabilities of 3′ terminal pentamers of primers as a function of G + C base composition. For example, primers #9–12 each have a G + C composition of 2/5 in the 3′ end of their primer sequence. Only the 3′ end pentamer sequence is shown for each primer (dots represent bases hidden from view). Stabilities are measured as ΔG (free energy) values.

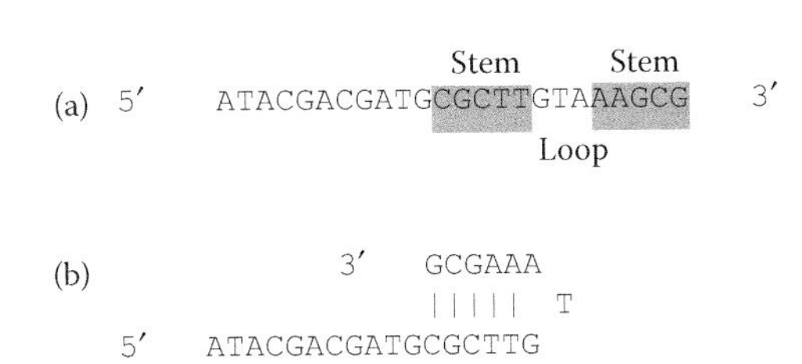

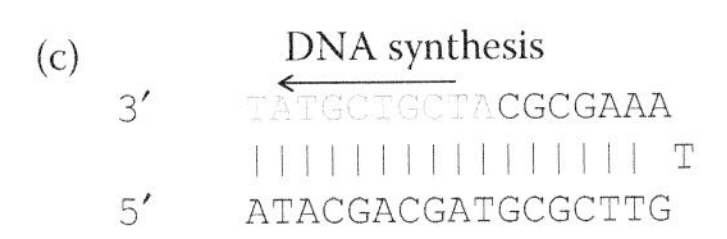

Figure 8.6. Potential consequence of a primer hairpin in PCR. (a) Primer containing a palindromic sequence (two gray rectangles). (b) Primer folds over on itself to form a hairpin involving the 3′end. This hairpin is characterized by a 3-bp loop, 5-bp stem, and a $\Delta G_{hairpin} = -5.0$ kcal/mol. (c) Because this hairpin forms a primer–template junction, self-priming can occur.

like primer dimers, hairpins are likely to form and subsequently self-prime DNA synthesis only when the reaction mixture is at permissible (low) temperatures (i.e., at the time when all PCR reagents are mixed together and before the first denaturation step). In such a scenario, the hairpin-forming primer can be rendered nonfunctional in the PCR, because most if not all the hairpin-forming primer will form PCR artifacts and not target products. Not all primer hairpins are problematic, as hairpins that only affect the 5′ end or internal part of the primer will likely not affect PCR efficiency because they do not contain primer–template junctions and therefore cannot self-prime (Rychlik 1995).

As with primer dimers, the stability of duplex formation is the key to whether or not synthesis of PCR artifacts can occur due to primer hairpins. Thus, the stability of a hairpin duplex on primers can be evaluated using ΔG but a term characterizing the free energy of the loop (ΔG_{loop}) must be included in the calculation because the loop is a destabilizing influence on the duplex (Rychlik 1995). Approximate free energies for loops ranging in sizes from 3 to 8 bases are the following: three base loop = 5.2 kcal/mol, four base loop = 4.5 kcal/mol, five base loop = 4.4 kcal/mol, six base loop = 4.3 kcal/mol, seven base loop = 4.1 kcal/mol, and eight base loop = 4.1 kcal/mol (Rychlik 1995). Thus, ΔG of the hairpin duplex ($\Delta G_{hairpin}$) can readily be calculated using Equation 8.1 but with an ΔG_{loop} term included

$$\Delta G_{hairpin} = \sum \Delta G_i + \Delta G_{loop}, \qquad (8.2)$$

where i is each nearest-neighbor group.

Again, this approximation of hairpin stability can be considered as an "uncorrected" estimate because we are omitting the initiation terms from the equation (see SantaLucia 1998 for full equation). Using Equation 8.2, Table 8.1, and the ΔG_{loop} for a 3-base loop, the stability of the hairpin in Figure 8.6b is estimated to be

$$\Delta G_{hairpin} = \Delta G(GC/CG) + \Delta G(CG/GC) + \Delta G(GA/CT) + \Delta G(AA/TT) + \Delta G_{loop}$$

$$\Delta G_{hairpin} = -3.1 + (-3.6) + (-1.6) + (-1.9) + 5.2$$

$$\Delta G_{hairpin} = -5.0 \text{ kcal/mol}$$

How bad is a hairpin with this level of stability? In general, negative $\Delta G_{hairpin}$ values should be avoided but primers with stabilities as high as $\Delta G_{hairpin} = -3.0$ kcal/mol may function properly in PCR (Rychlik 1995). Observing that the hairpin structure in Figure 8.6b involves the 3′ end and exhibits duplex stability exceeding the −3.0 kcal/mol threshold, this primer should not be used in PCR. Because PCR artifacts due to hairpins form prior to the initial denaturation step in PCR (like primer dimers), the hot start method (Chou et al. 1992) may be effective at preventing the initial formation of these products and thus allow the reaction to produce only target products. However, designing a hairpin-free primer offers a simpler approach.

Rule 6: Primers Should Have a Tm 50–65°C The melting temperature or "Tm" of a primer is a commonly used measure of the stability between the primer and its template when they are hybridized together via hydrogen bonding into a duplex molecule. The Tm is defined as the temperature at which half of the potential binding sites are bound with primer (Palumbi 1996). Knowing the Tm of the primers allows you to predict the optimal annealing temperature for thermocycling. Because low annealing temperatures can decrease primer specificity thereby leading to the production of PCR artifacts (i.e., a mis-priming problem), higher annealing temperatures should be used (e.g., ≥50°C) when possible. Innis and Gelfand (1990) suggested that primers should be designed to have Tm's in the 55–80°C range. However, in practice the most commonly used Tm's are in the 50–65°C range.

The most accurate way to estimate the Tm of a primer involves using nearest-neighbor thermodynamic theory. Many primer design software packages use this approach to estimate the Tm of primers. However, this method is difficult to do by hand such as we did earlier when estimating the ΔG for short primer–primer DNA duplexes. A less accurate "rule of thumb" method for estimating the Tm of a primer simply considers the length of a primer and weighs the four bases differently to reflect their general stabilities in nearest-neighbor interactions. Suggs et al. (1981), Thein and Wallace (1986), and Palumbi (1996) provided this simple equation for estimating the Tm of a primer:

$$\text{Tm} = 2°\text{C} \times (\text{A} + \text{T}) + 4°\text{C} \times (\text{G} + \text{C}) \quad (8.3)$$

This simple formula for estimating the Tm of a primer, which only requires counts of each of the four bases in a primer, has been widely used by researchers.

Rule 7: Each Primer in a Pair Should Have Tm < 5°C of Each Other It is critically important for a pair of primers to have similar melting temperatures because the primer with the higher Tm may mis-prime if the annealing temperature is set too low. Alternatively, if the annealing temperature is matched to the primer with the higher Tm, then the temperature may be too high for the lower Tm primer causing it to not anneal at all. Therefore, the rule of thumb is that the two primers should have Tm values within 5°C of each other (Innis and Gelfand 1990).

Rule 8: Avoid Mismatched Bases at the 3′ End of the Primer If a primer is annealed to the correct template locus but one or more bases are mismatched, then the consequences on PCR yield vary from having little effect to a more drastic effect. In a study that focused on the effects of such mismatches on PCR yield, Kwok et al. (1990) found that single mismatches had no significant effect on PCR yield as long as they were located internally within the primer-template duplex and did not involve the sensitive 3′ end of the primer. In contrast, their findings indicated that if two of the final four bases in the 3′ end were mismatched, then PCR yield was substantially lowered, especially if the terminal 3′ base was mismatched. In cases where only the 3′ terminal base was mismatched, PCR yield varied depending on mismatch: a 100-fold reduction resulted from A:G, G:A, and C:C mismatches, a 20-fold reduction from a A:A mismatch, but all other mismatches largely had little effect on yield under normal PCR conditions (Kwok et al. 1990). One surprising finding by the authors concerned the discovery that T:G, T:C, and T:T mismatches had little effect on PCR yield. This suggests that designing a primer with a "T" at the 3′ terminal base may be a good strategy if the primer is to be used in cross-species studies (Palumbi 1996). If the primer is embedded within a coding sequence, then a common strategy is to make sure that the 3′ terminal base corresponds to a 2nd codon position site (e.g., Backström et al. 2008). Owing to the lack of variability observed at 2nd codon sites, this strategy virtually guarantees that a mismatch involving the 3′ terminal base will not happen.

8.1.2 Final Comments about Primer Design Rules

Anytime new primers are designed without the aid of a primer selection software program, the candidate primers should be carefully evaluated to see how well they meet the aforementioned rules of design. The use of bioinformatics software tools can greatly facilitate the analysis of primers. For example, the *OligoAnalyzer* 3.1 tool by Integrated DNA Technologies® (http://www.idtdna.com/calc/analyzer; Owczarzy et al. 2008), which I routinely use, is very useful for analyzing characteristics of primers, including homodimers, heterodimers, hairpins, 3′ end pentamers, Tm, and G + C content. Kibbe (2007) also provides a web-based bioinformatics tool for examining the properties of primers.

When using primer selection software, occasionally the program will not be able to suggest many or any possible primer pairs even when the default settings are used. In these cases, changing some of the settings can lessen the stringency of the search, which usually increases the output of suggested primer pairs. The first rules to relax are (in order): rules #1 (primer length), #2 (G + C%), and #6 (Tm). For example, if you first search for primers that are 19–22 bases long, have a 40%–60% G + C content, and have Tm's 55–60°C, but are unable to find any candidate primers as a result, then it is time to broaden the search. In the second search, the aforementioned parameters could be expanded to include primers of: 18–25 bases, G + C content 30%–60%, and Tm's 50–65°C. If it is likely that the primers and template will match each other perfectly, then these relaxed criteria should still enable you to select good primers. If on the other hand, you suspect that mismatches might exist between the primers and template, then primer length and Tm should remain at the higher end of the search ranges to ensure a high level of specificity in PCR.

8.1.3 Testing New Primers in the Lab

Whenever new PCR primers are designed, they must be bench tested in the laboratory. The testing procedure amounts to performing PCR reactions and varying the annealing temperature until the thermal "window" in which the primers are able to amplify the correct product is found. Sometimes this thermal window at the annealing step spans 5–10°C (or more) and sometimes it is <1°C. This annealing temperature parameter can only be empirically determined for each untested primer pair and therefore it is a trial and error process. However, to reduce the number of test PCRs that must be performed, manufacturers of many thermocyclers have incorporated a thermal gradient feature during the annealing step. For example, if a researcher is evaluating annealing temperatures that range between 50 and 60°C, the PCR tubes containing identical reagents are placed across the 12 wells in the thermocycler's metal heating block where the temperature gradient is generated. This gradient feature not only helps the researcher identify the optimal annealing temperature for each primer pair, but it can also obviously reduce the amount of time expended for testing new primers. Occasionally, some new primers never perform well in PCR and so they should either be redesigned or discarded. However, assuming that the template sequences used to make the primers were of appropriate quality and primer design software used (or manually designed), then most if not all new primer pairs should generate the expected PCR products. Table 8.2 shows the success rates for new sets of primers from a variety of studies that used single genomic templates together with primer design software to develop new PCR-based anonymous loci. These data show that success rates are generally high, which varied between 47% and 86% (Table 8.2). Note that little if any PCR optimization (e.g., experimenting with different annealing temperatures) was done in some of these studies and therefore it is likely that at least some of the success rates shown in Table 8.2 represent underestimates of the true number of good-functioning primers.

8.2 PRIMER AND PROBE DESIGN APPROACHES

The rapidly progressing genomics and bioinformatics fields are removing the barriers to developing phylogenomic loci. In recent years, a number of published studies have proffered a flurry of new methods for designing new loci. These newer methods not only allow researchers to target particular loci without resorting to challenging and time-consuming genomic cloning approaches, but some methods are capable of generating hundreds or thousands of loci for organismal phylogenomic studies. Many of these newer approaches involve the use of Perl or Python scripts as part of

TABLE 8.2

PCR success rates for sets of anonymous loci primers involving studies of vertebrates

Organisms	# Primer sets tested	# Primer sets with successful PCR	% Success	Study
Birds	17	8	47	Amaral et al. (2012)
Lizards	15	12	80	Bertozzi et al. (2012)
Snakes	17	12	71	Bertozzi et al. (2012)
Birds	35	30	86	Jennings and Edwards (2005)
Lizards	77	50	65	Rosenblum et al. (2007)
Turtles	96	73	76	Thomson et al. (2008)

NOTE: A pair of primers is considered functional if the pair can successfully amplify the correct locus in a DNA sample from the same species (i.e., a single amplification band of the expected size in an agarose gel is sufficient evidence). Primers in all listed studies were designed using primer design software.

their pipelines. Thus, in order to implement these procedures with the least amount of troubles, it is helpful to have some knowledge about basic programming in the Perl and Python languages. Before a particular locus development method can be chosen, however, the researcher must first decide which type of locus or loci will be needed. Locus design methods can be divided into two basic approaches: (1) *single template approaches* and (2) *multiple homologous templates approaches.*

8.2.1 Single Template Approaches for Developing PCR-Based Loci

For studies involving highly similar sequences such as those involving a single gene family, a single species, or a complex of closely related species, a *single template approach* is generally used. This is because these types of sequences generally show low levels of nucleotide variation and hence mismatches at primer binding sites are expected to be too few to adversely impact a PCR. Thus, only a single representative sequence is needed to serve as the template from which primer design software can find appropriate forward and reverse primers.

Various approaches have been used for obtaining a single template sequence to design new primers. Until recently, most studies relied on genomic cloning to make new loci. However, simpler and more efficient methods have been developed since the dawn of the genomics era, which make use of existing genomic resources (e.g., clone libraries), partial genome sequences generated from NGS platforms, or use lab-free computer searches of whole genome sequences available on public databases. We will examine these methods below beginning with a discussion of genomic cloning methods. Although cloning methods have been superseded by newer genomics-based methods, important insights have emerged from that work that can benefit locus development endeavors in the future.

8.2.1.1 Single Template Methods Using Genomic Cloning Methods

Although this method is no longer used to generate phylogenomic loci, it is being presented here for two reasons. First, I believe it is of historical interest to see how novel loci (e.g., anonymous loci) were developed prior to the time of genomics and NGS. Secondly, there are some important implications from these early cloning-based phylogenomic studies that are instructive to modern genome-enabled and NGS approaches to loci development, particularly pertaining to anonymous loci.

The following represents a simplified description of the gene cloning process. Interested readers should consult Sambrook et al. (1989) for detailed protocols. The first major step in cloning is to construct a genomic library from an individual of the study species. To construct a genomic library, the following steps must be performed: (1) obtain a sample of purified genomic DNA from an individual of the reference species; (2) break the genomic DNA into shorter fragments using sonication or restriction enzymes; (3) run the fragmented DNA through an agarose gel and excise a gel slice that contains a subset of the DNA fragments (e.g., fragments in 1–2 kb range); (4) purify the size-selected DNA; (5) ligate the fragments into plasmid cloning vectors; (6) transform the recombinant vectors into *E. coli* bacteria; and (7) screen the bacterial colonies for the "clones" containing recombinant vectors. At this point in the cloning process, the next step will depend on the type of locus that the researcher wishes to develop. If the researcher would like to design primers that amplify a particular gene or gene family, then a radioactively labeled oligonucleotide probe must be used to hybridize only to the recombinant plasmids that contain a copy of that gene. The recombinant clones retained from this hybridization procedure are then selected for sequencing, which will hopefully yield the template + flanking sequences for the locus of interest. The final step is simple: the obtained template sequence is input into primer design software, which then outputs the primer sequences.

Alternatively, if the researcher wants to develop anonymous loci, then a random subset of the recombinant plasmids are selected for sequencing. Once a number of different plasmid inserts have been sequenced, the researcher can design a set of new anonymous loci simply by using primer design software to produce one primer pair nested within each insert sequence. However, if these anonymous loci are destined for use in phylogenomic analyses that assume each locus has the properties of being selectively neutral, genealogically independent from other sampled loci, and single-copy, then additional filtering steps are needed prior to the final primer design step.

To evaluate whether or not a given insert sequence meets the neutrality assumption, two approaches can be tried. First, a BLAST search of the Genbank database can be used to query the genome of the most closely related species to see if a candidate anonymous locus matches a known conserved part of the genome such as a protein-coding gene (Jennings and Edwards 2005). Candidate anonymous loci that show no "hits" with the reference genome are tentatively considered to be from noncoding regions of the genome and hence are deemed more likely to meet the neutrality assumption. Another strategy is to use bioinformatics software to scan each sequence for an "ORF" that spans hundreds of bases because sequences that have such long ORFs may represent protein-coding regions. Sequences with long ORFs can be discarded as a precautionary measure.

The independent loci assumption is thought to be largely satisfied if anonymous loci are obtained from random genomic locations. This is because such loci would be found on different chromosomes or, if found on the same chromosome, they would likely be found far enough apart from each other that their gene trees would effectively be independent of each other. Evidence that this approach works as desired comes from the study of Jennings and Edwards (2005), who used blunt-end genomic cloning methods to develop 30 anonymous loci for Australian grass finches. When these loci are mapped to the chromosomes of the zebra finch, a grass finch species that had its genome fully sequenced (Warren et al. 2010), we can see that these loci are scattered across the genome as expected (Figure 8.7). Although some loci appear to be close together (e.g., on Chromosomes 1, 4, 4A, 7, and 19), the minimum physical distances between neighboring loci is on the order of hundreds of thousands of bases and, in some cases, millions of bases apart. Given these rather large physical distances (see Table 3.1), it is likely that all sampled loci have independent gene trees.

Lastly, in an attempt to satisfy the single-copy locus assumption, Karl and Avise (1993) used laboratory procedures to identify and then eliminate from consideration all vector insert sequences containing repetitive DNA. This procedure was deemed necessary because it was originally believed that anonymous loci based on genomic sequences containing replicative transposons would likely violate the single-copy assumption (i.e., PCR would coamplify many copies of a single locus or different copies in different individuals). Since that time, many published methods for developing anonymous loci include a filtering step to eliminate sequences containing unwanted repetitive DNA (e.g., Chen and Li 2001; Shaffer and Thomson 2007; Thomson et al. 2008; Bertozzi et al. 2012; Lemmon and Lemmon 2012).

Owing to the high copy numbers of some transposable elements in many eukaryote genomes, there is certainly sufficient reason to be concerned about the possible adverse effects of repetitive DNA on locus development. However, these measures also can eliminate large parts of the genome from locus development consideration thereby limiting the number of loci for phylogenomic studies. Thus, an important question to ask is: *will anonymous loci with repetitive DNA likely violate the single-copy assumption?*

We can evaluate this question using the set of anonymous loci for the grass finches from the study of Jennings and Edwards (2005) because these authors did not include a repetitive DNA filtering step during loci development. First, we can get an idea about which of these loci are comprised of transposable elements by using the *CENSOR* software tool (Kohany et al. 2006) found on the online database *Repbase* (http://www.girinst.org/censor/index.php; Jurka et al. 2005). The results revealed that 10 of the 30 loci showed no evidence of containing any repetitive DNA. Of the remaining 20 loci, 16 contained a single transposon, three loci had two transposons embedded within their sequences, and one locus has three transposons. The transposable elements found amongst these loci consisted of four DNA transposons, 13 LTR retrotransposons, and eight non-LTR retrotransposons. As you will recall from Chapter 2, retrotransposons are represented by many large gene families, which can comprise large fractions of eukaryote genomes. Despite the threat posed by retrotransposons, several lines of evidence suggest that they do not pose problems with these particular anonymous loci. First, sequencing of multiple clones per PCR product did not reveal more than two alleles per individual (Jennings and Edwards 2005). Secondly, when sequences of each locus were used in a BLAST search against

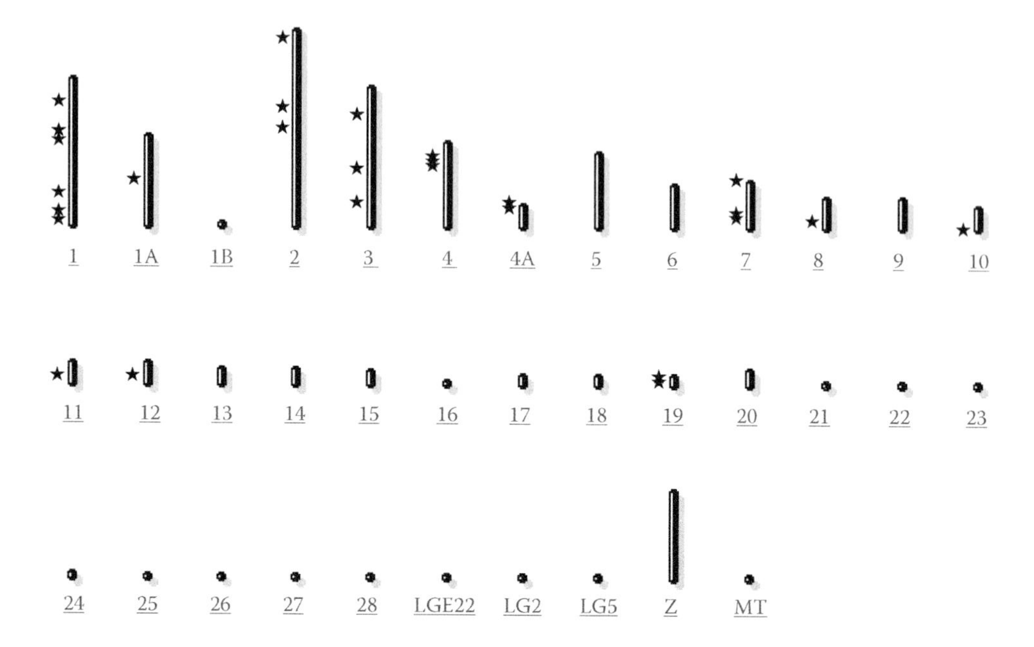

Figure 8.7. Chromosomal locations of 27 anonymous loci in the genome of the zebra finch. Loci were obtained from the genome of a closely related species of Australian grass finch (see Jennings and Edwards 2005). Each star indicates the chromosomal position of an anonymous locus. This figure was generated using the zebra finch genome (annotation release 101) on the NCBI genome browser site (www.ncbi.nlm.nih.gov/genome/browse/).

the zebra finch genome, 28/30 loci had only a single strongly supported genomic location. Results of *in silico* PCR of these loci against the zebra finch genome on the UCSC Genome Bioinformatics site (http://genome.ucsc.edu/cgi-bin/hgPcr) also did not reveal convincing evidence of any multiple copy loci. Thus, the vast majority of these loci appear to agree with the single-copy assumption despite the prevalence of transposable elements in their sequences. How can we explain this anomalous finding? Further reflection about the nature of replicative transposable elements and PCR principles reveals a simple explanation and shows why we need not worry about the possible adverse effects of repetitive DNA on the design of PCR-amplifiable anonymous loci.

To see why it may not be a problem to develop anonymous loci containing transposable elements, first consider a retroelement copy that resides in a particular chromosomal location of the zebra finch genome. If we were to place *both* PCR primers *within* this element as shown in Figure 8.8a, then it is possible that multiple copies of this element would subsequently be amplified in a single PCR reaction or different copies amplified from different individual genomes among PCR reactions—both scenarios would violate the single-copy assumption. However, recall that these genomic parasites are believed to insert themselves into random genomic locations—such as inside noncoding DNA, other transposable elements, coding regions, and introns. This means that the genomic landscape flanking each inserted retrotransposon is expected to be random with respect to the adjacent retrotransposon. Thus, if one primer is placed inside the element and the other primer is positioned *outside* of the element in the flanking region as shown in Figure 8.8b, then a subsequent PCR would not only amplify a single-copy of the element from a given genome, but also the same "orthologous" element would always be amplified in any closely related genome as well. Thus, as long as both PCR primers are not placed inside the *same* element, then the PCR is expected to generate only a single amplicon and thus satisfy the single-copy assumption in anonymous loci studies. The key here is that although each individual primer site might have multiple locations in a genome (especially for the primer within the retroelement), there is very likely only a single genomic location where *both* priming sites oppose each other on separate DNA strands and are capable of generating an amplicon that matches the

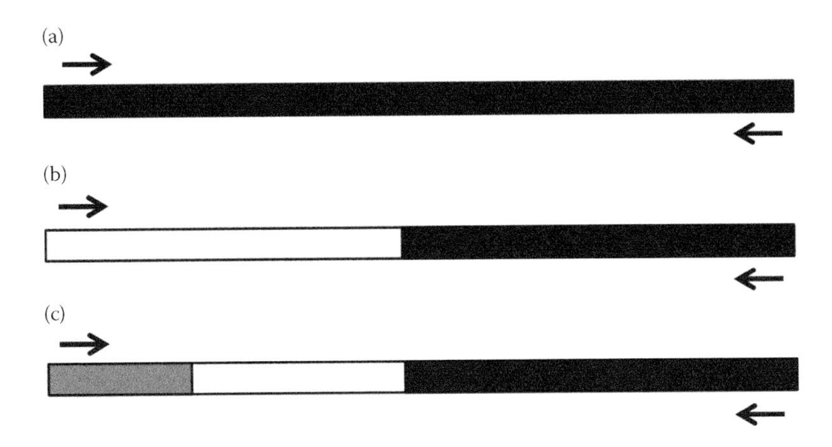

Figure 8.8 Forward and reverse PCR primers in relation to transposons embedded in a genome. (a) Forward and reverse primers (arrows) are positioned within a single transposon (black bar); (b) forward primer (above the white bar) is located outside the transposon while the reverse primer is inside (below the black bar); and (c) forward primer is located in one transposon (gray bar) and reverse primer is inside an adjacent but different transposon (black bar). The two shown transposons have independent insertion histories.

expected size. Given this principle, we can envision a third scenario shown in Figure 8.8c in which the forward and reverse primers are both inside *different* transposable elements; that is, both elements were inserted into the genomic DNA independently of each other regardless of their identity (i.e., independent insertion histories). Thus, like the scenario in Figure 8.8b, this third scenario is not expected to pose a problem in PCR. The scenarios illustrated in Figure 8.8b and 8.8c can explain the empirical results of Thomson et al. (P. 522, 2008) who observed no qualitative differences in PCR performance or DNA sequence characteristics between their apparently repeat-free anonymous loci versus a single locus that contained repetitive DNA. Returning to the grass finch example, of the 20 anonymous loci that contained one or more transposons, we see that only a single anonymous locus corresponds to the scenario in Figure 8.8a. In contrast, 19 loci matched the scenario in Figure 8.8b. Because a number of loci contained two or three distinct transposons, scenario "C" could have easily been observed if both primers were positioned in two separate transposons, but, by accident, this did not occur.

There is another good reason to expect that most anonymous loci will not violate the single-copy assumption. Recall from Chapter 2 that most retrotransposons are free to accumulate mutations from the moment they become inserted into the host genome. This means it is only a matter of time before primer-binding sites in one transposable element (or any other nonfunctional DNA not influenced by sites under selection) become mutated to the point that the primer cannot perform in PCR. Thus, primers developed for one neutral locus will likely only function for that same locus in other closely related genomes.

These findings suggest that most anonymous loci obtained from a genomic library will likely meet the single-copy assumption regardless of whether they contain transposable elements or not. This result has important implications, as we will see below when we consider NGS and bioinformatics approaches to anonymous loci development. Instead of the traditional practice of discarding candidate loci sequences that contain repetitive DNA, it may actually be advantageous to develop anonymous loci containing transposable elements. First, a larger number of genealogically independent anonymous loci can potentially be developed from a genome. Secondly, because most retroelements are no longer functional once they are inserted into a host genome, they will accumulate mutations that are not weeded out by natural selection and therefore they can be considered to meet the assumption of neutrality. The only issue with anonymous loci containing transposable elements would be their possible high GC content, which may result in elevated substitution rates at remaining CpG sites. However, a simple method for correcting this problem is to delete CpG sites from the multiple sequence alignment.

This would result in minimal loss of data and the remaining sites would likely have a single homogeneous substitution rate that reflects the natural spontaneous rate (Yang 2002; Jennings and Edwards 2005; Costa et al. 2016).

If a researcher still wants to eliminate possible multicopy loci from a sample of new loci, then a new and much simpler method can be used to remove these potentially poor quality loci. Instead of using *RepeatMasker* (Smit et al. 2004; Tarailo-Graovac and Chen 2009) to identify and discard template sequences containing repetitive DNA (which would likely be many sequences), a better strategy would be to use *RepeatMasker* (http://www.repeatmasker.org/; Smit et al. 2004) or the CENSOR software tool (Kohany et al. 2006) found on the *Repbase* site (http://www.girinst.org/censor/index.php; Jurka et al. 2005) to check each template sequence *after* the primer design step to determine whether or not the forward and reverse primers reside within the *same* transposable element. Any template sequence in which both primers are nested in the same element would be eliminated or redesigned prior to purchase and testing of new primers. This filtering step should ensure that all new loci meet the single-copy assumption. If the entire genome sequence of a closely related species is available, then *in silico* PCR and BLAST can be used to verify the single-copy nature of newly developed loci.

8.2.1.2 Single Template Methods Using Available Genomics Resources

During the early years of the genomics era the number of fully sequenced genomes in Genbank dramatically increased. At this time, which was just before NGS approaches to sequencing genomes entered the scene, the primary method for sequencing genomes consisted of shotgun cloning of genomic DNA, sequencing (Sanger) the ends of cloned inserts to a coverage at least 6× of the genome being sequenced, and then using bioinformatics tools to assemble the genome sequences (Venter et al. 2001). Forward and reverse vector (e.g., M13) sequencing primers were used to collect only the ends of the insert sequence because often the inserts were too long to be fully sequenced—particularly when the cloning vectors were bacterial artificial chromosomes (BAC) vectors, which could accommodate enormous 100–500 kb long chromosomal fragments. These genomic libraries were not only useful for sequencing entire genomes and obtaining particular chromosomal regions of interest to investigators (e.g., for gene evolution studies), but they could also be used for development of large numbers of genomic loci (Edwards et al. 2005; Shaffer and Thomson 2007; Thomson et al. 2008). By co-opting these genomics resources, researchers could have at their disposal the DNA templates for easily designing dozens or hundreds of different genomic loci without having to resort to genomic cloning methods.

For example, Shaffer and Thomson (2007) demonstrated the efficacy of this approach by designing a set of new anonymous loci for turtles using a BAC-end sequence library for the painted turtle (*Chrysemys picta*). "BAC-end" sequences are acquired simply by using the BAC-sequencing primers to sequence (via the Sanger method) the first 700–1,000 bases of the insert DNA molecule from one or both ends of the insert. Since BAC-end "reads" are generated from randomly selected BAC clones, they represent the sequences of anonymous chromosomal fragments and are, as such, from anonymous locations in the genome. To maximize the chance that new loci are single-copy and do not consist of repetitive genomic elements, Shaffer and Thomson (2007) used bioinformatics tools (*RepeatMasker*; Smit et al. 2004) to first filter away BAC-end sequences containing repetitive elements. Note, for reasons just discussed with the grass finch example, this step may not drastically improve the quality of the loci and thus it may be omitted. In the final step, primer design software is used to design a pair of primers from each retained template sequence followed by testing of the new primers on a panel of species that span the desired phylogenetic distances (e.g., intraspecific populations, family, order, etc.).

8.2.1.3 Single Template Methods Using NGS Partial Genome Data

More recently, NGS-based approaches for developing anonymous loci have been developed, which represent significant advances in loci development. Bertozzi et al. (2012) used 454 sequencing to generate a large number of random genomic sequences from low coverage libraries for a lizard and a snake species. After obtaining their raw sequencing data, the authors used a custom bioinformatics workflow to isolate high-quality sequence data, which could be used for candidate anonymous loci. This workflow included a number of filtering steps designed to discard sequences containing low quality scores, repetitive DNA, and known protein-coding regions. In the final step, the candidate loci sequences were fed into primer design software. In another NGS-based study, Lemmon and Lemmon (2012) presented a similar workflow for generating new anonymous loci but used an Illumina platform for generating raw sequence reads. Although both of these NGS studies included a filtering step to remove sequences containing repetitive DNA, this step can be replaced with the newer filtering strategy already suggested—i.e., eliminating candidate loci that have both primers within the same transposable element (see Figure 8.8a). These NGS-based approaches represent the best current methods for developing large numbers (tens to hundreds) of anonymous loci in species that have not yet had their genomes fully sequenced.

8.2.1.4 Single Template Methods Using Whole Genome Sequences

The previously described loci design methods involve a combination of laboratory work to generate candidate loci sequences (via Sanger or NGS platforms) followed by the use of bioinformatics tools to filter the sequences until a final set of single templates are acquired for the final primer design step. This loci development workflow can be made far simpler if a complete and annotated genome sequence is available for one of the target study organisms, as this would mean omitting the expensive and time-consuming laboratory step.

Chen and Li (2001) took advantage of the newly sequenced human genome when they pioneered an *in silico* method of anonymous locus development. Using human genome data and a computer-based workflow, these workers designed PCR primers to amplify 53 single-copy, presumably neutral, and independent anonymous loci. Once they obtained their set of primers, they used PCR and Sanger sequencing to obtain homologous sequences from the genomes of chimpanzee, gorilla, and orangutan to complement their human sequences. Their anonymous locus development workflow consisted of the following steps: (1) use a computer to find and retain all intergenic sequences; (2) from each intergenic region, retain one 2–20 kb segment if it is at least 5 kb from known functional genes in both directions—this step attempts to find sequences that are not strongly influenced by the possible effects of genetic hitchhiking or background selection (i.e., to satisfy the neutrality assumption); (3) use *RepeatMasker* (Smit et al. 2004) to mask repetitive DNA in each retained 2–20 kb segment (i.e., to satisfy the single-copy locus assumption); (4) retain each block of unmasked sequence ≥800 bp for use as a template sequence; and (5) design one PCR primer pair for each retained template sequence.

The main drawback with this approach is that only a miniscule number of organisms have had their genomes fully sequenced and even fewer have had their genomes annotated. Therefore, this approach has limited utility at the present time. However, in the future when more and more complete genomes become available, these *in silico*-based methods will become popular for designing large numbers of anonymous loci (and other types of loci).

8.2.2 Multiple Homologous Template Approaches for Designing PCR-Based and Anchor Loci

Primers designed using only a single template sequence usually work well for intraspecific studies or for studies involving a group of closely related species. However, when a researcher uses these same primers in attempts to amplify orthologous sequences in more evolutionarily distant taxa such as those in other genera, families, etc., then PCR will often fail. Such failures can be attributed to the fact that the primers were developed without consideration about the degree of evolutionary conservatism of the primer annealing sites.

As was discussed in Chapters 2 and 3, many genomic sites are undoubtedly free from evolutionary constraints (i.e., are not conserved) because they have no known function (e.g., intergenic DNA) or they cannot function (e.g., most retrotransposons) and hence there is nothing for natural selection to maintain (Graur et al. 2013). Thus, these sites are able to accumulate mutations (indels or substitutions) without any consequence for the organism. An implication of this is that the longer the time of divergence between two species, the more likely that the primer annealing sites will become mutated to the point that the primers will no longer function properly in cross-species PCR experiments.

A number of empirical studies have noticed this fall off in cross-species amplification success as the phylogenetic distance increases between primers and DNA being tested. Rosenblum et al. (2007), who developed 77 anonymous loci for the eastern fence lizard, observed that amplification success in cross-species PCR declined with increasing genetic distance. Similarly, Thomson et al. (2008) observed a similar drop in PCR success while testing 96 anonymous loci primer pairs on various turtle species; in particular, 50% of primer pairs had failed in PCR when divergences between the primers and species being tested moved into the 70–130 million year divergence range (see Figure 5 in Thomson et al. 2008). These authors further pointed out that for earlier microsatellite studies involving birds and mammals (Primmer et al. 1996) and fishes (Carreras-Carbonell et al. 2008), that this 50% failure mark (i.e., 50% of the loci fail to amplify) was reached when distances were only in the first tens of millions of years (Thomson et al. 2008). As to why anonymous loci for turtles are apparently able to perform well for cross-species PCRs spanning longer divergences than for birds, mammals, and fishes (i.e., >70 million years vs. 10s of millions of years, respectively), Thomson et al. (2008) hypothesized that this could be attributed to slower substitution rates in turtles. The microsatellite study on turtles by FitzSimmons et al. (1995) provides evidence supporting the hypothesis of Thomson et al. (2008), but this problem needs further study.

The aforementioned studies show that for some types of loci such as anonymous loci and microsatellites it is only a matter of time before one or more primer annealing sites become mutated in a manner that precludes successful PCR amplification. This also means that attempts to use primers developed for one species (or species group) on another species must be a trial and error process in the lab—the chance of success being inversely correlated with degree of evolutionary divergence between the sequences.

Of course, not all types of primers are vulnerable to the problem of primer sites decay because the degree of evolutionary conservatism of primer sites varies among types of loci. Universal primers are positioned on highly conserved sites of genomes, which can allow researchers to obtain homologous sequences for taxa spanning wide phylogenetic distances such as across genera, families, and orders. Universal primers have been developed for many nuclear loci such as exon and EPIC loci where the primer annealing sites are located within highly conserved portions of exons.

8.2.2.1 Designing Universal Primers by Comparative Sequence Analysis

In order to identify evolutionarily conserved sites suitable for primer placement, at least two homologous sequences must be aligned and compared site by site. Importantly, these sequences must span the evolutionary distances of the taxa for which the primers will be applied (Shaffer and Thomson 2007). Although two sequences may be sufficient for designing a pair of universal primers, the researcher can obtain a better understanding about within site variation at candidate primer locations by examining larger numbers of sequences in a multiple sequence alignment.

A visual inspection of a sequence alignment consisting of homologous sequences obtained from a number of different species can reveal blocks of nucleotide sites that are more evolutionarily conserved than other regions. An example of this type of analysis is shown in Figure 8.9, which shows an alignment of 15 sequences representing various hypothetical species of a study group. The block of highly conserved sites contained within the yellow square in Figure 8.9 can be used to design a new universal primer. Let's now take a closer look at the process of designing a universal primer.

First, note that the sequences in Figure 8.9 are shown in a $5' \rightarrow 3'$ direction and that the target sequence is to the right (not shown). Thus, we are looking for a suitable location for a new "forward" primer. Notice that the sites (columns) on the left and right sides include many variable sites. The left side is particularly messy as it contains several sites with indels, which have resulted in the placement of a several uncertain gaps in the alignment. On the right side, the sites appear to be well aligned and gap-free but they still contain a number of variable sites. However, the region near the middle that includes sites #1463–1499 is the block that contains the fewest variable sites and so this stretch of sequence seems to represent the best option for designing a new forward primer.

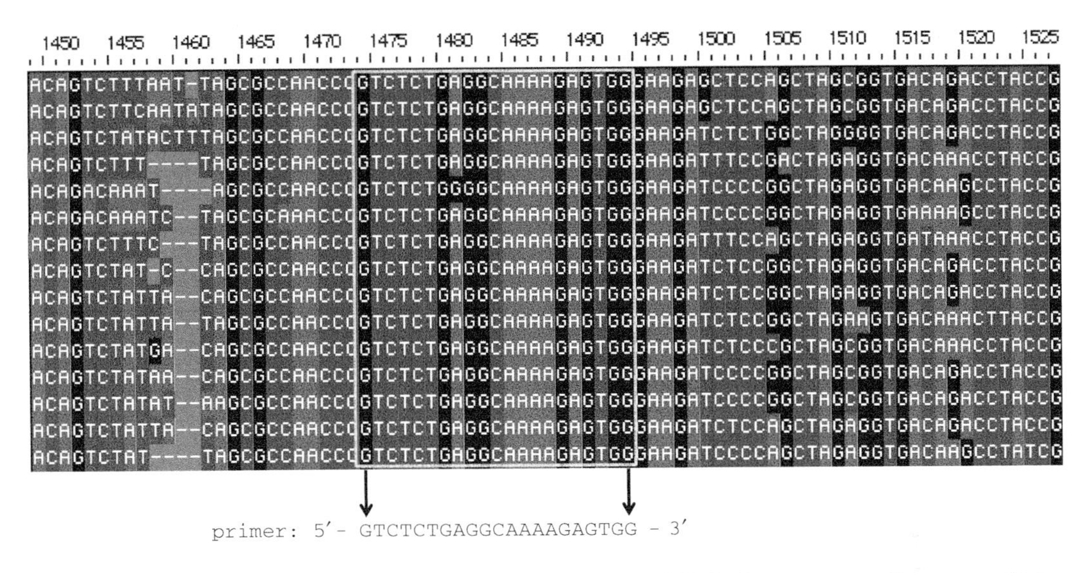

Figure 8.9. Design of a nondegenerate universal primer within a conserved block of DNA sequences. Shown is a multiple sequence alignment for 15 hypothetical species in a study group. Sites #1474–1494 (area within yellow rectangle) were used to design a new forward primer, which is shown below the alignment. Note that site #1481 is the only observed variable site within the primer design region. Sequences were aligned by eye using the software Se–Al (Rambaut 2007) and the 5′ → 3′ polarity of the sequences is from left to right.

Accordingly, sites #1474–1494 (delineated by yellow) were selected for the new primer. Notice that within this particular stretch of sequence there is only a single variable site (site #1481; A/G transition). This variable site will likely be acceptable to be included because single variable sites in the middle or at the 5′ end of a primer generally do not adversely affect PCR. Because this variable site consists of only two different bases among all sequences and only one sequence shows a "G" nucleotide, we can just use the most prevalent base at this position ("A"). The newly designed primer is shown below the alignment in Figure 8.9. Using the online program *Oligoanalyzer* 3.1 (http://www.idtdna.com/calc/analyzer; Owczarzy et al. 2008) we can check the physical characteristics of this primer:

21 bp long (satisfies Rule 1)

G + C% = 52.4% (satisfies Rule 2)

No stable homodimers at 3′ end (satisfies Rule 3)

$\Delta G_{pentamer} = -7.96$ kcal/mol (satisfies Rule 4)

$\Delta G_{hairpin} = -1.05$ kcal/mol (satisfies Rule 5)

Tm = 56°C (satisfies Rule 6)

No mismatches at 3′ end of primer (satisfies Rule 8)

If a "reverse" primer is also designed to pair with this forward primer, then Rule 3 (no stable heterodimers) and Rule 7 (Tm between primers <5°C) can also be evaluated. Although it may be tempting to add one additional base to the 3′ end of the primer (site #1495), the effect of adding this extra "G" base would be to increase the "stickiness" of the 3′ end of the primer (i.e., decrease $\Delta G_{pentamer}$ to −9.43 kcal/mol), which would violate Rule 4. Thus, it is preferable to not include this extra base.

In the aforementioned example, nearly all sites within the chosen primer sequence were invariable (conserved). However, it is often not possible to find such a highly conserved stretch of sites in which a new primer can be placed—remember that we need to find a contiguous stretch of sites that is *at least* 18 bp long in order to design a new primer. Figure 8.10 shows a different stretch of bases from the genomes of the same 15 hypothetical species used in the Figure 8.9 example. First, notice that the left and right sides again do not contain blocks of invariant sites long enough to place a primer. The least variable region,

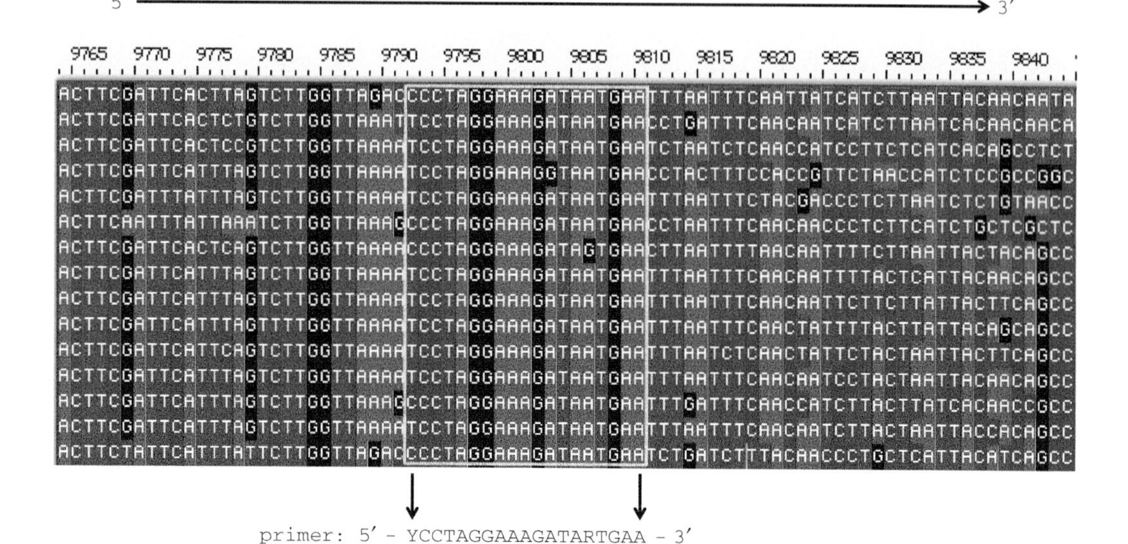

Figure 8.10. Design of a degenerate universal primer within a conserved block of DNA sequences. Shown is a multiple sequence alignment for 15 hypothetical species in a study group. Sites #9792–9810 (area within yellow rectangle) were used to design a new forward primer, which is shown below the alignment. This primer contains two degenerate sites ("Y" and "R") and therefore it is a 4-fold degenerate primer. Sequences were aligned by eye using the software Se–Al (Rambaut 2007) and the 5′ → 3′ polarity of the sequences is from left to right.

which includes sites #9792–9810, is found in the center (Figure 8.10). This particular 19 bp stretch of sequence, which is outlined in yellow, might suffice for placing a forward primer, but notice that it includes three variable sites (#9792, 9803, and 9806). However, because each of these three sites exhibits simple transition type substitutions (T/C, A/G, and A/G, respectively), it may be possible to design the primer using the most prevalent base at each of the variable sites (i.e., "T," "A," and "A," respectively). However, a primer that is only 19 bp long may not perform well in PCR if it has this many mismatches with the genomic DNA templates (longer primers can better tolerate mismatches).

One effective strategy for accommodating variable sites in a candidate primer location is to design a special type of universal primer called a *degenerate primer*. A degenerate primer is, in reality, a mixture of primers that are usually the same length but exhibit different bases at one or more sites. The rationale for using a mixture of primers is that during a PCR reaction at least one of the primers in the mixture will be the correct match (or close enough) with the target template for amplification to occur. Some mismatched bases can be tolerated as long as there are not too many of them and provided that they are at the 5′ end or in the center of the primer; mismatches located at the 3′ of the primer should be avoided. Thus, we can design a degenerate primer using the sequences in Figure 8.10 by designating each of the three variable sites with the appropriate IUPAC ambiguity code (see Table 6.4). This means that site #9792 is now indicated as "Y" to represent a "T" or "C" base; site #9803 as "R" to represent an "A" or "G"; and site #9806 as "R" to represent an "A" or "G." Because there are two possible bases at each degenerate site, this would be known as an "8-fold" degenerate primer ($2 \times 2 \times 2$) meaning that the actual primer mixture used in PCR would consist of eight different primer sequences representing all possible sequences. Similarly, if another primer had four degenerate sites consisting of "R," "Y," "D," and "N" codes at each degenerate site (Table 6.4), then this degenerate primer would exhibit $2 \times 2 \times 3 \times 4 = 48$-fold level of degeneracy and hence contain 48 unique primers! On the other hand, a primer with no degenerate sites would be 0-fold degenerate and thus consist of a single primer.

When designing degenerate primers it is critically important to remember that a tradeoff exists here: the more degenerate the primer (i.e., larger numbers of variant primers), the greater the chance of amplifying nontarget products including paralogous sequences (Chenuil et al. 2010). For example, Chenuil et al. (2010) found that their EPIC loci primers, which were developed for obtaining orthologous sequences throughout the metazoan tree of life, often performed poorly when the primers exhibited greater than 8-fold level of degeneracy. Primers of varying levels of degeneracy can still perform well in PCR, but it is advisable to minimize the levels of degeneracy in a primer whenever possible. In the example shown in Figure 8.10 we might elect to not make site #9803 a degenerate site because a single mismatch in the middle of the primer will likely not affect the primer's performance. Thus, the preferable strategy is to minimize levels of degeneracy in this primer, which gives us the 4-fold degenerate primer shown at the bottom of Figure 8.10. Using the *Oligoanalyzer* 3.1 tool (http://www.idtdna.com/calc/analyzer; Owczarzy et al. 2008) we can examine this primer's physical characteristics:

- 19 bp long (satisfies Rule 1)
- G + C% = 36.8% (satisfies Rule 2)
- No stable homodimers at 3′ end (satisfies Rule 3)
- $\Delta G_{pentamer} = -6.82$ to -6.95 kcal/mol (satisfies Rule 4)
- $\Delta G_{hairpin} = -0.02$ kcal/mol (satisfies Rule 5)
- Tm = 44.4–47.8°C (violates Rule 6)
- No mismatches at 3′ end of primer (satisfies Rule 8).

Again, if a "reverse" primer is also designed to pair with this forward primer, then Rule 3 (no stable heterodimers) and Rule 7 (Tm between primers <5°C) can also be evaluated. Note that because there is a mixture of primers involved in one degenerate primer (four different primer sequences), some of the primer's characteristics (above) are presented as minimum and maximum values. Also, you may have noticed that the 5′ terminal base located at site #9792 is variable and thus maybe you wondered if it could be omitted from the primer to further reduce the level of degeneracy. Doing this, however, would lower Tm of the primer down to the 42.4–44.6°C range, which might be too low for successful PCR. Remember that the Tm difference between forward and reverse primers should be <5°C (Rule 7). Primers with Tm values in this range can work well but it is best to strive for making primers with a Tm at least 50°C (Rule 6). These fine-tuning steps during the primer design process, which include deciding the exact span of sites to include and levels of degeneracy, represent a critical part of universal primer design.

8.2.2.2 Multiple Homologous Template Approaches Using Whole Genome Sequences

In an effort to address the longstanding paucity of available DNA sequence loci for phylogenomic studies concerned with inferring higher level organismal phylogenies, a number of innovative new genome-enabled methodologies for developing large numbers of exon, EPIC, and anonymous loci have been proffered in recent years. Similar to the single template whole genome methods, these *in silico*-based methods take advantage of existing whole genome sequences in databases such as ENSEMBL to find appropriate genomic regions for loci development. The main difference between the single and multiple template approaches is that the latter requires at least two whole genome sequences so that conserved primer sites can be identified.

An example of this approach is illustrated in the study by Li et al. (2007) who constructed a bioinformatics pipeline to find a large number of exon loci that could be useful in higher-level studies of ray-finned fishes, a clade that comprises about half of all extant vertebrate species. Figure 8.11 shows the bioinformatics pipeline developed by Li et al. (2007). First, the genome of the reference species (zebrafish) is downloaded from the ENSEMBL database before it is searched for all likely exonic segments of DNA, which consist of ORFs >800 bp. The resulting ORFs (probable exons) are then BLAST searched against the reference genome in an effort to filter out those ORFs that are not single-copy. The single-copy exons are then BLAST searched against a second query genome to narrow the list of exons to only those that are shared (conserved) between both genomes. The next step is to align homologous exon sequences and design primers.

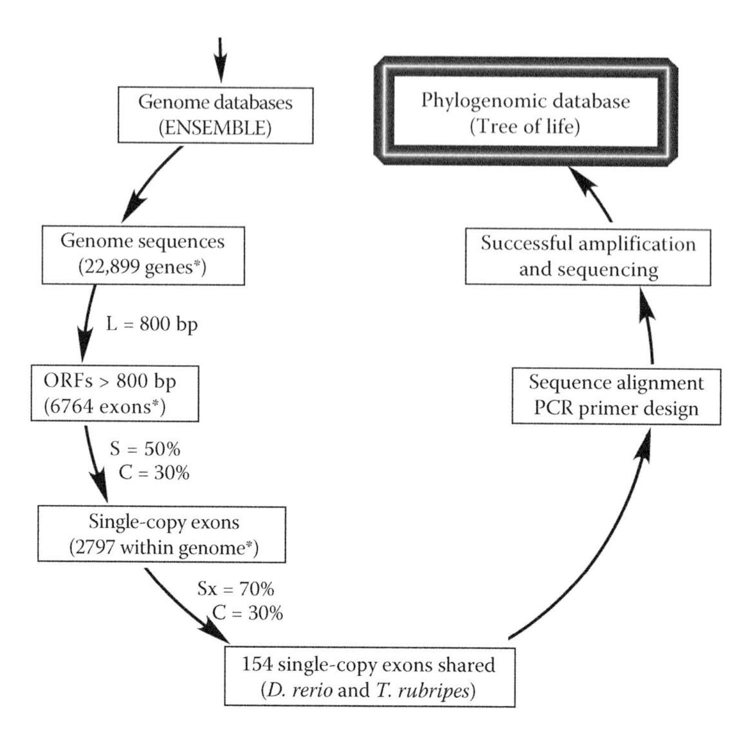

Figure 8.11. Bioinformatics pipeline for *in silico* development of large numbers of exon loci using one annotated whole genome sequence and at least one other whole genome sequence. (Reprinted from Li, C. et al. 2007. *BMC Evol Biol* 7:44. With permission.)

Interestingly, this last step is the only "low tech" step as the primer design is done by eye—similar to that described earlier for designing universal primers. If the sequences are divergent enough from each other, yet still alignable, then the primers will likely require one or more degenerate sites. If additional homologous sequences obtained from other species can be included in the alignment, then additional variation can be accounted for in the primers (i.e., the degeneracy of the primers is increased) making them even more universal in their ability to amplify the target locus on widely divergent species. Thus, even though this locus design approach can be largely accomplished using complete genome data and bioinformatics software, the actual design of the primers still relies on the investigator's knowledge about the rules of primer design. Comparable methods for designing large numbers of exon loci that can be used on phylogenetically diverse taxa come from the studies by Townsend et al. (2008) and Portik et al. (2012). Whole genome-based methods have also been devised for developing large numbers of EPIC loci (Backström et al. 2008; Chenuil et al. 2010; Li et al. 2010) and anonymous loci (Peng et al. 2009; Wenzel and Piertney 2015).

8.2.2.3 Designing Anchor Loci Probes Using Whole Genome Sequences

If you are planning to undertake a phylogenomic study that will require RNA bait sets for in-solution hybrid selection, then you should first check to see if RNA probe sets that could be used in your study already exist. For example, there are a number of UCE- and AE-anchored loci probe sets already available for vertebrates (e.g., Faircloth et al. 2012, 2013; Lemmon et al. 2012; Prum et al. 2015).

If suitable RNA bait sets do not yet exist for your study group, then you will need to develop your own probe set. Probes used for custom RNA bait kits are designed using complete genome data. Once designed, the new bait set can be produced from a company that offers this service. For example, custom RNA bait kits can be ordered from Agilent, Inc. (SureSelect kits) and from Macroarray (MYbaits kits).

How are RNA bait sets made? For highly conserved loci such as UCE- and AE-anchored loci, multiple genomes (at least two), which bracket the evolutionary divergences in the clade of interest, are first aligned to each other so that highly conserved regions can be identified as candidate loci regions. From this pool of candidate regions, probe sequences are designed in the most-conserved stretches of sites and possibly in less-conserved flanking sequences (see below). Because the types of RNA probes used in in-solution hybrid selection reactions are usually 120 bp long, which is shorter than typical phylogenomic loci, a number of additional probes are often designed in order to catch all adapter-ligated library fragments that contain some portion of a target locus sequence. Thus, for each candidate locus region, the initial probe is usually placed in the center of the most-conserved stretch of sites (e.g., middle of a UCE) and then additional probes are designed in the regions on both sides of the first probe. This practice of using multiple probes per probe region is termed *tiling*. Different tiling schemes have been used such as placing probes in a nonoverlapping or sequential manner to cover each protein-coding exon (Gnirke et al. 2009). Alternatively, and more commonly, probes are placed in overlapping configurations to some degree. For example, in some UCE-anchored loci studies (e.g., Faircloth et al. 2012, 2015), a tiling density of 2× was used whenever UCEs were >180 bp long; that is, 120 bp-long probes were tiled such that adjacent probes overlapped each other by 60 bp. In another UCE-anchored loci study, Faircloth et al. (2013) used a 4× tiling density. A much denser tiling strategy was employed in the first AE-anchored loci study (Lemmon et al. 2012), as the authors positioned a new probe every five bp along each distinct probe region. This meant that each probe region was represented by as many as 25 largely overlapping probes. In a more recent study using AE-anchored loci, Prum et al. (2015) used a 1.5× tiling density. Although probe-tiling density may be optimized for different groups of organisms, the enormous multilocus datasets generated in each of the aforementioned studies attests to the efficacy of using even a nonoptimized tiling density.

For additional details about how UCE-anchored probe sets are made see Faircloth et al. (2012, 2013, 2015) and the website (http://ultraconserved.org). Additionally, software called UCE-Probe-Design-Program or "UPDP" has been developed to help researchers design UCE probe sets (see http://github.com/faircloth-lab/uce-probe-design). For more information about how AE-anchored loci are developed see Lemmon et al. (2012), Prum et al. (2015), and the following website for additional information (http://anchoredphylogeny.com).

REFERENCES

Abd-Elsalam, K. A. 2003. Bioinformatic tools and guideline for PCR primer design. *Afr J Biotechnol* 2:91–95.

Amaral, F. R., S. V. Edwards, and C. Y. Miyaki. 2012. Eight anonymous nuclear loci for the squamate antbird (*Myrmeciza squamosa*), cross-amplifiable in other species of typical antbirds (Aves, Thamnophilidae). *Conserv Genet Resour* 4:645–647.

Backström, N., S. Fagerberg, and H. Ellegren. 2008. Genomics of natural bird populations: A gene-based set of reference markers evenly spread across the avian genome. *Mol Ecol* 17:964–980.

Bertozzi, T., K. L. Sanders, M. J. Sistrom, and M. G. Gardner. 2012. Anonymous nuclear loci in non-model organisms: making the most of high throughput genome surveys. *Bioinformatics* 28:1807–1810.

Breslauer, K. J., R. Frank, H. Blöcker, and L. A. Marky. 1986. Predicting DNA duplex stability from the base sequence. *Proc Natl Acad Sci USA* 83:3746–3750.

Carreras-Carbonell, J., E. Macpherson, and M. Pascual. 2008. Utility of pairwise mtDNA genetic distances for predicting cross-species microsatellite amplification and polymorphism success in fishes. *Conserv Genet* 9:181–190.

Chen, F. C. and W.-H. Li. 2001. Genomic divergences between humans and other hominoids and the effective population size of the common ancestor of humans and chimpanzees. *Am J Hum Genet* 68:444–456.

Chenuil, A., T. B. Hoareau, E. Egea et al. 2010. An efficient method to find potentially universal population genetic markers, applied to metazoans. *BMC Evol Biol* 10:276.

Chou, Q., M. Russell, D. E. Birch, J. Raymond, and W. Bloch. 1992. Prevention of pre-PCR mis-priming and primer dimerization improves low-copy-number amplifications. *Nucleic Acids Res* 20:1717–1723.

Costa, I. R., F. Prosdocimi, and W. B. Jennings. 2016. In silico phylogenomics using complete genomes: A case study on the evolution of hominoids. *Genome Res* 26:1257–1267.

Edwards, S. V., W. B. Jennings, and A. M. Shedlock. 2005. Phylogenetics of modern birds in the era of genomics. *Proc R Soc Lond B Biol Sci* 272:979–992.

Faircloth, B. C., M. G. Branstetter, N. D. White, and S. G. Brady. 2015. Target enrichment of ultraconserved elements from arthropods provides a genomic perspective on relationships among Hymenoptera. *Mol Ecol Resour* 15:489–501.

Faircloth, B. C., J. E. McCormack, N. G. Crawford, M. G. Harvey, R. T. Brumfield, and T. C. Glenn. 2012. Ultraconserved elements anchor thousands of genetic markers spanning multiple evolutionary timescales. *Syst Biol* 61:717–726.

Faircloth, B. C., L. Sorenson, F. Santini, and M. E. Alfaro. 2013. A phylogenomic perspective on the radiation of ray-finned fishes based upon targeted sequencing of ultraconserved elements (UCEs). *PLoS One* 8:p.e65923.

FitzSimmons, N. N., C. Moritz, and S. S. Moore. 1995. Conservation and dynamics of microsatellite loci over 300 million years of marine turtle evolution. *Mol Biol Evol* 12:432–440.

Folmer, O., M. Black, W. Hoeh, R. Lutz, and R. Vrijenhoek, R. 1994. DNA primers for amplification of mitochondrial cytochrome c oxidase subunit I from diverse metazoan invertebrates. *Mol Mar Biol Biotechnol* 3:294.

Freier, S. M., R. Kierzek, J. A. Jaeger et al. 1986. Improved free-energy parameters for predictions of RNA duplex stability. *Proc Natl Acad Sci USA* 83:9373–9377.

Gibbs, Josiah Willard. Wikipedia Page. http://en.wikipedia.org/wiki/Josiah_Willard_Gibbs (accessed September 5, 2015).

Gnirke, A., A. Melnikov, J. Maguire et al. 2009. Solution hybrid selection with ultra-long oligonucleotides for massively parallel targeted sequencing. *Nature Biotechnol* 27:182–189.

Graur, D., Y. Zheng, N. Price, R. B. Azevedo, R. A. Zufall, and E. Elhaik. 2013. On the immortality of television sets: "function" in the human genome according to the evolution-free gospel of ENCODE. *Genome Biol Evol* 5:578–590.

Haas, S., M. Vingron, A. Poustka, and S. Wiemann. 1998. Primer design for large scale sequencing. *Nucleic Acids Res* 26:3006–3012.

Hebert, P. D. N., A. Cywinska, S. L. Ball, and J. R. deWaard. 2003. Biological identifications through DNA barcodes. *Proc R Soc Lond B Biol Sci* 270:313–321.

Innis, M. A. and D. H. Gelfand. 1990. Chapter 1. Optimization of PCRs. In *PCR Protocols: A Guide to Methods and Applications*, eds. M. A. Innis, D. H. Gelfand, J. J. Sninsky, and T. J. White, 3–12. New York: Academic Press.

Innis, M. A., D. H. Gelfand, J. J. Sninsky, and T. J. White, eds. 1990. *PCR Protocols: A Guide to Methods and Applications*. New York: Academic Press.

Jennings, W. B. and S. V. Edwards. 2005. Speciational history of Australian Grass Finches (*Poephila*) inferred from thirty gene trees. *Evolution* 59:2033–2047.

Jurka, J., V. V. Kapitonov, A. Pavlicek, P. Klonowski, O. Kohany, and J. Walichiewicz. 2005. Repbase update; a database of eukaryotic repetitive elements. *Cytogenet Genome Res* 110:462–467.

Karl, S. A. and J. C. Avise. 1993. PCR-based assays of mendelian polymorphisms from anonymous single-copy nuclear DNA: Techniques and applications for population genetics. *Mol Biol Evol* 10:342–361.

Kibbe, W. A. 2007. OligoCalc: An online oligonucleotide properties calculator. *Nucleic Acids Res* 35:W43-W46.

Kocher, T. D., W. K. Thomas, A. Meyer et al. 1989. Dynamics of mitochondrial DNA evolution in animals: Amplification and sequencing with conserved primers. *Proc Natl Acad Sci USA* 86:6196–6200.

Kohany, O., A. J. Gentles, L. Hankus, and J. Jurka. 2006. Annotation, submission and screening of repetitive elements in Repbase: RepbaseSubmitter and censor. *BMC Bioinf* 7:474.

Koressaar, T. and M. Remm. 2007. Enhancements and modifications of primer design program Primer3. *Bioinformatics* 23:1289–1291.

Kwok, S., D. E. Kellogg, N. McKinney, D. Spasic, C. Levenson, and J. J. Sninsky. 1990. Effects of primer–template mismatches on the polymerase chain reaction: Human immunodeficiency virus type 1 model studies. *Nucleic Acids Res* 18:999–1005.

Lemmon, A. R., S. A. Emme, and E. M. Lemmon. 2012. Anchored hybrid enrichment for massively high-throughput phylogenomics. *Syst Biol* 61:727–744.

Lemmon, A. R. and E. M. Lemmon. 2012. High-throughput identification of informative nuclear loci for shallow-scale phylogenetics and phylogeography. *Syst Biol* 61:745–761.

Li, C., G. Ortí, G. Zhang, and G. Lu. 2007. A practical approach to phylogenomics: The phylogeny of ray-finned fish (Actinopterygii) as a case study. *BMC Evol Biol* 7:44.

Li, C., J.-J. M. Riethoven, and L. Ma. 2010. Exon-primed intron-crossing (EPIC) markers for non-model teleost fishes. *BMC Evol Biol* 10:90.

Lincoln, S. E., M. J. Daly, and E. S. Lander. 1991. *PRIMER*: A computer program for automatically selecting PCR primers. MIT Center for Genome Research and Whitehead Institute for Biomedical Research, Nine Cambridge Center, Cambridge, Massachusetts,

2142. ftp://ftp.broadinstitute.org/distribution/software/Primer0.5/readme.txt (accessed September 17, 2016).

Owczarzy, R., A. V. Tataurov, Y. Wu et al. 2008. IDT SciTools: A suite for analysis and design of nucleic acid oligomers. *Nucleic Acids Res* 36:W163–W169.

Palumbi, S. R. 1996. Chapter 7. Nucleic acids II: The polymerase chain reaction. In *Molecular Systematics*, 2nd edition, eds. D. M. Hillis, C. Moritz, and B. K. Mable, 205–247. Sunderland: Sinauer.

Peng, Z., N. Elango, D. E. Wildman, and V. Y. Soojin. 2009. Primate phylogenomics: Developing numerous nuclear non-coding, non-repetitive markers for ecological and phylogenetic applications and analysis of evolutionary rate variation. *BMC Genomics* 10:247.

Portik, D. M., P. L. Wood Jr., J. L. Grismer, E. L. Stanley, and T. R. Jackman. 2012. Identification of 104 rapidly-evolving nuclear protein-coding markers for amplification across scaled reptiles using genomic resources. *Conserv Genet Resour* 4:1–10.

Primmer, C. R., A. P. Møller, and H. Ellegren. 1996. A wide-range survey of cross-species microsatellite amplification in birds. *Mol Ecol* 5:365–378.

Prum, R. O., J. S. Berv, A. Dornburg et al. 2015. A comprehensive phylogeny of birds (Aves) using targeted next-generation DNA sequencing. *Nature* 526:569.

Rambaut, A. 2007. *Se-Al, version 2.0 a11*. Edinburgh: University of Edinburgh.

Rosenblum, E. B., N. M. Belfiore, and C. Moritz. 2007. Anonymous nuclear markers for the eastern fence lizard, *Sceloporus undulatus*. *Mol Ecol Notes* 7:113–116.

Rozen, S. and H. Skaletsky. 1999. Primer3 on the WWW for general users and for biologist programmers. In *Bioinformatics methods and protocols*, ed. S. Misener, 365–386. New York: Humana Press.

Rychlik, W. 1995. Selection of primers for polymerase chain reaction. *Mol Biotechnol* 3:129–134.

Rychlik, W. and R. E. Rhoads. 1989. A computer program for choosing optimal oligonucleotides for filter hybridization, sequencing and *in vitro* amplification of DNA. *Nucleic Acids Res* 17:8543–8551.

Sambrook J., E. F. Fritsch, and T. Maniatis. 1989. *Molecular Cloning: A Laboratory Manual*, 2nd edition. Cold Spring Harbor: Cold Spring Harbor Laboratory Press.

SantaLucia, J., Jr. 1998. A unified view of polymer, dumbbell, and oligonucleotide DNA nearest-neighbor thermodynamics. *Proc Natl Acad Sci USA* 95:1460–1465.

Shaffer, H. B. and R. C. Thomson. 2007. Delimiting species in recent radiations. *Syst Biol* 56:896–906.

Smit, A., R. Hubley, and P. Green. 2004. *RepeatMasker Open-3.0*. Available from http://www.Repeatmasker.org.

Suggs, S. V., T. Hirose, E. H. Miyake, M. J. Kawashima, K. I. Johnson, and R. B. Wallace. 1981. In *Developmental Biology Using Purified Genes*, ed. D. D. Brown, 23:683–693. New York: Academic Press.

Tarailo-Graovac, M. and N. Chen. 2009. Using RepeatMasker to identify repetitive elements in genomic sequences. *Curr Protoc Bioinformatics* 25:4.10.1–4.10.14.

Thein, S. L. and R. B. Wallace. 1986. The use of synthetic oligonucleotides as specific hybridization probes in the diagnosis of genetic disorders. In *Human Genetic Diseases: A Practical Approach*, ed. K. E. Davis, 33–50. Herndon: IRL Press.

Thomson, R. C., A. M. Shedlock, S. V. Edwards, and H. B. Shaffer. 2008. Developing markers for multilocus phylogenetics in non-model organisms: A test case with turtles. *Mol Phylogenet Evol* 49:514–525.

Thomson, R. C., I. J. Wang, and J. R. Johnson. 2010. Genome-enabled development of DNA markers for ecology, evolution and conservation. *Mol Ecol* 19:2184–2195.

Townsend, T. M., R. E. Alegre, S. T. Kelley, J. J. Wiens, and T. W. Reeder. 2008. Rapid development of multiple nuclear loci for phylogenetic analysis using genomic resources: An example from squamate reptiles. *Mol Phylogenet Evol* 47:129–142.

Untergasser, A., I. Cutcutache, T. Koressaar et al. 2012. Primer3 – new capabilities and interfaces. *Nucleic Acids Res* 40:e115.

Venter, J. C., M. D. Adams, E. W. Myers et al. 2001. The sequence of the human genome. *Science* 291:1304–1351.

Warren, W. C., D. F. Clayton, H. Ellegren et al. 2010. The genome of a songbird. *Nature* 464:757–762.

Watson, J. D., T. A. Baker, S. P. Bell, A. Gann, M. Levine, and R. Losick. 2014. *Molecular Biology of the Gene*, 7th edition. New York: Pearson Education, Inc.

Wenzel, M. A. and S. B. Piertney. 2015. In silico identification and characterisation of 17 polymorphic anonymous non-coding sequence markers (ANMs) for red grouse (*Lagopus lagopus scotica*). *Conserv Genet Resour* 7:319–323.

Yang, Z. 2002. Likelihood and Bayes estimation of ancestral population sizes in hominoids using data from multiple loci. *Genetics* 162:1811–1823.

CHAPTER NINE

Future of Phylogenomic Data Acquisition

Prior to the arrival of molecular biology kits and outsourcing of Sanger sequencing, researchers had to be skilled in all aspects of phylogenomic data acquisition—from DNA extractions to preparing sequencing gels for an automated sequencing machine. This arduous and low throughput workflow meant that it was not unusual to wait many months or even years before the desired dataset was in hand. Thus, the welcome arrival of outsourcing for the Sanger sequencing step greatly simplified the process of obtaining DNA sequence data. Outsourcing is having a similar effect on NGS. During the early years of NGS, researchers needed to prepare their own sequencing libraries before outsourcing the sequencing step (e.g., "buying a lane" on an Illumina sequencer). Now, many NGS facilities offer to perform library preparations—for nontarget partial genome sequencing, target capture (e.g., UCE-anchored loci), or whole genome sequencing—as well as the final sequencing step. Thus, researchers who opt to outsource all NGS work only need to perform DNA extractions in order to generate enormous datasets. Whether researchers prepare their own libraries or not does not change the fact that it now only takes weeks to obtain large datasets via the NGS route. By spending less time in the laboratory obtaining data, researchers can invest more time in the analysis of their datasets. Indeed, the long-term trend in phylogenomics has been a reduction in time for amassing sequence data, which has had the favorable consequence of allowing researchers more time to conduct analyses. Can this trend be extended further? Given the fast pace of genomics and biotechnology, we should ponder what the future holds for DNA sequence data acquisition.

9.1 THE IMPENDING FLOOD OF GENOMES

Jarvis et al. (2014) published a massive phylogenomic study of birds based on 48 assembled and annotated genomes that spanned the avian tree of life. This was hailed as a landmark achievement in tree of life studies (Callaway 2014; Zhang et al. 2014). However, this feat required many collaborators from different laboratories, access to substantial funding, and years of hard work. Despite the advent of NGS technologies more than a decade ago, obtaining large genomes such as those for vertebrates still remains a difficult and expensive task. That only 259 vertebrate genomes had been sequenced and assembled up to and including the year 2014 (OBrien et al. 2014) bespeaks the challenges of acquiring whole genome sequences.

For a phylogenomic study, the current high costs of genomes means that a tradeoff exists between the numbers of genomes that can be sequenced versus numbers of species that can be sampled. Prum et al. (2015) conducted a large-scale study of the avian tree of life with this tradeoff in mind. Instead of acquiring whole genome sequences for each sampled species—as was done by Jarvis et al. (2014), these authors used a target capture method to obtain 259 anchored loci from 198 avian species distributed throughout the avian tree of life. The main advantage of the approach advocated by Prum et al. (2015) is that it allows for more extensive taxon sampling, which is expected to improve the accuracy of gene trees (Graybeal 1998; Heath et al. 2008; Townsend and Lopez-Giraldez 2010). Thus, subsampling genomes for many loci using target capture methods is currently the best (or only) strategy for phylogenomic studies involving many species. However, at some point in the future the cost of obtaining full genomes may fall low enough to trigger a major change in how DNA sequence datasets are acquired.

How much does a new genome sequence cost? In early 2014, Illumina announced that it was making a sequencing platform called the HiSeq X Ten that is expected to sequence human genomes for under 1,000 USD (Hayden 2014). This cost estimate is being made possible by improvements to Illumina's existing NGS technology. There has also been some speculation that the cost of outsourcing a genome (for any species) will soon be in the low thousands of dollars (Hayden 2014). While it is impossible to know how low the cost of full genome sequencing will ultimately reach, it is not far-fetched to believe that the cost will eventually plunge down into the low hundreds of dollars if not lower. Human medicine is a powerful driver of genomics and biotechnology and thus the cost of obtaining full genome sequences will likely continue to fall as a result of competition among sequencing technologies.

The appearance of a novel genome sequencing platform, which has low labor and reagent costs combined with software that automatically assembles genomes, could make the $100–$400 genome a reality at some point in coming years. In this scenario, a researcher could send dozens or more of purified DNA samples to a genome sequencing facility where all sequencing and genome assemblies are performed. Later, perhaps within days or 1–2 weeks, the researcher would be able to download all genome data files and begin phylogenomic analyses. Even if this futuristic scenario does not become reality in the near future, a large number of vertebrate genomes may still soon be available. OBrien et al. (2014) predicted that the number of sequenced vertebrate genomes will grow to more than 10,000 within 5–10 years. This estimate was based on assumptions that existing NGS technologies and bioinformatics tools for assembling genomes may see modest improvements and therefore it does not take into consideration what might happen if a new and higher performing NGS platform becomes available. This impending flood of genomes is going to vigorously stimulate a large number of new full genome-based studies of the vertebrate tree of life, which, in turn, will enable *in silico* phylogenomics to come of age (Costa et al. 2016).

9.2 *IN SILICO* ACQUISITION OF PHYLOGENOMIC DATASETS

As we saw in Chapters 7 and 8, available genome sequences are enabling researchers to easily design large numbers of phylogenomic loci. Once sets of target capture probes or PCR primers are designed, laboratory methods are required to obtain DNA sequence data from the genomes of all sampled individuals. Although this hybrid *in silico*-laboratory workflow for obtaining datasets represents a quantum-leap improvement over earlier methods, the ultimate approach to acquiring a dataset is to use 100% *in silico* methods whereby all data are directly obtained from computer files containing complete genome sequences. This approach, however, requires complete genome sequences for each individual or species in the study, which explains why there have been few such studies to date.

Rokas et al. (2003) conducted a completely *in silico* phylogenomic study involving eight species of yeasts whose genomes had already been sequenced. Using only a computer, these authors extracted DNA sequences representing 106 orthologous loci from each of the genomes before performing analyses of yeast phylogeny. In similar fashion, Faircloth et al. (2012) and McCormack et al. (2012) took advantage of existing full genome data in order to generate large datasets consisting of UCE-anchored loci for mammals. Jarvis et al. (2014) also used computer-based methods to construct datasets based on UCE-anchored loci but their data were derived from genomes that they themselves had earlier sequenced and assembled. More recently, Costa et al. (2016) conducted a fully *in silico*-based study of hominoid evolution, which was based on a large number of anonymous and AE-anchor loci extracted from available genome sequences.

The 100% *in silico* approach to data acquisition offers several advantages over other methods. First, given complete genome data for each individual or species in a study, the amount of time required to construct multiple loci datasets is on the order of days or less. A second advantage, at least over some of the alternative methods, is that selected loci can be verified as being single-copy in genomes. A third advantage is that *a priori* physical distance thresholds between loci located on the same chromosomes can be used to ensure that each locus is genealogically independent

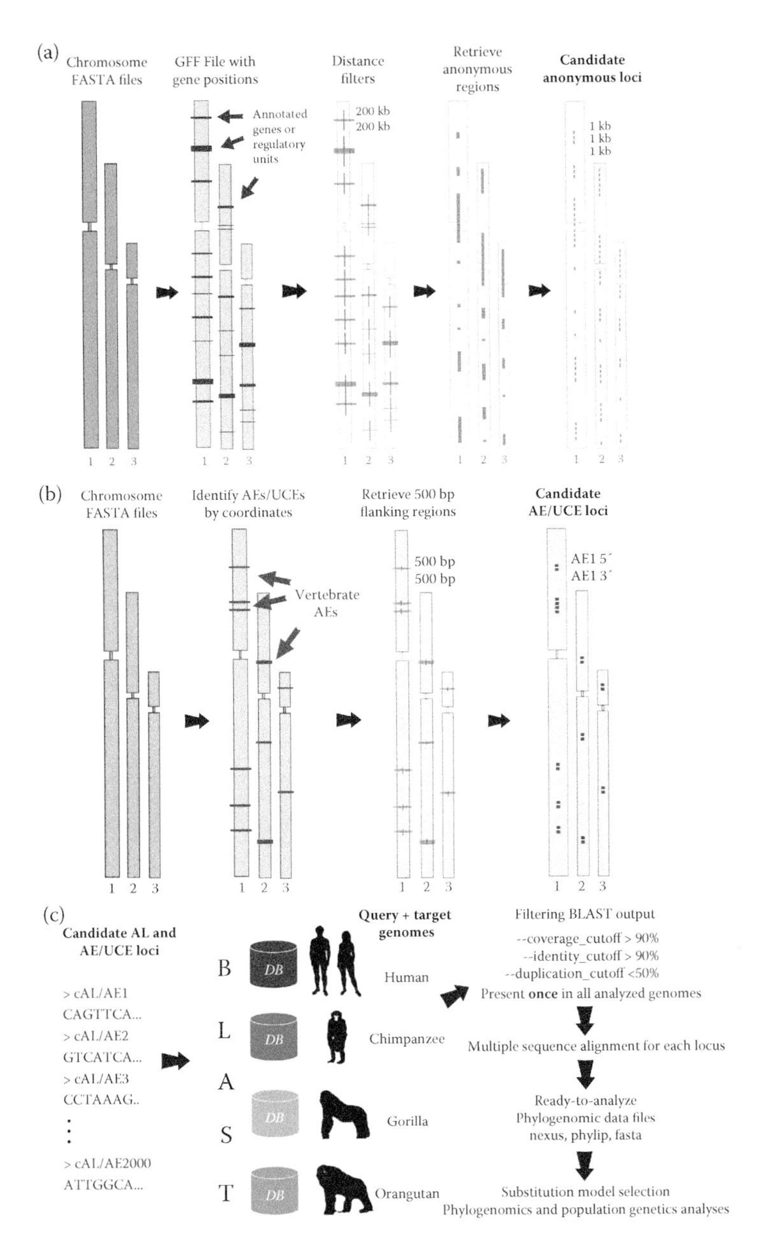

Figure 9.1 A software pipeline for acquiring ready-to-analyze phylogenomic datasets consisting of anonymous or anchor (AE/UCE) loci (Costa et al. 2016, fig. 1). The legend reads, "ALFIE software pipeline. (a) Anonymous loci (AL) finding module: User inputs complete genome sequences in a FASTA format and a general feature format (GFF) file for the query genome. Program first applies a user-defined 'distance filter,' which removes all known functional elements + flanking sequences of user-specified lengths (purple color blocks). Remaining (presumably neutral) intergenic regions (orange color blocks), called candidate ALs, are retrieved and cut into consecutive segments of user-defined length and saved in FASTA files. (b) Anchor loci (AE/UCE) finding module: User inputs genome sequences in FASTA format. Program finds locations of target AEs/UCEs in a reference human genome with a coordinate file that currently contains 512 vertebrate AEs (included in package). Module retrieves flanking regions with user-defined length (e.g., 500 bp). User also specifies distance (in base pairs) between flanking sequences and their AEs/UCEs. Paired flanking sequences (i.e., candidate AE/UCE loci) are saved in FASTA files. (c) Downstream analyses: AL or AE/UCE candidate loci are used as query sequences in BLAST searches against target genomes. Single-copy loci are retained and subsequently aligned. A user-specified distance filter retains loci that are likely independent from other sampled loci. Each pair of AE/UCE flanking sequences is concatenated to form independent loci. Lastly, ALFIE outputs ready-to-analyze data sets." (Reprinted from Costa, I. R., F. Prosdocimi, and W. B. Jennings. 2016. *Genome Res* 26:1257–1267.)

from other sampled loci (Sachidanandam et al. 2001; O'Neill et al. 2013; Leaché et al. 2015; Costa et al. 2016; see Table 3.1 for additional examples). Finally, the computer-based process of generating DNA sequence datasets can be automated from start to finish: software can be used to extract target loci sequences from all sampled genomes, perform multiple sequence alignments, and output ready-to-analyze datasets.

The recent work by Costa et al. (2016) provides us with an example of how phylogenomic data acquisition can be fully automated. In this study, the authors developed a software pipeline called *ALFIE*, which is for Anonymous/ Anchor Loci FIndEr (Figure 9.1). After the user inputs full genome data for each individual or species and specifies values for various distance filters (or thresholds), desired locus length, etc., the program seamlessly performs the following steps: (1) extracts the maximum number of single-copy, presumably neutral, and independent anonymous loci from an annotated genome (Figure 9.1a) or obtains a set of predefined AE or UCE loci from an unannotated genome sequence (Figure 9.1b); (2) extracts orthologous sequences for all anonymous or AE/UCE loci from other complete genomes; (3) constructs multiple sequence alignments for each locus; and (4) outputs ready-to-analyze datasets in various commonly used formats such as NEXUS, PHYLIP, and FASTA (Figure 9.1c). These authors bench tested this software using complete genome data from the well-studied extant hominoids (humans, chimpanzees, gorillas, and orangutans). In less than 3 hours, this software output a 1.2 Mb-sized dataset consisting of 292 anonymous loci (each ~1 kb long) while only 13 minutes was needed to output a dataset consisting of 242 AE loci of similar lengths. Although this type of study can only be done using fully sequenced genomes for all study individuals, this example shows the exciting full potential of *in silico* data acquisition.

REFERENCES

Callaway, E. 2014. Flock of geneticists redraws bird family tree. *Nature* 516:297.

Costa, I. R., F. Prosdocimi, and W. B. Jennings. 2016. In silico phylogenomics using complete genomes: A case study on the evolution of hominoids. *Genome Res* 26:1257–1267.

Faircloth, B. C., J. E. McCormack, N. G. Crawford, M. G. Harvey, R. T. Brumfield, and T. C. Glenn. 2012. Ultraconserved elements anchor thousands of genetic markers spanning multiple evolutionary timescales. *Syst Biol* 61:717–726.

Graybeal, A. 1998. Is it better to add taxa or characters to a difficult phylogenetic problem? *Syst Biol* 47:9–17.

Hayden, E. C. 2014. Is the $1,000 genome for real? *Nature*. doi: 10.1038/nature.2014.14530.

Heath, T. A., S. M. Hedtke, and D. M. Hillis. 2008. Taxon sampling and the accuracy of phylogenetic analyses. *J Syst Evol* 46:239–257.

Jarvis, E. D., S. Mirarab, A. J. Aberer et al. 2014. Whole-genome analyses resolve early branches in the tree of life of modern birds. *Science* 346:1320–1331.

Leaché, A. D., A. S. Chavez, L. N. Jones, J. A. Grummer, A. D. Gottscho, and C. W. Linkem. 2015. Phylogenomics of phrynosomatid lizards: Conflicting signals from sequence capture versus restriction site associated DNA sequencing. *Genome Biol Evol* 7:706–719.

McCormack, J. E., B. C. Faircloth, N. G. Crawford, P. A. Gowaty, R. T. Brumfield, and T. C. Glenn. 2012. Ultraconserved elements are novel phylogenomic markers that resolve placental mammal phylogeny when combined with species-tree analysis. *Genome Res* 22:746–754.

OBrien, S. J., D. Haussler, and O. Ryder. 2014. The birds of Genome 10K. *GigaScience* 3:32.

O'Neill, E. M., R. Schwartz, C. T. Bullock et al. 2013. Parallel tagged amplicon sequencing reveals major lineages and phylogenetic structure in the North American tiger salamander (*Ambystoma tigrinum*) species complex. *Mol Ecol* 22:111–129.

Prum, R. O., J. S. Berv, A. Dornburg et al. 2015. A comprehensive phylogeny of birds (Aves) using targeted next-generation DNA sequencing. *Nature* 526:569.

Rokas, A., B. L. Williams, N. King, and S. B. Carroll. 2003. Genome-scale approaches to resolving incongruence in molecular phylogenies. *Nature* 425:798–804.

Sachidanandam, R., D. Weissman, S. C. Schmidt et al. 2001. A map of human genome sequence variation containing 1.42 million single nucleotide polymorphisms. *Nature* 409:928–933.

Townsend, J. P. and F. Lopez-Giraldez. 2010. Optimal selection of gene and ingroup taxon sampling for resolving phylogenetic relationships. *Syst Biol*. doi: 10.1093/sysbio/syq025.

Zhang, G., E. D. Jarvis, and M. T. P. Gilbert. 2014. A flock of genomes. *Science* 346:1308–1309.

INDEX

D

E

F

G

H

I

N

O

P

Q

R

S

T

U

V

W

X